FLORA C
BHUTAN

INCLUDING A RECORD OF PLANTS FROM SIKKIM

VOLUME 1   PART 3

A.J.C. GRIERSON & D.G. LONG

ILLUSTRATIONS BY MARY BATES

ROYAL BOTANIC GARDEN, EDINBURGH.
1987.

Published by the Royal Botanic Garden, Edinburgh, for the Overseas Development Administration, Eland House, Stag Place, London, and the Royal Government of Bhutan.

© Royal Botanic Garden, Edinburgh, 1987.

ISBN 0 9504270 6 3

Typeset and printed by Outline, Edinburgh. Marketing Agents: Wheldon & Wesley Ltd., Lytton Lodge, Codicote, Hitchin, Herts, SG4 8TE, England.

Contents

		Page
Additional bibliography		466

ANGIOSPERMAE (continued)

65	Hamamelidaceae	468	76	Leguminosae	607
66	Crassulaceae	471	77	Podostemaceae	738
67	Saxifragaceae	485	78	Oxalidaceae	739
68	Parnassiaceae	515	79	Geraniaceae	744
69	Hydrangeaceae	517	80	Tropaeolaceae	749
70	Grossulariaceae	522	81	Zygophyllaceae	750
71	Philadelphaceae	525	82	Balanitaceae	750
72	Iteaceae	527	83	Linaceae	751
73	Pittosporaceae	528	84	Euphorbiaceae	754
74	Rosaceae	529	85	Daphniphyllaceae	813
75	Connaraceae	607			

Index of botanical names	815
Index of common names	832

ADDITIONAL BIBLIOGRAPHY 140−177

(For bibliography 1−129 see Vol. 1 Part 1 pages 37−43; for bibliography 130−139 see Vol. 1 Part 2 page 190)

140. Baehni, C., Bonner, C.E.B. & Vautier, S. 1951. Plantes recoltées par le Dr Wyss-Dunant au cours de l'Expedition suisse a l'Himalaya en 1949. *Candollea* 13: 213−236
141. Balakrishnan, N.P. 1981. *Flora of Jowai and vicinity, Meghalaya.* Vol. 1. Calcutta.
142. Beauverd, G. 1909. Notes sur une collection de plantes de l'Himalaya. *Bull. Soc. Bot. Geneve* Ser. 2, 1: 104−107.
143. Bennet, S.S.R. & Sharma, B.K. 1983. Indian Ginseng. *Indian Forester* 109: 840−845.
144. Burkill, I.H. 1908. Notes on the pollination of Flowers in India. Note No. 5 — Some autumn observations in the Sikkim Himalaya. *J. Proc. Asiatic Soc. Bengal (NS)* 4(4): 179−195.
145. Chang, D.H.S. 1983. The Tibetan plateau in Relation to the Vegetation of China. *Ann. Missouri Bot. Gard.* 70: 564−570.
146. Das, S.N., Janardhanan, K.P. & Roy, S.C. 1983. Some observations on the ethnobotany of the tribes of Totopara and adjoining areas in Jalpaiguri District, West Bengal. *J. Econ. Tax. Bot.* 4(2): 453−474.
147. Devitt, J. 1986. Forestry in Bhutan. *Unasylva* 38: 46−51.
148. Dogra, P.D. 1985. Conifers of India and their Wild Gene Resources in Relation to Tree Breeding. *Indian Forester* 111: 935−955.
149. Gansser, A. 1983. *Geology of the Bhutan Himalaya.* Basel, Boston, Stuttgart.
150. Ghosh, R.B. 1982. Taxonomic and botanical glimpses on the Primulas of Sikkim — A census. *J. Econ. Tax. Bot.* 3: 923−930.
151. Gopal, B. & Meher-Homji, V.M. 1983. Temperate broad-leaved evergreen forests of the Indian subcontinent. In Ovington, J.D. (ed), *Ecosystems of the World 10. Temperate Broad-leaved evergreen forests,* pp 125−134. Amsterdam, Oxford, New York.
152. Grey-Wilson, C. 1984. Alpine ecology in the Barun Khola, Nepal. *Kew Magazine* 1(1): 30−35.
153. Grierson, A.J.C. 1984. Notes relating to the Flora of Bhutan: X. *Notes RBG Edinburgh* 42(1): 107−111.
154. Grierson, A.J.C. & Long, D.G. 1984. *Flora of Bhutan* Vol 1. Part 2. Edinburgh.
155. Hansen, B. 1983. The Flora of Bhutan (Book Review). *Notes RBG Edinburgh* 41(1): 132.
156. Haridasan, K. & Rao, R.R. 1985. *Forest Flora of Meghalaya* Vol. 1. Dehra Dun.
157. Herklots, G.A.C. 1972. *Vegetables in South-east Asia.* London.
158. Kochhar, S.L. 1981. *Economic Botany in the Tropics.* Delhi.
159. Lachungpa, S.T. 1985. Sikkim's Alpine Flowers. *Himalayan Plant J.* 3: 6−10.

160. Long, D.G. 1983. Notes relating to the Flora of Bhutan: VI. A new species of Erycibe (Convolvulaceae). *Notes RBG Edinburgh* 41(1): 127–131.
161. Long, D.G. 1984. Notes relating to the Flora of Bhutan: VIII. Lauraceae. *Notes RBG Edinburgh* 41(3): 505–525.
162. Long, D.G. 1984. Notes relating to the Flora of Bhutan: IX. Corydalis (Fumariaceae). *Notes RBG Edinburgh* 42(1): 87–106.
163. Majumdar, N.C., Krishna, B. & Biswas, M.C. 1984. Vegetation of Neora Valley and adjacent regions in Kalimpong Forest Division, West Bengal. *J. Econ. Tax. Bot.* 5(5): 1013–1025.
164. Millais, E.G. 1984. Rhododendrons in West Bengal and Sikkim. *Rhododendrons* 1983–84: 12–17.
165. Nakao, S. & Nishioka, K. 1984. *Flowers of Bhutan*. Tokyo.
166. Nayar, M.P. 1984. *Key Works to the taxonomy of Flowering Plants of India*. Vol. 1. Botanical Survey of India, Calcutta.
166a. Polhill, R.M. & Raven, P.H. (eds) 1981. *Advances in legume systematics*. Vols 1 & 2. London.
167. Polunin, O. & Stainton, A. 1984. *Flowers of the Himalaya*. Delhi.
168. Pradhan, U.C. 1985. Lloyd Botanic Garden - Darjeeling. *Himalayan Plant J.* 3: 38–41.
169. Pradhan, U.C. & Rai, B.M. 1983–5. Summer Flora of the Singalila Mountains. *Himalayan Plant J.* 1: 52–54; 2: 10–11, 20–22; 3: 24–25.
170. Robson, N.K.B. & Long, D.G. 1983. Notes relating to the Flora of Bhutan: VII. Notes on Hypericum L. *Notes RBG Edinburgh* 41(1): 133–139.
171. Safui, B., Chandra, S. & Bhattacharya, A. 1985. Some additions to the Flora of Jalpaiguri district, West Bengal. *J. Econ. Tax. Bot.* 7: 1–4.
172. Sargent, C., Sargent, O. & Parsell, R. 1985. The forests of Bhutan: a vital resource for the Himalayas. *J. Trop. Ecology* 1: 265–286.
173. Sargent, C., Sargent, O., Parsell, R., Clark, J. & Dorji, T. 1985. *The Forest Resource of Bhutan*. I.T.E., Abbots Ripton.
174. Sikdar, J.K. & Samanta, D.N. 1984. Herbaceous flora (excluding Cyperaceae, Poaceae and Pteridophytes) of Jalpaiguri district, West Bengal — A check list. *J. Econ. Tax. Bot.* 4(2): 525–538.
175. Sikdar, J.K. 1984. Contributions to the flora of Baikunthapur Forest Division, Jalpaiguri District (West Bengal). *J. Econ. Tax. Bot.* 5(3): 505–532.
176. Singh, U., Wadhwani, A.M. & Johri, B.M. 1983. *Dictionary of Economic Plants in India*. Ed. 2. Delhi.
177. Yonzone, G.S., Yonzone, D.K.N. & Tamang, K.K. 1984. Medicinal plants of Darjeeling district. *J. Econ. Tax. Bot.* 5(3): 605–616.

Family 65. HAMAMELIDACEAE

by D.G. Long

Evergreen or deciduous trees or shrubs, with simple or stellate hairs. Leaves alternate, simple, pinnately or palmately veined, stipulate. Flowers free or connate in heads, spikes or racemes, unisexual (plants monoecious) or bisexual, actinomorphic. Calyx with tube adnate to ovary, lobes 5 or absent. Petals 4 − 6, free, or absent. Stamens 5 or 10−14 or many, free. Lobed disc often present. Ovary inferior or semi-inferior, 2-celled; styles 2, free, spreading; ovules 1 or 6−8 or many per cell, axile. Fruit a woody capsule; seeds 2−many, winged or not.

1. Evergreen trees with entire leaves (sometimes lobed when juvenile); venation palmate at base, not close and parallel above; stipules coriaceous, obovate-oblong, obtuse .... **2. Exbucklandia**
+ Evergreen or deciduous trees or shrubs with serrate leaves; venation pinnate throughout or palmate at base but close and parallel above; stipules membranous, elliptic, acute or subulate 2

2. Deciduous shrub or small tree; venation palmate at base, close and parallel above; flowers bisexual in catkin-like spikes .... **1. Corylopsis**
+ Tall evergreen tree; venation pinnate, distant; flowers unisexual in solitary or racemosely arranged heads .... **3. Altingia**

1. CORYLOPSIS Siebold & Zuccarini

Deciduous shrubs or small trees with stellate hairs. Leaves palmately veined at base, with close, parallel veins above, serrate; stipules

FIG. 34. **Hamamelidaceae** and **Crassulaceae**. **Hamamelidaceae**. a−d, *Altingia excelsa:* a, shoot with male and female heads; b, male head in section; c, female head; d, fruiting head. e−i, *Exbucklandia populnea:* e, shoot with female heads; f, male head; g, male flower in section; h, female flower; i, fruiting head. j−p, *Corylopsis himalayana:* j, inflorescence and leaf; k, flower; l, floral bract; m, stamens and nectaries; n, ovary; o, fruiting inflorescence; p, capsule. **Crassulaceae**. q−t, *Sedum multicaule:* q, habit; r, flower; s, petal and antipetalous stamen; t, dehisced capsule. u−x, *Rhodiola bupleuroides:* u, flowering stem with portion of rhizome; v, male flower from above; w, female flower in section; x, capsule. y−ab, *Kalanchoe integra:* y, portions of shoot and inflorescence; z, corolla dissected; aa, calyx and carpels; ab, fruit. Scale: y × ⅓; e, j, q, u × ½; a × ⅔; f, i × ¾; d, o × 1; g, 1, z, aa × 1½; h, ab × 2; b, k × 2½; c, t × 3; m, r, x × 4; s, v, w × 5; n × 6.

65. HAMAMELIDACEAE

membranous, caducous. Flowers precocious in bracteate catkin-like spikes, bisexual, not connate. Calyx lobes 5, short. Petals 5. Stamens 5, alternating with 5 nectaries. Cells of ovary with 1 ovule. Capsules free, subglobose, woody, 2-seeded, surrounded by persistent calyx. Seeds all fertile, not winged.

1. C. himalayana Griff. Sha: *Grong Grongmo Shing.* Fig. 34 j–p

Shrub or small tree to 7m, shoots stellate-tomentose. Leaves ovate, 7–10×5–7cm, shortly acuminate, base cordate, finely serrate (with excurrent vein-ends), palmately 7–9-veined at base, with 7–10 pairs of close, parallel lateral veins above, veins prominent and tomentose beneath; petioles 1.5–2cm; stipules elliptic 1.5–2cm, acute, cordate. Spikes 2.5–5cm, fragrant. Lower bracts (bud scales) ovate, 1.4–1.7cm, glabrous; upper (floral) bracts ovate, 5–6mm, tomentose. Calyx lobes triangular, 0.8mm. Petals yellow, obovate, 7–9mm. Capsules 5–6mm diameter.

Bhutan: S—Chukka and Deothang districts, **C**—Ha, Thimphu (cultivated), Tongsa and Tashigang districts. Amongst shrubs on open hillsides, and at margins of Cool broad-leaved forests, 1800–2560m. February–April.

Sometimes cultivated as an ornamental shrub.

2. EXBUCKLANDIA R. W. Brown

Large evergreen trees, glabrous or with few simple hairs. Leaves palmately 5–7-veined at base, entire (or 3–5-lobed in juvenile plants); stipules coriaceous, solitary or paired and coherent, caducous. Flowers coherent in globose, pedunculate, axillary heads; heads with only female or with bisexual flowers. Bisexual flowers: calyx absent; petals 4–6, linear; stamens 10–14, borne on lobed disc; ovary with 6–8 ovules in each cell. Female flowers: calyx with 5 truncate rim-like lobes, petals and stamens absent. Capsules in globose woody heads, each 2-celled, 4-valved, 4–8 seeds in each cell, upper 4–5 seeds sterile, unwinged, lower 1–2 seeds fertile with oblong wing.

1. E. populnea (Griff.) R. W. Brown; *Bucklandia populnea* Griffith, *Symingtonia populnea* (Griff.) van Steenis. Dz: *Chenju Shing;* Sha: *Lem Shing;* Nep: *Pipli* Fig. 34 e–i

Tree 15–33m. Leaves coriaceous, broadly ovate, 10–27×7–20cm, acuminate, base truncate or shallowly cordate, usually glabrous; petioles 2–8cm; stipules obliquely obovate-oblong, 2–4×1–2cm, obtuse. Flower heads 2–4 per axil, 5–12-flowered, on peduncles 1–2.5cm, females 8–10mm diameter, bisexual heads 15–20mm diameter, stamens creamy-white. Fruiting heads 1.7–2.2cm diameter; fertile seeds with wing c 9×2mm.

Bhutan: S—Phuntsholing, Chukka, Gaylegphug and Deothang districts, **C**—Punakha and Mongar districts, **N**—Upper Mo Chu district; **Sikkim.** Warm broad-leaved and Evergreen oak forests, 1260–2200m. November–March.

An ornamental tree with valuable timber used in building construction (34). Regenerates freely on disturbed soil of road-cuttings and land-slips, and useful in stabilizing hillsides against erosion.

3. ALTINGIA Noronha

Large evergreen trees; hairs stellate and simple. Leaves pinnately veined throughout, serrate; stipules minute, caducous. Flowers coherent in axillary, pedunculate heads, unisexual (plants monoecious), male heads arranged racemosely, females solitary, below males. Male flowers: sepals, petals and ovary absent, stamens numerous. Female flowers connate by calyx tubes; calyx lobes, petals and stamens absent; disc minutely lobed; ovary with many ovules in each cell. Fruiting heads globose, woody; capsules 2-celled, 4-valved, with many seeds in each cell, upper seeds sterile, wingless, lower 1 or few fertile, with narrow marginal wing.

1. A. excelsa Noronha. Nep: *Seti.* Fig. 34 a–d

Tall tree 20–40(–50)m. Leaves ovate-lanceolate, 8–17×4–5.5cm, acuminate, base rounded or broadly cuneate at base, margins glandular-serrulate, glabrous but with tufts of hair in vein-axils beneath; petioles 2–3.5cm, glandular near apex; stipules subulate c 4.5mm, early caducous. Male racemes 2–3.5cm, yellow, many-flowered. Female heads globose, 5–8mm diameter, 10–14-flowered, pinkish. Fruiting heads 1.8–2.2cm diameter; fertile seeds 5–6mm.

Bhutan: S—Sarbhang and Deothang districts, **C**—Tongsa, Mongar and Tashigang districts. Warm broad-leaved forests, 890–1830m. March–April.

A very valuable timber species often attaining a huge size; unknown in W Bhutan and Sikkim. Bark contains a resin *'Burmese Storax'* used as perfume and incense (126).

Family 66. CRASSULACEAE

by A. J. C. Grierson

Succulent herbs, sometimes woody at base. Leaves alternate, opposite or subverticillate, sometimes forming a basal rosette, usually simple, sometimes compound, pinnately veined, exstipulate. Flowers in cymes, corymbs

66. CRASSULACEAE

or panicles, actinomorphic, (3−4)5(−6)-merous, bisexual or unisexual. Sepals shortly united at base into a tube. Petals free or ± connate, sometimes forming a tubular corolla. Stamens as many or twice as many as petals, free or united with them. Carpels superior or semi-inferior, as numerous as petals, free or united at base, 1-celled. Nectar glands scale-like, alternating with carpels and inserted at their base. Fruit a cluster of follicles often surrounded by persistent perianth parts. Seeds few or numerous.

1. Large herbs 30−200cm; petals deeply connate into a tubular corolla .... **1. Kalanchoe**
+ Small to medium sized herbs rarely up to 100cm; petals free or connate only at base 2

2. Flowering stems arising from a terminally produced rosette of leaves 1−2cm; stamens as numerous as petals .... **2. Sinocrassula**
+ Flowering stems not arising from a terminally produced rosette of leaves; rosettes, if present, laterally produced, with leaves 5mm or less; stamens twice as many as petals (except in *Sedum correptum, S. perpusillum* and *S. fischeri*) 3

3. Slender herbs with opposite leaves; flowers borne in axils of most leaves .... **3. Crassula**
+ Usually more robust herbs with alternate or opposite leaves but if slender then leaves alternate; flowers in terminal cymes or corymbs or in axils of upper leaves only 4

4. Plants with stout creeping or suberect rhizomes 3−30mm thick; flowering stems arising from axils of brownish scale leaves at rhizome apex .... **4. Rhodiola**
+ Plants with slender rhizomes not more than 2mm thick, without scale leaves, or plants without rhizomes .... **5. Sedum**

1. KALANCHOE Adanson

Erect or ascending perennials. Leaves simple, trifoliate or pinnate. Flowers in terminal corymbose or paniculate cymes. Calyx shortly or deeply divided into 4 lobes. Corolla tubular, 4-lobed, segments spreading or recurved. Stamens 8, borne in 2 series on corolla tube. Carpels free or slightly connate at base.

1. Flowers pendulous; calyx divided to above middle .... **1. K. pinnata**
+ Flowers erect or spreading; calyx divided almost to base ..... **2. K. integra**

1. KALANCHOE

1. K. pinnata (Lamarck) Persoon; *Bryophyllum calycinum* Salisbury, *B. pinnatum* (Lamarck) Oken

Stems 0.3−2m, somewhat woody at base. Leaves simple or 3−5-foliate, thickly fleshy, leaves or leaflets ovate-oblong 5−10×2.5−5cm, obtuse, base rounded, margin crenately serrate, glaucous beneath; petioles 2−5cm. Panicles 10−30cm, flowers pendulous. Calyx 2.5−3cm, crimson streaked with green, lobes triangular 0.75−1cm. Corolla 3.5−4cm, constricted in lower third then bulbous at base, lobes reddish, 1−1.5cm, exserted from calyx mouth. Stamens inserted in lower quarter of corolla, anthers slightly exserted. Carpels 0.75−1cm, connate at base, styles 2.5−3.5cm.

Bhutan: S—Gaylegphug and Sarbhang districts, **C**—Tashigang district. Cultivated in gardens and sometimes becoming naturalised, 300−1050m. March−May.

Native of Tropical Africa.

2. K. integra (Medikus) Kuntze; *K. spathulata* DC., *Cotyledon spathulata* Poiret. Fig. 34 y−ab

Similar to *K. pinnata* but leaves always simple, ovate or spathulate, 10−15(−25)×3.5−6.5(−12)cm, ± shallowly crenate-serrate; petiole 2−5(−7)cm; inflorescence cymose, corymbose or paniculate, flowers ± erect; calyx c 0.75cm, lobes triangular, united at base; corolla yellow, tube c 1.5cm constricted at middle, expanded at base, lobes ovate 5−10×6mm, acuminate; stamens inserted at middle of corolla, anthers not exserted.

Bhutan: S—Deothang district (near Deothang), **C**—Tashigang district (Ghunkarah); **Sikkim.** Naturalised on hot dry hillsides, 600−1200m. October−November.

Native of Tropical Africa.

2. SINOCRASSULA Berger

Perennial herbs with creeping rhizomes, branches rooting and forming terminal rosettes of leaves. Flowering stems erect, leafy and surrounded at base by rosette of leaves (if fallen, leaf scars usually visible). Leaves alternate, simple, glabrous. Inflorescence corymbose or paniculate. Flowers bisexual, 5-merous. Sepals basally connate. Petals free, longer than sepals. Stamens 5, alternating with petals. Nectar scales oblong or squarish. Carpels basally connate, style short, stigma minute capitate.

1. S. indica (Decaisne) Berger; *Crassula indica* Decaisne, *Sedum indicum* (Decaisne) Hamet

Stems 10−25(−35)cm, glabrous. Leaves obovate 1−2.5×0.5−1cm, acute or obtuse, mucronate, base attenuate, margin entire. Calyx reddish, tube c 1.5mm, lobes triangular c 2.5×1mm, acuminate. Petals greenish

crimson, narrowly ovate, c 3×1mm, finely acuminate. Stamens c 2mm. Carpels 2.5–3mm, outwardly curved above.
Bhutan: C—Thimphu, Punakha, Tongsa and Tashigang districts; **Sikkim.** Rock faces and mossy tree trunks, 1450–2700m. August.

3. CRASSULA L.

Small slender annuals with thread-like stems. Leaves simple, opposite, connate at base. Flowers axillary, solitary or cymose. Sepals and petals 5, both ± free at base. Stamens 5. Carpels 5, free.

1. C. schimperi Fischer & Meyer; *C. pentandra* (Edgeworth) Schoenland, *Tillaea pentandra* Edgeworth

Stems 2–7cm, glabrous. Leaves linear-lanceolate 3–6×0.5–1mm, acuminate, base narrowed. Sepals lanceolate 1–1.5mm, acuminate. Petals c 1mm, ± transparent. Stamens c 0.5mm. Follicles c 0.75mm, each 2-seeded.

Bhutan: C—Thimphu district (Dotena). Under overhanging cliff, 2550m. July–September.

4. RHODIOLA L.

Perennial herbs with slender or thick suberect rhizomes bearing scales at apex, from the axils of which rise the simple annual stems bearing leaves and flowers. Radical leaves sometimes present; cauline leaves usually alternate, simple, usually sessile, herbaceous or fleshy, often glaucous, rarely leaves subverticillate. Inflorescence terminal, flowers several or numerous in dense or spreading cymes, sometimes solitary. Flowers bisexual or unisexual (then plants dioecious), (3–)4–5(–6)-merous. Calyx shortly tubular at base, lobes ± erect. Petals free, borne on margin of calyx tube, white, red, yellow or greenish. Stamens twice as many as petals, those opposite petals epipetalous; stamens or staminodes absent in female flowers. Carpels superior or semi-inferior, ± as long as petals, ± connate at base, pistillodes present in male flowers, shorter than petals. Nectar scales oblong or squarish borne at base between carpels.

1. Leaves verticillate, ovate-elliptic to suborbicular .... **Species 1 & 2**
+ Leaves alternate (sometimes crowded in upper part of stem), variously shaped 2

2. Leaves deeply toothed or lobed often half-way to midrib
   3. R. chrysanthemifolia
+ Leaves entire or shallowly toothed 3

3. Leaves obovate to suborbicular with thick pale coloured margins, crowded in upper part of stem .... **4. R. marginata**
+ Leaves variously shaped, rarely suborbicular in *R. bupleuroides*, without thick pale coloured margins, ± evenly distributed along stem 4

4. Rhizomes stout mostly 1.5−3cm thick, radical leaves always absent; stems usually stout, 10cm or more tall (sometimes shorter and with slender rhizomes in *R. bupleuroides* but then flowers dark reddish brown, and in *R. coccinea* but then rhizomes densely surrounded by dead stems of previous years and flowers usually solitary) 5
+ Rhizomes mostly slender 0.5−1(−1.5)cm thick; stems mostly slender usually less than 10cm tall (sometimes up to 25cm and with stouter rhizomes in *R. hobsonii* but then radical leaves usually present) 10

5. Dead flowering stems of previous years persisting on rhizome 6
+ Dead flowering stems deciduous 8

6. Flowering stems slender 2−5cm, usually 1-flowered, those of previous years densely surrounding rhizome, growth cushion-like **5. R. coccinea**
+ Flower stems stout 10cm or more tall; flowers several or numerous, growth not cushion-like 7

7. Inflorescence of subumbellately arranged cymes; stems glandular pubescent .... **6. R. himalensis**
+ Inflorescence of compact corymbose cymes; stems glabrous
Species 7 & 8

8. Flowers bisexual; leaves linear, toothed in upper half .. **9. R. wallichiana**
+ Flowers unisexual; leaves lanceolate, oblanceolate, elliptic to suborbicular, if ± linear then entire 9

9. Petals purplish-red .... **10. R. bupleuroides**
+ Petals greenish-yellow .... **11. R. sherriffii**

10. Flowers bisexual; rhizomes bearing stem and leaf remains at apex 11
+ Flowers unisexual; rhizomes scaly at apex but usually without stem and leaf remains, stems rarely persisting in *R. cretinii* 13

11. Radical leaves 2−5×0.5−1cm usually present; anthers dorsifixed
12. R. hobsonii
+ Radical leaves smaller or absent; anthers basifixed 12

12. Stems prostrate or decumbent .... ± .... **13. R. humile**
+ Stems ± erect .... **Species 14 & 15**

66. CRASSULACEAE

13. Flowers solitary .... **16. R. ludlowii**
\+   Flowers several or numerous in corymbose cymes .... **Species 17 & 18**

1. R. stapfii (Hamet) Fu; *Sedum stapfii* Hamet

Rhizomes c 8mm thick, erect. Flowering stem solitary, erect 1.0−2.5cm, glabrous, deciduous. Leaves usually 4−6, verticillate at or near top of stem, ovate-elliptic 8−15×4−8mm, obtuse or acute, base rounded; petiole 2−3.5mm. Cymes usually 3−6-flowered, bracteate, pedicels 0.5−1.5cm. Flowers unisexual, usually 5-merous. Calyx tube 1−1.5mm, lobes triangular 2.5−3.5×1−2mm. Petals oblong-ovate, 2−3×1−1.75mm, entire or crenulate near apex. Stamens shorter than petals. Carpels 4.5−5mm, longer than petals, outwardly curved above.

Bhutan: C—Thimphu district (Tremo La to Sharna), **N**—Upper Mo Chu district (Chhew La); **Sikkim; Chumbi.** 4265m. August−September.

2. R. prainii (Hamet) Ohba; *Sedum prainii* Hamet

Similar to *R. stapfii* but larger, rhizomes 1−2cm thick; stems up to 11cm, deciduous; leaves usually 4, verticillate in lower half of stem, broadly ovate to suborbicular, 1−5×0.75−5.5cm, obtuse or subacute, base rounded, truncate or shallowly cordate, margin minutely papillate, petiole 0.75−4cm; inflorescence corymbose, 12−18-flowered, bracteate; flowers bisexual; petals oblong or ovate, 4−6×1.5−3mm, minutely serrulate at margin, white or pale pink; carpels 3.5−5mm, shorter than petals, styles c 0.5mm, ± straight.

Sikkim: locality unknown. Boulders and rock crevices. September.

3. R. chrysanthemifolia (Léveille) Fu; *Sedum chrysanthemifolium* Léveille, *S. trifidum* Hook. f. & Thomson p.p., *S. linearifolium* Royle var. *ovatisepalum* Hamet, *S. ovatisepalum* (Hamet) Ohba

Rhizomes 5−10mm thick, often creeping. Flowering stems 8−30cm, deciduous. Leaves clustered near stem apex, oblanceolate or obovate in outline, 2−8×0.5−2cm, obtuse, base attenuate, subpetiolate, pinnatifid with 1−7 pairs of obtuse teeth or lobes up to 8mm, glabrous. Inflorescence a ± densely corymbose cyme, 10−30(−40)-flowered. Flowers bisexual, usually 5-merous. Calyx tube 0.5−1mm, lobes triangular 1.5−3.5× 1−1.5mm, acuminate. Petals linear-lanceolate 6−7.5×1mm, white, pink or crimson. Stamens ± as long as petals. Carpels 5−7mm, style 1−2mm, slightly curved outwards.

Bhutan: C—Ha, Thimphu, Punakha and Bumthang districts, **N**—Upper Mo Chu district; **Sikkim.** Mossy rocks and hillsides, 3000−3800m. July−September.

4. R. marginata Grierson

Rhizomes erect, 5−15mm diameter. Stems erect 0.5−2.5cm, densely papillate, naked below but with 5−8 leaves aggregated near apex. Leaves obovate, suborbicular or flabellate, 3−7mm long and broad, rounded or subacute, base abruptly attenuate, margin thick and pale coloured 0.4−0.5mm broad, glabrous above, papillate beneath; petioles up to 7mm. Flowers 4-merous, unisexual, 3−20 subsessile in axils of leaf-like bracts. Calyx lobes elliptic c 2.5×1mm, obtuse, pale margined. Petals reddish, ovate, slightly shorter than calyx lobes. Stamens c 1mm, epipetalous ones subulate c 0.5mm. Female flowers unknown.

Bhutan: N—Upper Mo Chu district (Laya) and Upper Bumthang Chu district (Pangotang). Rock crevices, 3800−4480m. June−September.

5. R. coccinea (Royle) Borissova; *?R. asiatica* D. Don, *Sedum coccineum* Royle, *?S. asiaticum* (D. Don) DC., *S. quadrifidum* sensu F.B.I. p.p. non Pallas

Dwarf alpine cushion plant. Rhizomes 1−3cm thick, branching, densely surrounded by dark brown persistent flowering stems from previous years. Stems 1−5cm, straw coloured, brownish or purplish, smooth or papillate-scabrid. Leaves linear-elliptic 3−7×0.5−1mm, obtuse. Flowers 1(−3), usually 4-merous, unisexual. Calyx tube 0.5−1mm, lobes linear-ovate 1.5−2.5×0.5−1mm. Petals pink, white or yellow, narrowly oblong-ovate, 2.5−3.5×0.5−1.5mm. Stamens usually slightly longer than petals. Carpels 4−6mm, style indistinct.

Bhutan: C—Thimphu district, **N**—Upper Pho Chu, Upper Bumthang Chu and Upper Kulong Chu districts; **Sikkim.** On peaty soil and mossy boulders, 3950−5000m. June−August.

The correct name for this plant may be *R. asiatica* D. Don but this requires lectotypification.

Two subspecies are recognised: the commoner subsp. **coccinea** which has smooth straw-coloured or brownish flowering stems and subsp. **scabrida** (Franchet) Ohba in which the stems are purplish and papillate scabrid. The latter is known from Sikkim (Chakalung La).

Plants intermediate between *R. coccinea* and *R. nobilis* (Franchet) Fu, a species described from W China, are known from Upper Bumthang Chu district (Dhur Chu) and Upper Kulong Chu district (Me La). They have more elongated rhizomes with less dense persistent flowering stems, broader (2−4mm) leaves and cymes of about 5 flowers.

6. R. himalensis (D. Don) Fu; *Sedum himalense* D. Don, *S. quadrifidum* Pallas var. *himalense* (D. Don) Froderstrom

Rhizomes elongate, suberect, 2−3cm thick. Flowering stems 20−30cm tall, 2−3mm broad, persistent, minutely glandular pubescent. Leaves oblanceolate or obovate, 5−20×2.5−5mm, acute, base attenuate, margin subentire or denticulate, papillate at least near margin on upper surface. Inflorescences more dense in male than female plants, cymes subumbellately arranged in loose corymbs. Flowers usually 5-merous, unisexual. Calyx tube 2−3.5mm, lobes triangular ± as long. Petals dark reddish purple, oblong-ovate 2.5−4.5×1.5mm. Stamens slightly shorter than petals. Carpels 4.5−5.5mm, styles c 0.5mm outwardly curved.

Bhutan: C—Thimphu and Bumthang districts, **N**—Upper Mo Chu, Upper Bumthang Chu and Upper Kulong Chu districts; **Sikkim**. In screes and dwarf Rhododendron scrub, 3800−4265m. June−August.

7. R. fastigiata (Hook. f. & Thomson) Fu; *Sedum fastigiatum* Hook. f. & Thomson, *S. quadrifidum* sensu F.B.I. p.p. non Pallas, *S. venustum* Praeger p.p.

Similar to *R. himalensis* but smaller, stems 6−15cm, glabrous; leaves linear-elliptic, 8−12×1−1.5mm, subobtuse, base attenuate, margin entire; flowers 6−15 forming compact corymbose cymes; calyx tube c 0.5mm, lobes ovate-triangular 2.25−3×0.5−1.5mm; petals yellowish white, narrowly linear-obovate 3.5−6×1.25−1.75mm; stamens slightly longer than petals; carpels 6−8mm; style c 1mm, slightly curved outwards.

Bhutan: C—Bumthang district (Dhur Chu), **N**—Upper Mo Chu district; **Sikkim:** Kankola, Lama Kangra and Dzalep La. 4265−4570m. June−August.

8. R. crenulata (Hook. f. & Thomson) Ohba; *Sedum crenulatum* Hook. f. & Thomson, *S. rotundatum* Hemsley

Similar to *R. himalensis* but flowering stems up to 20cm, 7−9mm thick, older ones turning reddish brown, glabrous; leaves ± densely arranged in upper parts of stem, broadly elliptic 1−3×0.5−1.5cm, obtuse or subacute, base cuneate, ± sessile, subentire to crenate; flowers 20−40 forming compact corymbs surrounded by upper leaves; calyx tube 1.5−2.5mm, lobes narrowly ovate 3−5×0.5−1mm; petals dark violet, linear elliptic 5−6×1mm; stamens ± as long as petals; carpels 7−10mm, style 2−3mm, straight.

Bhutan: C—Bumthang district (Dhur Chu), **N**—Upper Mo Chu (Shingche La) and Upper Bumthang Chu (Tolegang) districts; **Sikkim**. On cliffs and screes, 4570−4880m. May−July.

R. imbricata Edgeworth *(Sedum rhodiola* sensu F.B.I. non DC.) has been doubtfully recorded from Sikkim. This West Himalayan species resembles *R. crenulata* but has pale coloured persistent stems, more narrowly elliptic leaves and yellow petals.

9. R. wallichiana (Hooker) Fu; *Sedum wallichianum* Hooker, *S. crassipes* Hook. f. & Thomson, *S. asiaticum* sensu F.B.I. p.p. non (D. Don) DC.

Rhizomes large, 2−3cm thick. Stems 15−30cm, glabrous, deciduous. Leaves linear 10−30×1−3mm, acuminate, base attenuate, margin with 2−3 pairs of small teeth in upper half. Flowers in dense compact corymbs, bisexual, usually 5-merous. Calyx tube c 1mm, lobes triangular 5−8× 1.5mm, acuminate. Petals greenish-yellow, often flushed pink, elliptic, 7−10×1.5−2.5mm. Stamens ± as long as petals. Carpels 9−14mm, styles 2.5−4mm slightly spreading.

Sikkim: Jongri, Ningbil, Changu, Yak La etc.; **Chumbi.** Moss covered boulders, 3650−3950m. July−September.

10. R. bupleuroides (Hook. f. & Thomson) Fu; *Sedum bupleuroides* Hook. f. & Thomson, *S. elongatum* Hook. f. & Thomson non Ledebour, *S. discolor* Franchet, *S. hookeri* Balakrishnan, *S. cooperi* Praeger non Clemenceau, *S. bhutanense* Praeger, *S. bhutanicum* Praeger, *S. phariense* Ohba, *R. purpureoviridis* (Praeger) Fu subsp. *phariensis* (Ohba) Ohba. Fig. 34 u−x

Rhizomes elongate or stout, (0.5−)1−3cm thick, scale leaves persistent. Stems 5−45(−100)cm, leafy throughout, glabrous, deciduous. Leaves usually elliptic but sometimes obovate or suborbicular (0.5−)1−4(−9)× 0.5−2(−4)cm, obtuse or acute, base attenuate, rounded or cordate, margin entire or remotely serrate. Flowers 7−many forming a ± loose corymb, unisexual, usually 5-merous. Calyx purplish, tube 1−2mm, lobes lanceolate 1−6×0.5−1mm. Petals dark purplish-red, oblong or ovate 1.5−4× 0.5−1.5mm usually somewhat larger (3−4mm) in male than female flowers (1.5−3mm). Stamens ± as long as petals. Carpels 3.5−9mm, longer than petals, style very short, outwardly curved.

Bhutan: C—Thimphu district, N—Upper Mo Chu, Upper Pho Chu, Upper Bumthang Chu, Upper Kuru Chu and Upper Kulong Chu districts; **Sikkim.** Among rocks and screes, 3800−4570m. June−August.

11. R. sherriffii Ohba

Similar to *R. bupleuroides* but leaves always oblanceolate, 2−4×0.2−0.75cm, acuminate, base attenuate, margin entire, ± inrolled, dark green above, pale beneath, papillate especially at margins above; inflorescence a compact corymbose cyme 2−4cm diameter, pedicels densely papillate; calyx tube 0.5−1mm, lobes oblong 3−4.5×1mm; petals greenish-yellow or reddish, oblanceolate, 4−6×1−1.5mm; stamens longer than petals, anthers red; carpels 6−12mm, almost free at base, styles c 0.5mm.

Bhutan: C—Tongsa district (Black Mountain), N—Upper Mo Chu, Upper Pho Chu, Upper Bumthang Chu and Upper Kulong Chu districts; **Sikkim.** On peat and mossy boulders, 3650−4265m. May−July.

12. R. hobsonii (Hamet) Fu; *Sedum hobsonii* Hamet, *S. praegerianum* W. W. Smith, *S. mirabile* Ohba

Rhizomes 5−15(−20)mm thick. Flowering stems slender, 5−15cm, ± erect, usually persistent. Radical leaves commonly present, elliptic or spathulate, 1−2×0.5−1cm, acute, base narrowing into petiole 1−4cm, margin entire. Cauline leaves elliptic to obovate 8−12×2.5−8mm, acute, base attenuate, subpetiolate. Flowers bisexual, usually 5-merous, few (3−12) forming arching cymes. Calyx tube c 2mm, lobes triangular c 4×1.5mm, acuminate. Petals pink, elliptic c 7×2.75mm, acute or acuminate. Stamens usually shorter than petals, anthers dorsifixed, connective bearing a minute hair-like appendage c 0.2mm. Carpels 5−7mm, styles 1.5−2mm, straight.

Bhutan: C—Ha, Thimphu and Bumthang districts, N—Upper Mo Chu and Upper Bumthang Chu districts; **Sikkim.** Rock crevices and mossy boulders, 2550−3950m. July−September.

13. R. humilis (Hook. f. & Thomson) Fu; *Sedum humile* Hook. f. & Thomson, *S. levii* Hamet, *S. barnesianum* Praeger

Rhizomes erect, 5−15mm thick, densely covered above by remains of stems and petioles from radical leaves of previous years. Flowering stems prostrate or decumbent, 2.5−6cm, deciduous. Radical leaves narrowly elliptic 3−8×1−2mm, subacute, base attenuate into petiole c 1cm. Cauline leaves elliptic or oblanceolate, 5−9×1−2mm, acute, base shortly attenuate, entire. Flowers 1−4 often solitary, bisexual. Calyx reddish, tube 0.5−1.5mm, lobes ovate 2.5−3.5×1.5−2mm, ± acuminate. Petals white at first becoming pink, ovate 5−6.5×1.5−2mm, subacute. Stamens shorter than petals. Carpels 4−7mm, styles c 2mm, spreading.

Bhutan: N—Upper Mo Chu district (Laya), Upper Pho Chu district (Gafoo La) and Upper Kulong Chu district (Me La); **Sikkim; Chumbi.** Gravelly hill slopes and on boulders, 3950−4880m. July−September.

14. R. smithii (Hamet) Fu; *Sedum smithii* Hamet

Similar to *R. humilis* but rhizomes without foliar remains; flowering stems erect, 2−3(−7)cm, deciduous; leaves elliptic or oblanceolate, 5−12×2−3mm, obtuse, base attenuate; flowers 3−10 forming small rounded heads; calyx reddish, tube c 1mm, lobes linear-ovate 2.5−4×1−1.5mm; petals pink, linear-lanceolate 3.5−6mm, ± obtuse; carpels 4−6mm, style 1−1.5mm.

Sikkim: Lhonak; **Chumbi.** Rock faces, 4570m. July−September.

15. R. amabilis (Ohba) Ohba; *Sedum amabile* Ohba

Similar to *R. humilis* but rhizomes without foliar remains; flowering stems erect or ascending 5−12cm, usually deciduous; leaves numerous, rather densely arranged throughout stem, oblong or linear-lanceolate, 6−10(−12)×1−1.5mm, obtuse or acute, base scarcely narrowed, minutely papillate near apex; inflorescence 1−5(−10)-flowered forming a compact cyme; calyx tinged red, tube c 0.5mm, lobes triangular 3−5×1.5mm, acuminate; petals white or red at apex, ovate-elliptic 5−7.5×1.5−2mm, acute or acuminate; stamens slightly shorter than petals; carpels 5−6mm, styles c 2mm outwardly curved.

Bhutan: C—Mongar and Tashigang district boundary (Donga La), **N**—Upper Kuru Chu and Kulong Chu district boundary (Panga La). On rocks on dry hillsides, 2740–3800m. August–September.

16. R. ludlowii Ohba

Rhizomes slender c 2.5mm thick. Stems erect or ascending up to 12cm, deciduous. Leaves rather densely clustered at apex of stem, oblanceolate, 10–15×2–5mm, rounded or subacute, base attenuate, subpetiolate, margin entire, glabrous. Flowers unisexual, solitary, 5-merous. Calyx lobes ovate-lanceolate c 4.5×1.5–2mm. Petals white or pink, broadly elliptic c 7× 3.5mm. Stamens c 4.5mm. Carpels 5–6mm, style c 1mm.

Bhutan: C—Tongsa district (Rinchen Chu). On cliffs, 4570–4900m. July–August.

17. R. cretinii (Hamet) Ohba; *Sedum cretinii* Hamet

Rhizomes slender, 2–4mm thick, creeping and producing side branches. Flowering stems 4–7(–12)cm, erect, papillate especially above, usually deciduous. Leaves rather numerous. evenly distributed along stems, linear-elliptic 7–10×1.5–2.5mm, acute or acuminate, base narrowed, margin entire or with 3–5 small teeth on each side, glabrous. Inflorescence 3–20-flowered forming a dense corymb. Flowers unisexual, 4–5-merous. Calyx tube 0.5–1mm, lobes narrowly triangular 3–4.5×0.5–1mm. Petals greenish or creamy white often tinged with red, linear-elliptic 4–6×1mm. Stamens ± as long as petals. Carpels 6–7mm, styles c 1mm ± straight.

Bhutan: N—Upper Mo Chu district (Laya); **Sikkim.** In dry gravelly ravines, 4000–4880m. August–September.

18. R. atsaensis (Froderstrom) Ohba; *Sedum atsaense* Froderstrom

Similar to *R. cretinii* but rhizomes stouter, 5–10mm thick, less branched; flower stems up to 10cm, deciduous; leaves mostly clustered at stem apex, spathulate 10–25×2–5mm, rounded, base attenuate, margin entire; inflorescence 30–40-flowered, densely corymbose, surrounded by leaves and bracts; flowers 4-merous, unisexual; calyx tube 2–5mm, lobes spathulate, 2–9×1–3mm, rounded at apex; petals reddish-brown, oblong 2–7×0.5–1.25mm; stamens shorter than petals; carpels c 8mm, connate in lower half, ovules 2 in each.

Bhutan: C—Thimphu district (Tremo La); **Sikkim:** Chumighata. Screes, 4730–4880m. June–September.

5. SEDUM L.

The medicinal name *Tshenmar* appears to apply to several species of this genus.

66. CRASSULACEAE

Annual, biennial, perennial or monocarpic herbs, ± succulent; rootstocks usually thin, sometimes tuberous; stems erect or ascending. Radical leaves absent from flowering stems, cauline leaves alternate, opposite or whorled, simple, usually sessile, often with a short rounded appendage below point of insertion. Inflorescence loosely or densely cymose. Flowers bisexual. Sepals 4–6, sometimes with a short rounded appendage below point of insertion. Petals 4–6, usually somewhat longer than sepals, free or basally connate. Stamens 8–12, those opposite petals inserted on them. Carpels 4–6, basally connate or nearly free.

1. Leaves in upper part of stem opposite or in whorls of 3–4 2
+ Leaves alternate throughout 4

2. Leaves spathulate or oblanceolate, 1–2×0.2–0.7cm; flowers yellow
   1. S. triactina
+ Leaves ovate, obovate or suborbicular, 2–7×1–4cm; flowers white or pink 3

3. Stems 3–20cm; petals c 4mm, white .... **2. S. filipes**
+ Stems 15–30cm; petals c 7mm, pink .... **3. S. spectabile**

4. Annuals or small monocarpic plants; stems 2.5–4(–5)cm; leaves never papillate; petals 3–4mm .... **Species 4–6**
+ Perennials; stems (3–)5–12cm or more; petals 5–10mm (in *S. correptum* stems 2–4cm, petals 3–3.5mm but then leaves densely papillate) 5

5. Leaves ± appressed to stems, acuminate to a fine hair-like tip; flowers in dense cymose heads .... **Species 7 & 8**
+ Leaves ± spreading, obtuse or acuminate; flowers in loose cymes or racemes (rarely flowers in dense heads in *S. oreades* but then leaves obtuse or acute) 6

6. Flowers in long spreading spike-like cymes .... **9. S. multicaule**
+ Inflorescence not as above 7

7. Leaves densely papillate .... **10. S. correptum**
+ Leaves glabrous .... **Species 11 & 12**

1. S. triactina Berger; *S. verticillatum* (Hook. f. & Thomson) Hamet non L., *Triactina verticillata* Hook. f. & Thomson

Perennial. Stems weak, succulent, ascending 10–25cm, overwintering rosettes sometimes present on lower parts. Leaves usually opposite or in whorls of 3, sometimes also alternate below, oblanceolate or spathulate,

10−20×2−7mm, obtuse or emarginate, attenuate at base. Cymes loosely corymbose. Sepals 0.8−1×0.5mm, obtuse. Petals yellow, free, lanceolate 5−6×1mm, acuminate. Stamens 10. Carpels 3, ± erect at first, later divergent.

Bhutan: C—Thimphu, Tongsa and Bumthang districts, N—Upper Mo Chu and Upper Kulong Chu districts; **Sikkim.** On mossy rocks in forests, 3050−3350m. June−August.

2. S. filipes Hemsley; *S. pseudostapfii* Praeger

Biennial or monocarpic herb. Stems erect or ascending, 3−20cm, naked below, usually with a pair of opposite leaves near middle and with a whorl of 4 leaves near apex below inflorescence. Leaves ovate, obovate or suborbicular 2−4.5×1−4cm, obtuse, base rounded, narrowed to a broad petiole 0.75−4cm. Cymes densely corymbose. Sepals oblong c 1.5× 0.75mm, obtuse. Petals white, oblong-lanceolate c 4×1mm, acuminate. Stamens 10−12. Carpels erect at first, later spreading.

Bhutan: C—Punakha district (Kencho) and Mongar district (Unjar), N—Upper Mo Chu district (Gasa); **Sikkim:** Lachen, Cheuntong and Chateng. Among moss in forests, 2000−2400m. August.

3. S. spectabile Boreau; *Hylotelephium spectabile* (Boreau) Ohba

Stems 15−35cm. Leaves opposite or in whorls of 3, obovate 2.5−7×2−4cm, subacute, base cuneate, margin weakly crenate-toothed or subentire. Flowers in dense corymbs. Sepals lanceolate, 2−2.5×0.75mm. Petals elliptic c 7×1.75mm, pink. Stamens exserted 8−9mm.

Bhutan: cultivated e.g. in Thimphu, 2370m. September.
Native of Eastern China.

4. S. perpusillum Hook f. & Thomson

Stems usually simple or branched at base, ± erect, 2.5(−5)cm. Leaves oblanceolate or spathulate, 4−5×1−1.5mm, obtuse or subacute. Corymbs ± dense c 10-flowered. Sepals oblanceolate 3−4×0.75−1.5mm, ± acute, base appendaged. Petals lanceolate, ± as long as sepals, white, free to base. Stamens 5. Carpels connate only at base.

Sikkim: locality unknown. 650m.

5. S. fischeri Hamet

Similar to *S. perpusillum* but stems ascending, 2.5−5cm; cymes looser; sepals oblong, 2− 4×0.5−1.5mm, not appendaged at base; petals c 3−4.5mm, acuminate to an obtuse tip; stamens 5.

?Bhutan: ?Ha district; **Sikkim:** Tarkarpo and Kalaeree; **Chumbi.** 3350−4880m. August−September.

The record (111) of *S. przewalskii* Maximowicz from Sikkim probably refers to this species.

66. CRASSULACEAE

6. S. henrici-robertii Hamet
Similar to *S. perpusillum* but stems 1−3cm, branched below, ascending; leaves linear to subobovate 2.5−5mm; sepals c 3mm, acute, base appendaged; petals ± as long as sepals, somewhat obtuse, submucronate; stamens 10; carpels connate to above middle.
Chumbi: Phari and Takekung. July.

7. S. trullipetalum Hook. f. & Thomson
Stems decumbent 5−10cm. Leaves lanceolate, ± appressed to stem, 4−6×1−1.5mm, acuminate to a fine hair-like tip, margin entire. Flowers up to 20 in dense rounded heads. Sepals elliptic c 3×0.75mm, not appendaged at base. Petals yellow, ovate, 6−10×2mm, acute, tapering abruptly from middle into a narrow basal claw, free to base. Carpels connate to c 2mm from base.
Bhutan: C—Thimphu district (above Thimphu), N—Upper Bumthang Chu district (Lhabja, Kopub and Pangotang) and Upper Kulong Chu district (Me La); **Sikkim.** On mossy rocks, 3050−4000m. August−September.

8. S. gagei Hamet
Similar to *S. trullipetalum* but leaves minutely denticulate or ciliate-dentate at margin; sepals minutely denticulate; petals elliptic c 8−10×1.5−2mm, narrowly acuminate, free to base; carpels connate up to 2.5mm from base.
Bhutan: N—Upper Mo Chu district (Chhew La) and Upper Pho Chu district (Tranza); **Sikkim:** Giagong. On alpine rocks, 3950−5000m. August−September.

9. S. multicaule Lindley. Fig. 34 q−t
Perennial. Stems sprawling, usually branched below, 7−12cm. Leaves linear-lanceolate, 10−15×1−1.5mm, acuminate, margin entire. Cymes with spreading spike-like branches up to 5cm. Sepals oblong-lanceolate 5−6×1−1.5mm, acuminate, base unappendaged. Petals ± as long as sepals, acuminate, free to base. Stamens 10. Carpels connate 1.5−2mm from base, ± erect at first later spreading ± horizontally forming a star-shaped cluster 8−10mm across.
Bhutan: S—Chukka district, C—Ha to Mongar districts; **Sikkim.** On moist rocks. 1525−3200m. July−August.

10. S. correptum Froderstrom
Perennial herb. Stems 2−4cm erect, densely papillate. Leaves alternate, spathulate or narrowly obovate, 5−7×1−3mm, obtuse or subacute, base attenuate, margin entire, densely papillate on both surfaces. Peduncles slender bearing a few leaf-like bracts, flowers 1−5, ± racemose, pedicels

6–15mm. Calyx tubes c 0.25mm, lobes liner-lanceolate 1–1.5× 0.2–0.3mm. Petals greenish, oblanceolate 3–3.5×0.75–1mm. Stamens 5, alternating with petals, anthers red at first. Nectar glands club-shaped c 0.5mm. Carpels 2.5–3mm, styles c 0.5mm outwardly curved.

Bhutan: N—Upper Mo Chu district (Laya). Dry rocks and cliffs, 3500m. June.

11. S. griffithii Clarke; *S. pseudosubtile* Hara

Stems 3–12cm, fleshy. Leaves oblanceolate or spathulate, 10–30×1.5–2.5mm, lower ones acute or acuminate, upper ones especially among flowers obtuse. Cymes loosely corymbose. Sepals spathulate 3–4×0.5–0.75mm, obtuse, base with a rounded appendage. Petals lanceolate c 5×1.5mm, acuminate. Carpels ± erect in fruit.

Bhutan: S—Gaylegphug district (Gale Chu), **N**—Upper Mo Chu district (Khosa and Tamji) and Upper Kuru Chu district (Denchung); **Sikkim**. On rocks and steep banks by streams, 1200–2133m. April–May.

12. S. oreades (Decaisne) Hamet; *S. jaeschkei* Kurz

Similar to *S. griffithii* but stems less fleshy; leaves oblong-lanceolate c 6×1.5mm, obtuse or acute; flowers usually 2–5, ± racemose, sometimes crowded into a head; sepals lanceolate 5–6×1.5–2mm, acuminate, base with a short rounded appendage; petals yellow, obovate or oblanceolate, c 10×4mm, connate to 2.5mm from base; stamens 10 or sometimes 5 with 5 staminodes; carpels lanceolate 5–7mm, shortly connate at base.

Bhutan: C—Ha and Bumthang districts, **N**—Upper Mo Chu, Upper Bumthang Chu, Upper Kuru Chu and Upper Kulong Chu districts; **Sikkim**. Stony ground, screes and rock crevices on mountains, 3650–4300m. July–September.

Records of the Mexican *S. quevae* Hamet from Sikkim (112, 113) are presumably erroneous. *S. cavei* Hamet (113) has never been validly published and is of unknown identity.

Family 67. SAXIFRAGACEAE

by A. J. C. Grierson

Herbs. Basal leaves sometimes rosetted, stem leaves alternate, rarely opposite, simple, pinnate or ternate, exstipulate but sometimes with stipule-like sheathing petiole bases. Flowers actinomorphic, bisexual, or sometimes unisexual, solitary or in cymes, racemes or panicles. Calyx of 4–5 sepals, often united at base and adnate to ovary. Petals usually 5, free, sometimes absent. Stamens 5–10, free. Ovary superior or semi-inferior, 1–2-celled, styles 2, ± diverging. Fruit a capsule; seeds numerous on axile or parietal placentas.

67. SAXIFRAGACEAE

1. Leaves compound 2
+ Leaves simple 3

2. Leaves pinnate; flowers c 5mm across .... **1. Rodgersia**
+ Leaves usually ternately divided (leaflets sometimes pinnate); flowers 2–3mm across .... **2. Astilbe**

3. Petals absent (very rarely present in *Tiarella*) 4
+ Petals present, showy (rarely absent in *Saxifraga*) 5

4. Flowers in racemes; carpels unequal in size .... **3. Tiarella**
+ Flowers in cymes, sometimes ± crowded; carpels of equal size
   4. Chrysosplenium

5. Plants with thick creeping rhizomes; leaves medium-sized or large 4–25cm long .... **5. Bergenia**
+ Plants often tufted or caespitose, sometimes creeping at base but then with slender rhizomes; leaves small seldom attaining 4cm long
   6. Saxifraga

1. RODGERSIA Gray

Robust rhizomatous herbs. Leaves odd pinnate. Inflorescence paniculate, branches ± racemose. Calyx 5-lobed, usually accrescent. Petals absent. Stamens 10, exserted. Ovary semi-inferior.

1. R. nepalensis Cullen. Fig. 35 a&b

Stems c 1m, villous and glandular particularly above. Basal leaves up to 60cm, leaflets 7–11, oblong-elliptic, 12–21×4–7cm, acuminate, base rounded, margin serrate, glabrous above, villous with long brown hairs on veins beneath; stem leaves similar but smaller, the uppermost one often 3-foliate. Panicles 30–40cm, villous and glandular pubescent. Calyx tube obconical, 3–4mm, glandular, lobes triangular, 3.5–5×2.5mm, acute, greenish white. Stamens 5–8mm. Capsules 7–10mm, tapering above into the divergent styles.

Sikkim: near Toong. On scree, 3000m. June–September.

FIG. 35. **Saxifragaceae.** a & b, *Rodgersia nepalensis:* a, part of leaf and inflorescence; b, flowers with two calyx lobes bent forward. c–e, *Chrysosplenium griffithii:* c, habit; d, flower from above; e, flower in section. f–h, *Saxifraga andersonii:* f, habit; g, upper surface of leaf; h, flower with some calyx lobes, petals and stamens removed. i–l, *Saxifraga hemisphaerica:* i, habit; j, leaf; k, male flower with two calyx lobes and 4 stamens removed; l, female flower with calyx lobes removed; m, *Saxifraga moorcroftiana:* habit. n & o, *Saxifraga brunonis:* n, habit; o, leaf. p, *Saxifraga tangutica:* habit. q–s, *Tiarella polyphylla:* q, part of flowering stem; r, flower; s, fruiting carpels, t & u, *Astilbe rubra:* t, part of leaf and inflorescence; u, flower. Scale: a × ¼; t × ⅓; m, n × ½; c × ⅔; q × ¾; p × 1; f × 1⅓; b × 2½; i, o, s, u × 3; d, h, r × 4; e, g, j, 1 × 5; k × 6.

2. ASTILBE D. Don

Erect rhizomatous perennial herbs. Leaves alternate, bi-or triternate; petiole bases sheathing, stipule-like. Flowers small in elongate terminal panicles, branches narrowly racemose. Calyx shortly adnate to base of ovary, lobes (4−)5. Petals (4−)5 or absent. Stamens 5 or (8−)10. Carpels connate below, tapering above into short styles. Mature carpels spreading and dehiscing ventrally.

1. A. rivularis D. Don. Sha: *Tonsar Gugay;* Nep: *Buro Okhate, Buru Okhati*
Stems 0.6−2mm sparsely brown villous. Leaves up to 45cm long and broad; leaflets ovate or elliptic 4−12×2−7cm, acuminate, base rounded or cordate, margin doubly serrate, appressed brown hirsute on veins beneath; petioles bearing long (up to 7mm) brown hairs especially at axils of leaflets, sheathing bases up to 5mm broad. Peduncles pale brownish pubescent. Calyx divided almost to base into lanceolate teeth 1.5−2mm, white or reddish. Petals absent. Stamens 5, opposite sepals, c 3mm. Ovary semi-inferior. Capsules ovoid c 5mm. Seeds ellipsoid c 0.5mm, tapering at either end into a tail up to twice as long.

Bhutan: S—Deothang district (Tshilingor and Narfong), **C**—Bumthang district (near Bumthang) and Tashigang district (near Tashi Yangtsi); **Sikkim.** Hillsides and valley bottoms, 1850−2900m. July−October.

Foliage eaten by cows and goats, flowers by people (131).

2. A. rubra Hook. f. & Thomson. Fig. 35 t&u
Similar to *A. rivularis* but stems seldom more than 1m, densely brown hirsute especially above; leaflets ovate, 3−7×1.5−4cm, sparsely pilose along veins beneath; petioles less densely brown hirsute; peduncles brown tomentose; calyx c 1.5mm; petals linear 4−5×0.1−0.2mm, purplish; stamens usually 10, c 2−3mm.

Bhutan: C—Thimphu district (Dotena, Taba, etc) and Bumthang district (near Bumthang); **Chumbi.** Edges of forests and streamsides, 2200−2900m. June−July.

3. TIARELLA L.

Erect perennial herbs. Leaves simple, lobed; basal leaves with large sheathing petiole base. Racemes terminal, ebracteate. Calyx 5-lobed, adnate to ovary at base. Petals absent, rarely 5. Stamens 10. Ovary superior, carpels 2, unequal, each attenuate into a slender minutely capitate style; ovules numerous on 2 subbasal placentas. Ripe carpels very unequal, dehiscing ventrally.

1. T. polyphylla D. Don. Fig. 35 q−s

Stems 20 −35cm, finely glandular pubescent. Leaves broadly ovate, up to 6cm long and broad, obtuse, base cordate, margin shallowly crenate-serrate, ± coarsely hirsute on upper surface, finely glandular pubescent beneath; petioles of basal leaves up to 15cm, ± hirsute, basal sheath c 1×0.5cm, brown; stem leaves similar, petioles shortly fimbriate at base. Flowers 10−20, white or purplish. Calyx lobes ovate, c 3×1mm. Stamens and styles shortly exserted. Longer mature carpel c 10mm, shorter one c 6mm.

Bhutan: S—Chukka district, **C**—Thimphu, Punakha and Mongar districts, **N**—Upper Kulong Chu district; **Sikkim**. Shady streamsides in forests, 2000−3200m. April−July.

4. CHRYSOSPLENIUM L.

Fleshy or slender perennial herbs, often rhizomatous. Leaves opposite or alternate, simple, toothed or lobed, petiolate, exstipulate. Flowers terminal, solitary or several in bracteate cymes, bracts leaf-like. Calyx tube adnate to ovary, 4-lobed. Petals absent. Disc ± well developed, nectariferous. Stamens 8, inserted at margin of disc. Ovary semi-inferior. Capsules 2-lobed, flattened, valves spreading at maturity.

1. Leaves opposite .... **1. C. nepalense**
+ Leaves alternate 2

2. Stems, leaves and petioles pubescent with fine brownish hairs
   Species 2 & 3
+ Stems and leaves glabrous (or with a few scattered short thick hairs in *C. griffithii, C. forrestii* and *C. nudicaulis*) 3

3. Stems bearing elliptic scale leaves in lower half .... **4. C. carnosum**
+ Stems without scale leaves 4

4. Stems slender, thread-like, bearing leaves in lower half .... **5. C. tenellum**
+ Stems ± robust without leaves in lower half .... **Species 6−8**

1. C. nepalense D. Don

Glabrous rhizomatous herb. Stems 5−15cm, ± erect. Leaves opposite, broadly ovate or suborbicular, 2−12mm long and broad, obtuse, base truncate or abruptly narrowed to petiole up to 1cm, margins with 4−7 low crenate teeth on either side. Flowers greenish yellow in loose leafy cymes. Calyx lobes rounded 1−1.25mm. Filaments c 0.5mm. Capsules 3−4mm, lobes divergent.

Bhutan: S—Gaylegphug district (Chabley Khola), **C**—Ha, Thimphu and Tongsa districts, **N**—Upper Mo Chu and Upper Kulong Chu districts; **Sikkim**. Streamsides and damp hollows, 1900−2750m. April−June.

2. C. adoxoides (Griffith) Maximowicz; *C. lanuginosum* Hook. f. & Thomson

Lanate rhizomatous herb. Stems erect, reddish, 7−20cm, densely brown lanate especially when young. Basal leaves broadly ovate or oblong, 1.5−3.5×1.5−3cm, obtuse, base rounded or shallowly cordate, margin with 5−7 shallow crenate teeth on either side, lanate especially on upper surface; petioles 1−3cm, lanate; stem leaves few, smaller than basal leaves. Inflorescence diffusely branched, flowers greenish, c 4mm across. Calyx lobes 1.5−2mm. Filaments c 0.5mm.

Bhutan: C—Punakha, Tongsa and Tashigang districts; **Sikkim**. Mossy banks in forests, 2550−2750m. January−May.

3. C. singalilense Hara

Similar to *C. adoxoides* but smaller, stems 2−7cm, sparsely pubescent; basal leaves ovate or suborbicular, 2.5−5×3−6mm, truncate or rounded at base, margin with up to 3 crenate teeth on each side, sparsely pubescent; petioles 3−5mm; stems leaves similar but smaller than basal; cymes (1−)2−6-flowered.

Sikkim: near Phalut. On mossy rocks, 3500m. June.

4. C. carnosum Hook. f. & Thomson

Glabrous tufted herb without rhizomes. Stems 3−15cm, erect, lower half bearing pale-coloured elliptic scale leaves 3−4×2−3mm, usually rounded and entire, rarely with 3 blunt teeth at apex, upper leaves ovate or obovate 10−12×6−8mm, rounded, margin with 2−4 blunt teeth on each side, base attenuate, subpetiolate. Cymes usually compact, bracts yellowish. Flowers dark red. Sepals rounded 1.5−2mm.

Bhutan: N—Upper Bumthang Chu district (Marlung), Upper Kuru Chu district (Gong La) and Upper Kulong Chu district (Me La); **Sikkim**. Moist gravel and screes, 4400−4570m. July−September.

5. C. tenellum Hook. f. & Thomson

Slender glabrous herb, rhizomatous, often becoming matted. Stems thread-like 2−5cm. Lower leaves suborbicular 7−12mm across, cordate at base, margin with 2−3 rounded crenate teeth on each side; petioles up to 1cm; upper leaves smaller. Cymes 1−5-flowered. Flowers 3.5−5mm across, greenish yellow. Sepals obovate c 1.5×3mm.

Bhutan: C—Thimphu district (between Nala and Tzatogang) (71). Moist rocks in Fir forest, 3200m. May.

6. C. griffithii Hook. f. & Thomson. Fig. 35 c−e

Tufted herb without rhizomes, almost glabrous, sometimes with a few short thick hairs at leaf and inflorescence axils. Stems 7−25(−30)cm, leafless below. Basal leaves suborbicular, 7−15×7−30mm, base cordate, margin with 3−5 lobes up to 7mm deep and broad on either side, lobes rounded or emarginate above; petioles 4−10cm. Stem leaves similar, inserted above middle of stem. Cymes loose. Flowers dark green, 4−6mm across. Sepals rounded, c 2.5mm.

Bhutan: C—Thimphu district (near Naha), N—Upper Mo Chu district (between Chamsa and Yabu Thang) and Upper Mangde Chu district; **Sikkim.** Moist gravel, 3200−4400m. May−June.

7. C. forrestii Diels; *C. alternifolium* sensu F.B.I. non L.

Similar to *C. griffithii* but ± thickly rhizomatous, surrounded at base by remains of petioles; leaves up to 5cm across, margin with 5−13 shallow lobes 1−3mm deep on either side; cymes ± compact, flowers subsessile; bracts bright yellow; flowers yellow.

Bhutan: C—Ha, Thimphu, Tongsa, Bumthang and Tashigang districts, N—Upper Mo Chu and Upper Kulong Chu districts; **Sikkim.** Steep banks in forest, 3350−4250m. May−July.

8. C. nudicaule Bunge. Med: *Yakima*

Similar to *C. griffithii* but thickly rhizomatous; stems smaller 2−10(−17)cm; radical leaves 8−20(−45)mm across with 3−7 lobes on each side, sinus 0.5−1mm, very narrow; true stem leaves absent or 1, bract leaves obovate 5−10mm across with 3−5 teeth; cymes compact, 3−7-flowered; flowers greenish, 4−5mm across.

Bhutan: C—Thimphu district (Barshong), N—Upper Mo Chu district (Lingshi La) and Upper Kulong Chu district (Me La); **Sikkim.** Among wet stones, 4100−4570m. May−August.

The above description relates to var. **intermedium** Hara which differs from the typical var. *nudicaule* in often having one cauline leaf and smaller flowers.

5. BERGENIA Moench

Perennial herbs with thick rhizomes. Leaves rosetted, simple, gland-dotted beneath, petiolate, base of petiole with sheathing flap on either side. Inflorescence paniculate, branches cymose. Calyx cup-shaped, adnate to ovary, lobes 5. Petals 5. Stamens 10, alternate ones slightly unequal in length. Ovary semi-inferior, carpels 2, diverging and dehiscing ventrally at maturity.

67. SAXIFRAGACEAE

1. B. purpurascens (Hook. f & Thomson) Engler; *Saxifraga purpurascens* Hook. f. & Thomson

Flowering stems 7–30cm, reddish brown, glandular-pubescent. Leaves elliptic or ovate-elliptic, 7–25×5–17cm, rounded, base cuneate or rounded, margin entire or shallowly sinuate, glabrous or ciliate near base; petioles 1.5–10cm, basal 1–4cm sheathing. Flowers 1–8, nodding. Calyx 8–14mm, dull crimson, lobes oblong, rounded. Petals obovate, 15–25×7–9mm, tapering to a basal claw 2–3mm, bright pink. Stamens 9–14mm, ± as long as styles. Capsules ellipsoid, c 1cm, tapering above into styles ± as long, base surrounded by persistent calyx.

Bhutan: C—Ha and Bumthang districts, **N**—Upper Mo Chu, Upper Pho Chu, Upper Bumthang Chu and Upper Kulong Chu districts; **Sikkim**. On rock ledges and cliffs, 3800–4550m. May–July.

2. B. ciliata (Haworth) Sternberg; *Saxifraga ligulata* Wall. Nep: *Pakhanbet*

Similar to *B. purpurascens* but leaves suborbicular or broadly obovate, 4–15×4–14cm, rounded at base and apex, margin finely denticulate and densely ciliate, leaves otherwise glabrous; flowers 1–20; calyx 7–12mm, green, lobes acute, denticulate near apex; petals obovate, 11–15× 7–13mm, white tinged pink; stamens 6–12mm.

Bhutan: S—Phuntsholing district (Kamji) and Deothang district (Chungkhar), **C**—Ha district (Chungu Gompa) and Mongar district (Khine Lhakang); **Sikkim; Arunachal Pradesh:** Nyam Jang Chu. Rocks and cliff faces, 1500–3050m. February–April.

The above description and records refer to forma **ligulata** (Wall.) Yeo.

6. SAXIFRAGA L.

Perennial herbs mostly small, sometimes dioecious. Basal leaves when present not infrequently rosulate, stem leaves alternate or opposite, usually entire, sometimes with minute pits near apex on upper surface. Flowers solitary or several, loosely racemose or corymbose, bisexual or unisexual. Calyx adnate to base of ovary, lobes 5. Petals 5, rarely absent, yellow or white, sometimes red or purple. Stamens 10, almost as long as petals in bisexual and male flowers, shorter than styles in female flowers. Ovary semi-inferior, carpels 2, united below, tapering upwards into 2 free styles, almost as long as petals in bisexual and female flowers, shorter than stamens in male flowers.

1. Leaves on vegetative part of shoot (those on flowering shoots may be different) with one or several minute pits at apex or near margin in upper part of leaf, pits usually surrounded by white calcareous secretions 2
+ Leaves without pits 8

2. Leaves with 3–7 pits near margin in upper part 3
+ Leaves with 1 ± apical pit 5

3. Flowers yellow .... **Species 1 & 2**
+ Flowers white or pink 4

4. Flowers several (3–6) on peduncles 2.5–8cm .... **Species 3 & 4**
+ Flowers solitary or up to 3 on peduncles 0.8–2cm .... **Species 5 & 6**

5. Leaves subacute, pit rarely producing calcareous secretion
 7. S. subsessiliflora
+ Leaves obtuse or ± truncate at apex, pits usually producing white calcareous secretions 6

6. Flowers yellow, on peduncles 0.5–1cm; loosely tufted plants
 8. S. flavida
+ Flowers white, sessile; densely caespitose plants 7

7. Leaves on vegetative parts of shoots opposite .... **9. S. georgei**
+ Leaves on vegetative parts of shoots alternate .... **Species 10–12**

8. Plants producing slender stolons from among basal leaves 9
+ Plants not producing stolons 11

9. Leaves orbicular; flowers with 2 petals much longer than the others
 13. S. stolonifera
+ Leaves oblanceolate; petals all ± equal in length 10

10. Leaf margins stiffly spinulose-ciliate, teeth sometimes gland-tipped
 Species 14 & 15
+ Leaf margins entire or minutely serrulate, ciliate with fine gland-tipped hairs .... **Species 16 & 17**

11. Leaves toothed or lobed, sometimes entire in *S. gageana* (but then flowers red and ovary flattened and lobed) and in *S. hispidula* (but then stems bearing a few scaly bulbils in lower part) 12
+ Leaves entire, sometimes with a fimbriate hyaline tip or stiffly ciliate margin 16

12. Petals yellow, sometimes white in *S. strigosa*; leaves mostly on stems
 Species 18 & 19
+ Petals white (sometimes yellow or red spotted) or entirely red; leaves mostly basal 13

13. Petals entirely red .... **Species 20 & 21**
+ Petals mainly or entirely white 14

14. Leaves ovate-oblong .... **Species 22−26**
+ Leaves suborbicular or reniform 15

15. Leaves glabrous beneath, teeth or lobes mostly blunt ... **27. S. asarifolia**
+ Leaves blackish glandular beneath, teeth coarse and acute
   28. S. granulifera

16. Leaves obovate with white fimbriate hyaline apex . **29. S. hemisphaerica**
+ Leaves variously shaped but without white hyaline apex 17

17. Petals pure white, not red-spotted .... **Species 30 & 31**
+ Petals yellow or rarely reddish or purple.................................... 18

18. Stems bearing scaly buds (bulbils) in axils of upper leaves
   Species 32 & 33
+ Stems without scaly buds 19

19. Flowers pendulous at first; very glandular plants (except *S. bergenioides*) 20
+ Flowers always erect 21

20. Petals purple; pedicels and calyx brown lanate .... **34. S. bergenioides**
+ Petals yellow; pedicels and calyx glandular pubescent .... **Species 35−37**

21. Leaf margins bearing stiff spinulose cilia, sometimes gland-tipped
   Species 38−40
+ Leaf margins without stiff spinulose cilia, if ciliate then hairs soft (sometimes with a few stiff ciliate hairs at apex of basal leaves in *S. umbellulata*) 22

22. Large plants, stems (7−)20−45cm tall; leaves 2−6×1−2cm; 23
+ Smaller plants, stems mostly less than 15cm; leaves usually less than 2×0.7cm 25

23. Flowers usually solitary or 2−4; petals 8−12×6−8mm .. **Species 41 & 42**
+ Flowers usually 5−10 or more; petals 7−10×3−5(−7)mm 24

24. Basal leaves usually absent; stems brownish lanate or sparsely and softly glandular... **Species 43 & 44**
+ Basal leaves usually present; stems densely covered with short thick glandular hairs .... **Species 45 & 46**

25. Small to medium-sized plants usually caespitose; stems 0−15(−20)cm; leaves 5−25×1−7mm; flowers solitary or several 26

+	Miniature plants with creeping stems; stems 0–5cm; leaves 2–5×1–3mm; flowers usually solitary (but up to 6 in *S. punctulata*) . 35
26.	Flowers several (5–10), rarely solitary but then stems 10cm or more . 27
+	Flowers solitary or few (2–3), stems less than 7cm (except *S. kinchingingae* with stems up to 12cm and rarely with 2 flowers) 29
27.	Pedicels densely brown lanate-tomentose, eglandular .. **47. S. tangutica**
+	Pedicels glabrous, sparsely or densely glandular-pubescent or eglandular-lanate, but not brown tomentose 28
28.	Basal leaves numerous, rosulate .... **48. S. umbellulata**
+	Basal leaves few or absent, not rosulate .... **Species 49–51**
29.	Flowers usually ± sessile amongst foliage; petals crimson or yellow within and red outside .... **Species 52 & 53**
+	Flowers exserted from foliage; petals yellow sometimes orange spotted within (rarely red in *S. contraria*) 30
30.	Leaves ovate, sessile throughout stem, upper ones, at least, cordate at base .... **54. S. cordigera**
+	Leaves variously shaped, lower ones, at least, usually petiolate but never cordate at base 31
31.	Leaves acuminate to a fine awn-like point .... **Species 55 & 56**
+	Leaves obtuse, acute or acuminate but without an awn-like point 32
32.	Stem leaves linear-lanceolate, scarcely 1mm broad, usually completely glabrous and eglandular .... **57. S. lepida**
+	Stem leaves variously shaped, 1–5mm broad, usually hairy or glandular 33
33.	Stems glabrous or (sometimes brown) shortly glandular in upper parts, without brown eglandular lanate hairs .... **Species 58–60**
+	Stems bearing long brown eglandular lanate hairs in upper parts 34
34.	Brown lanate hairs dense .... **Species 61 & 62**
+	Brown lanate hairs sparse, mostly at leaf axils .... **Species 63–66**
35.	Plants densely caespitose; leaves 0.7–1mm broad 36
+	Plants loosely or not at all caespitose; leaves 1–2.5mm broad 37
36.	Leaves elliptic, subterete, glabrous; petals green ..... **67. S. inconspicua**
+	Leaves linear-elliptic or oblanceolate, flat, ciliate at least as base; petals yellow .... **Species 68 & 69**

37. Plants loosely caespitose; leaves ± rosetted 38
+ Plants with slender stems; leaves scarcely rosetted 39

38. Flowers on pedicels up to 1.5cm .... **70. S. stella-aurea**
+ Flowers sessile .... **71. S. jacquemontiana**

39. Leaves opposite .... **72. S. contraria**
+ Leaves alternate 40

40. Stems and leaves glandular; petals white or yellow with red spots
 73. S. punctulata
+ Stems and leaves glabrous, eglandular; petals yellow (sometimes orange-spotted), green or absent 41

41. Petals yellow (sometimes orange-spotted) .... **74. S. engleriana**
+ Petals green or absent .... **75. S. microphylla**

1. S. thiantha H. Smith

Plants somewhat laxly caespitose; basal shoots up to 8cm. Leaves ± coriaceous, oblong, 6−7×1.5mm, obtuse, with 3−5 calcareous pits near apex, ciliate near base. Flowers solitary on leafy peduncles 2−3cm, reddish glandular-pubescent. Calyx 5−6mm reddish-glandular, lobes ovate, 3.5−4×2.5mm, obtuse, each with a calcareous pit at apex. Petals obovate, c 8×5−6mm, sulphur yellow.

Bhutan: C—Punakha district (Ritang and Tang Chu), N—Upper Bumthang Chu district (Waitang). On cliff faces, 4000−4570m. June−August.

The above description and records refer to the typical var. **thiantha**; var. **citrina** H. Smith with peduncles and calyces bearing black glands and broader (up to 8mm) lemon yellow petals is known from Tongsa district (Omta Tso) at 4650m.

2. S. sherriffii H. Smith

Similar to *S. thiantha* but more densely caespitose, basal shoots up to 7cm; leaves densely imbricate, oblong-obovate, c 4×2mm, subacute with 3−5 pits near apex, slightly recurved, ciliate near base, ± thick, concave within, convex or keeled on back; flowers 3−12, subumbellate on peduncles 3−4.5cm, glandular-pilose, bearing 5−7(−9) thin spathulate leaves; calyx 3−3.5mm, glandular-pubescent, lobes ovate-lanceolate, c 2×1mm, acute, without an apical pit; petals obovate, 5−6×3.5−5mm, narrowed into a basal claw, yellow.

Bhutan: N—Upper Bumthang Chu district (Pangotang). On cliffs and steep rocky banks, 3800m. May.

3. S. andersonii Engler. Fig. 35 f—h

Somewhat loosely cushion-forming, basal shoots 6—7cm. Leaves ± coriaceous, oblong-spathulate, 4—6×1.5—2mm, obtuse or subacute with 3—7 calcareous pits at apex, margin ciliate near base. Flowers 3—5 on leafy peduncles 2.5—5cm, glandular-pubescent, pedicels up to 4mm in flower (—10mm in fruit). Calyx c 4mm, lobes ovate c 2×1.5mm, glandular and with 1—3 apical pits on the inner surface. Petals white to pale pink, obovate, 4—5×2—3.5mm.

Bhutan: C—Thimphu and Tongsa districts, **N**—Upper Mo Chu, Upper Pho Chu, Upper Kuru Chu and Upper Kulong Chu districts; **Sikkim.** Open hillsides and screes, 3500—4570m. May—July.

4. S. stolitzkae Engler & Irmscher

Similar to *S. andersonii* but leaves lime-encrusted in the upper half from 7—11 marginal pits; flowers 3—6 on peduncles 7—8cm, glandular-pubescent; calyx c 4.5mm, purplish, glandular, lobes ovate c 3×1mm, without apical pits; petals obovate, c 5×2.5mm, pink.

Bhutan: C—Thimphu district (between Barshong and Naha). On damp cliffs, 3650m. May.

5. S. clivorum H. Smith

Densely caespitose, basal shoots up to 5cm. Leaves ± coriaceous, oblong, densely imbricate, 5—6×2—2.5mm, apex obtuse with 3(—5) calcareous pits at margin, glandular-ciliate near base. Flowers 1—3, subsessile or on peduncles up to 8mm, with a few narrower (c 0.75mm) leaves each with 1 apical pore. Calyx 3.5—4mm, lobes oblong-ovate, c 2×1.75mm, glandular-ciliate. Petals white, obovate, c 3.75×2.5—3mm.

Bhutan: C—Tongsa district (Dungshinggang); **Sikkim:** Guicha La. On open rocks and cliff faces, 3960—4570m. June.

6. S. vacillans H. Smith

Similar to *S. clivorum* but less densely caespitose, basal shoots up to 15cm; leaves ± herbaceous, opposite (or, in vigorously growing parts, alternate), oblong, 3.5—5×1.5—2mm, subacute, with 3—7 calcareous pits near apex, ciliate near base; flowers 1—2 on ± naked peduncles 1.5—2cm, glandular-pubescent; calyx c 3mm, lobes ovate-oblong c 1.5×1mm with a calcareous pit at the apex; petals obovate, 4—5×2—3mm, white.

Bhutan: C—Thimphu district (N of Barshong). On wet cliff faces, 4100m. May.

7. S. subsessiliflora Engler & Irmscher

Densely caespitose, basal shoots 3—7cm. Leaves densely imbricate, erect, concave on inner surface, convex or keeled on the back, oblong-obovate, 4—5×1.5—2mm with a pit near subacute apex, rarely secreting lime, margin minutely glandular-ciliate. Flower solitary, sessile. Calyx 3.5–4mm, lobes ovate, c 2.5×1.5mm, obtuse, glandular-ciliate. Petals oblong-obovate 3–5×1.5mm, white.

Bhutan: N—Upper Bumthang Chu district (Dole La); **Sikkim:** Bijan. 4570m. May–June.

8. S. flavida H. Smith

Somewhat loosely caespitose, basal shoots up to 7cm. Leaves ± spreading, rather thin, oblong-obovate, c 5×2mm, apex obtuse with a calcareous pit, margin sparsely glandular-ciliate near base. Flowers solitary on leafy glandular peduncles 0.5–1cm. Calyx c 5mm, glandular, lobes broadly ovate, c 2×2.5mm, glandular. Petals obovate, c 4×3–4mm, yellow.

Bhutan: C—Punakha district (Ritang). On cliff faces, 4400m. June.

9. S. georgei Anthony

Plants loosely caespitose forming mats, basal shoots 3–5cm. Leaves thinly coriaceous, opposite, spreading, oblong 3–4×1.5–2mm, obtuse, recurved and ± triquetrous at apex bearing a calcareous pit, glabrous. Flowers solitary, sessile. Calyx 3.5–4mm, glabrous or glandular-pubescent at base, lobes broadly ovate, c 2×2mm. Petals obovate, c 5×3mm, white.

Bhutan: C—Thimphu district, **N**—Upper Bumthang Chu and Upper Kulong Chu districts; **Sikkim.** On cliff faces, 3050–4570m. May–June.

10. S. pulvinaria H. Smith; *S. imbricata* Royle non Lamarck

Densely caespitose forming cushions or mats, basal shoots 3–6cm. Leaves closely imbricate, thickish, appressed, concave inside, convex or keeled on the back, oblong-ovate, 3.5–4×1–1.5mm, obtuse, orbicular or rhombic at apex with a calcareous pit, margin ciliate. Flowers solitary, sessile. Calyx c 3mm, lobes ovate, c 2×1mm, without calcareous pits, glandular-ciliate. Petals oblong-obovate, 4–5×2–3mm, white.

Bhutan: N—Upper Mo Chu district (Lingshi) and Upper Kuru Chu district (Gong La); **Sikkim:** Tangu, Zemu and Lhonak. Cliff faces and dry hill slopes, 4100–4570m. July–August.

11. S. matta-florida H. Smith

Similar to *S. pulvinaria* but calyx lobes 1.8–2×1.8–2mm, bearing a calcareous pit at the apex; petals larger 5–6×3.5–4.5mm.

Bhutan: N—Upper Kulong Chu district (Lao). On wet mossy cliffs, 3050m. May.

12. S. saxorum H. Smith

Similar to *S. pulvinaria* and *S. matta-florida* but leaves obtuse, apiculate, those on shoots glandular-ciliate near base, otherwise glabrous, but those subtending flowers glandular-pilose throughout; calyx lobes broadly ovate c

1.7×2−2.5mm with a calcareous pit at apex; petals obovate 3−4×2−2.8mm.

Bhutan: C—Tongsa district (Dungshinggang) and Sakden district (Orka La). On open rock and cliff faces. 3900−4260m. June.

13. S. stolonifera Curtis; *S. sarmentosa* Schreber

Stems 20−35cm, hirsute, bearing at base several sparsely hirsute stolons up to 50cm long. Leaves ± rosetted, suborbicular 2−7×2.5−10cm, dentate or shallowly lobed, sparsely hirsute above, purplish beneath; petioles up to 20cm, hirsute. Flowers in a loose raceme or panicle. Sepals ovate c 4×1.5mm, pubescent. Petals unequal, 3 upper ones elliptic c 5×2mm, white but pink in upper half with purple spots, yellow at base, 2 lower petals lanceolate 10−15×3−5mm, white.

Sikkim: Darjeeling. Apparently naturalised on stone walls, etc. May. Native of China.

14. S. mucronulata Royle; *S. flagellaris* Sternberg var. *mucronulata* (Royle) Clarke, *S. mucronulata* Royle subsp. *sikkimensis* (Hulten) Hara. Med: *Dreeta Sazin*

Stems simple 5−10(−15)cm, glandular-pubescent, bearing at base as many as 10 slender glandular-pubescent stolons up to 12cm. Basal leaves rosetted, oblanceolate, c 10×3mm, acuminate to a cartilaginous bristle, attenuate at base, margin stiffly serrulate with spinulose teeth some of them gland-tipped; stem leaves similar, decreasing in size above. Flowers 3−10 usually in a subumbellate raceme surrounded at base by linear bracts c 8×1mm, pedicels 3−10mm. Calyx 5−6mm, glandular, lobes ovate-lanceolate, c 4.5×1mm. Petals obovate, c 8×3mm, yellow.

Bhutan: C—Thimphu district (Barshong) and Punakha district (Jewphu), **N**—Upper Mo Chu district (Kangla Karchu La and Lingshi) and Upper Bumthang Chu district (Marlung); **Sikkim.** On moist gravel, 4570−5200m. June−August.

15. S. brunonis Seringe; *S. brunoniana* Sternberg. Fig. 35 n & o

Similar to *S. mucronulata* but more slender; stolons thread-like, red, glabrous; leaves pale green, glabrous except for spinulose-ciliate margin, teeth sometimes gland-tipped; flowers solitary or up to 4, loosely racemose on pedicels up to 5cm, sparsely glandular; calyx c 3mm, glabrous, reddish, lobes ovate c 2×1.5mm, acute; petals elliptic-oblanceolate, 7−8×2−3mm, yellow, sometimes orange spotted.

Bhutan: C—Thimphu and Bumthang districts, **N**—Upper Mo Chu, Upper Pho Chu, Upper Bumthang Chu and Upper Kuru Chu districts; **Sikkim.** On sandy river banks and mossy boulders, 3050−4880m. June−September.

16. S. pilifera Hook. f. & Thomson

Similar to *S. mucronulata* but leaves quite entire, finely glandular-puberulous and ciliate; flowers solitary or up to 7 on pedicels 3−10mm; calyx c 4mm, glandular-pubescent, lobes ovate c 2.5−3×1.5mm; petals obovate c 3×1.5−3mm, dark red.

Bhutan: N—Upper Mo Chu district (Yale La), Upper Pho Chu district (Gafoo La) and Upper Bumthang Chu district (Marlung); **Sikkim:** Lhonak and Sang La. On grassy rocks and sandy screes, 4265−4880m. July−September.

17. S. tentaculata Fischer

Similar to *S. mucronulata* but stems glabrous or sparsely glandular; leaves tending to be aggregated at base and apex of stem, oblanceolate, 5−10×3−4mm, entire or minutely serrulate, finely glandular-ciliate; flowers solitary or 2; calyx 3.5−4mm, lobes ovate, 2.5×1.5mm; petals oblanceolate or obovate, 5−6×2mm, yellow.

Bhutan/Chumbi frontier: between Phari and Tsethanka: **Sikkim:** Chakalung La. 4570−4880m. July−September.

18. S. hispidula D. Don

Caespitose; stems 3−15cm, leafy throughout, usually bearing a few scaly buds (bulbils) 0.5−10mm near base, pubescent with spreading white usually glandular hairs. Leaves ovate or oblong, 3−18×2−8mm, acute, base rounded, sessile, margin with 1 or 2 teeth on each side, sometimes entire, stiffly white pilose especially on upper surface. Flowers solitary or 2. Calyx c 5mm, glandular-pubescent especially at base, lobes ovate, c 3.5×2mm. Petals obovate 5−8×3−4mm, yellow.

Bhutan: C—Ha, Thimphu, Tongsa and Bumthang districts, **N**—Upper Mo Chu, Upper Pho Chu, Upper Bumthang Chu, Upper Kuru Chu and Upper Kulong Chu districts; **Sikkim.** On cliffs, rocks and in Fir forests, 3650−4570m. July−September.

The typical var. **hispidula** has entire leaves whereas in var. **doniana** Engler the leaves have toothed margins. In all the Bhutanese material seen so far at least some of the leaves have toothed margins. Both varieties are widespread in Sikkim.

19. S. strigosa Seringe; ?*S. bumthangensis* Wadhwa

Similar to *S. hispidula* but not caespitose, stems solitary or few, simple, leafy throughout or with largest leaves clustered and ± rosetted in lower part but above base, leaves diminishing in size above and often with scaly buds in axils; leaves oblanceolate or obovate, 1−2.5×0.5−1cm, acuminate, base attenuate, subpetiolate, margin serrate with 2−3 bristle-pointed teeth on either side, ± strigose with stiff appressed hairs especially on upper surface; flowers solitary or up to 6 on pedicels 1−4cm or the lower ones transformed into sessile clusters of bulbils; calyx c 3mm, lobes ovate-triangular, c 2 × 1mm; petals obovate, c 5.5.×2mm, yellow or white marked with yellow or red within.

Bhutan: C—Thimphu, Tongsa, Bumthang and Tashigang districts, N—Upper Bumthang Chu and Upper Kuru Chu districts; **Sikkim.** Streamsides, 2750–3200m. July–September.

20. S. gageana W. W. Smith

Stems slender 2–2.5(−5)cm, finely pubescent. Basal leaves spathulate or elliptic, c 5×3mm, acute or subacute, base attenuate into petiole 5–10mm, margin entire or with 1–2 teeth on each side, sparsely villous. Flowers solitary or up to 3 on pedicels 1–1.5cm. Calyx c 3.5mm, purplish, lobes oblong c 2×1mm. Petals obovate, c 3×1mm, bright red. Ovary flattened, disc-like, lobed at margin.

Sikkim: Ningbil and Chola range. 4400m. August.

21. S. rubriflora H. Smith

Similar to *S. gageana* but aculescent; leaves rosetted, obovate, 7–10×4–7mm, obtuse or subacute and with a few low broad teeth around upper margin, attenuate and subpetiolate at base, margin ciliate; flowers solitary, subsessile among leaves or on pedicels 2–7mm; calyx 3–3.5mm, reddish, glabrous, lobes ovate c 2×2mm; petals elliptic-oblanceolate c 2.5×1–1.5mm, red; stamens and ovaries red.

Bhutan: C—Tongsa district (Omta Tso). On earthy hill slopes, 4700m. August.

22. S. pallida Seringe; *S. micrantha* Edgeworth non Fischer, *S. himalaica* Balakrishnan

Stems 10–20(−30)cm, sparsely pubescent near base, ± densely glandular-pubescent above. Basal leaves ovate-oblong, 0.75–2.5(−4.5)× 0.5–1.5(−2)cm, obtuse or subacute, base truncate, rounded or shallowly cordate, margin shallowly crenate-serrate, sparsely lanate above, glabrous beneath; petioles 1–5cm. Flowers 3–5, racemose or paniculate with several few-flowered branches. Calyx 4–6mm, often red or brown, lobes ovate c 2×1.5mm, becoming reflexed. Petals obovate, 3–4×1.5–2mm, white. Anthers reddish-brown, filaments clavate.

Bhutan: C—Mongar district, N—Upper Mo Chu, Upper Pho Chu, Upper Mangde Chu and Upper Kulong Chu districts; **Sikkim.** Alpine turf and screes, 3050–4250m. June–July.

23. S. pluviarum W. W. Smith

Similar to *S. pallida* and apparently differing from it by the presence of 4–8 ovoid bulbils (c 1mm) in the axils of bracts borne on the pedicels.

Sikkim: Tosa, Chola Range. 4250–4570m.

67. SAXIFRAGACEAE

24. S. pseudopallida Engler & Irmscher
Similar to *S. pallida* but smaller, flower stems 4−10cm and with larger flowers; calyx 6−8mm, reddish; petals white, 6−10×3−5mm; styles and ovary reddish-brown.
Bhutan: N—Upper Mo Chu district (Yale La and Lingshi) and Upper Kulong Chu district (Me La); **Sikkim**. Alpine streams and open rocky banks, 3950−4570m. June−July.
Possibly not specifically distinct from *S. pallida*.

25. S. dungbooii Engler & Irmscher
Similar to *S. pallida* and especially *S. pseudopallida* but stems 8−12cm, brownish-lanate above; basal leaves rosetted, ovate-elliptic, 10−15×5−7mm, subacute, base narrowed into petiole up to 3cm, margins entire, glabrous; flowers 3−7; calyx lobes spreading, ovate-triangular, 2.5−3.5×1.5−2mm, glabrous; petals obovate-oblong, 4.5−7×2−3.5mm.
Sikkim: Yume Chu; **Chumbi:** Phari. 5100m. August.

26. S. melanocentra Franchet
Similar to *S. pallida* and more especially *S. pseudopallida* in plant and flower size but petals obovate, 6−8×4mm with 2 yellow or orange spots at the base of each within and sometimes becoming purplish outside; filaments subulate; carpels dark blackish-crimson.
Bhutan: N—Upper Mo Chu district (Sinchu La), Upper Bumthang Chu district (Pangotang and Marlung), Upper Kuru Chu district (Kang La) and Upper Kulong Chu district (Me La). Open sandy screes, 3950−4570m. June−August.
Possibly not specifically distinct from *S. pallida* and *S. pseudopallida*.

27. S. asarifolia Sternberg; *S. odontophylla* Hook. f. & Thomson non Sternberg
Rhizome c 5mm thick, often bearing foliar remains. Stems 10−20cm sparsely brownish-lanate. Basal leaves suborbicular or reniform, crenate with 7−9 broad rounded teeth, base deeply cordate, ciliate, under surface usually purplish; petioles 3−8cm, villous. Stem leaves similar, decreasing in size upwards and becoming more acutely toothed. Flowers 2−4(−10), loosely racemose, pedicels 1−5cm. Calyx 5−7mm, glandular-pubescent, lobes ovate 4−5×2mm, acute. Petals obovate, 8−10×6−7mm, white sometimes spotted reddish within. Anthers reddish or orange.
Bhutan: C—Tongsa district (near Thita Tso); **Sikkim**. Under Rhododendrons, 4100−4250m. August.

28. S. granulifera H. Smith
Similar to *S. asarifolia* but more slender, sparsely glandular-pubescent; leaves 3−15mm with 5−7 coarse sharp teeth, truncate or shallowly cordate

at base, blackish glandular-pubescent beneath; lower leaves with petioles up to 12mm; upper ones ± sessile and usually with ovoid bulbils (c 0.75mm) in the axils, flowers 1−5(−9) on pedicels up to 3cm; calyx c 4mm, glandular at base, lobes narrowly ovate, 2.5−3×1−1.5mm; petals oblanceolate c 8× 2.5mm, white.

Bhutan: N—Upper Pho Chu district (Leji) and Upper Kulong Chu district (Me La); **Sikkim:** Galing. On rocks and cliffs, 3050−3800m, June−July.

E. Himalayan records of *S. cernua* L. (111) probably refer to this species.

29. S. hemisphaerica Hook. f. & Thomson. Fig. 35 i−1

Caespitose, forming dense low cushions, dioecious; basal shoots 1−2(−4)cm. Leaves densely rosetted, ± concave, obovate, c 2.5× 0.75−1mm, obtuse, base cuneate, surrounded in upper part by a fimbriate white hyaline margin c 1mm. Flowers solitary, sessile among upper leaves. Calyx c 2mm, lobes fimbriate like leaves, c 1×0.7mm. Petals absent. Male flowers with exserted yellow stamens twice as long as calyx lobes and ovary shorter than them. Female flowers with staminodes shorter than calyx, styles thick, shortly exserted.

Bhutan: N—Upper Mo Chu district (Phile La), Upper Mangde Chu district and Upper Bumthang Chu district (Marlung); **Sikkim:** Chakalung. On rocks and boulder screes, 4265−4880m. July.

30. S. humilis Engler & Irmscher

Loosely caespitose. Stems 1−2cm, glandular-hairy, densely leafy. Leaves broadly oblanceolate, 7−15×2−3.5mm, subacute, base attenuate, lanate-ciliate, otherwise glabrous. Flowers solitary or rarely up to 4 on pedicels c 5mm, glandular-pubescent. Calyx c 5mm, lobes ovate c 2.5× 1.5mm, obtuse, glandular-ciliate. Petals obovate, 3−4×2−2.5mm, white.

Bhutan: C—Tongsa district (Omta Tso), **N**—Upper Bumthang Chu district (Marlung), Upper Kuru Chu district (Kang La) and Upper Kulong Chu district (Me La). Open hill slopes and screes, 4100−4720m. June−August.

31. S. coarctata W. W. Smith

Similar to *S. humilis* but leaves always densely imbricate, coriaceous, elliptic or spathulate up to 5mm, entire or 3-toothed at apex; flowers always solitary, sessile; sepals broadly ovate or suborbicular, 1−1.5mm; petals oblong or obovate 2−3×1.5mm.

Sikkim: Yumchho La. 4570m.

32. S. brachypoda D. Don; *S. fimbriata* Seringe, *S. gouldii* Fischer var. *eglandulosa* H. Smith

Caespitose; stems simple, 9—15mm, leafy throughout, often with a scaly bud in axil of upper leaves, glabrous below, glandular-pubescent above. Leaves stiff, lanceolate, 4—15×1—3mm, acuminate to sharp almost prickly point, base rounded, sessile, margin sparsely spinulose-ciliate, ± glossy on upper surface, pale beneath, 3-veined from base. Flowers often solitary but sometimes up to 6. Calyx 5—6mm, glandular, lobes ovate-triangular, c 4× 1.5mm. Petals obovate, 6—10×3—4mm, yellow, margins entire or minutely fimbriate.

Bhutan: C—Thimphu district (Barshong), N—Upper Pho Chu district (Gafoo La), Upper Bumthang Chu district (Ju La and Tolegang) and Upper Kulong Chu district (Me La); **Sikkim.** Moist rocks on open hillsides, 3650—4570m. July—August.

33. S. filicaulis Seringe

Densely caespitose with wiry, branching, erect or trailing stems 5—15cm, sparsely glandular-pubescent, often bearing scaly buds in leaf axils above. Leaves elliptic, 4—8×0.5—1mm, acuminate to a bristle-like point, narrowed to a ciliate stem-clasping petiolar base, margin inrolled and bearing 1—2 cilia on each side. Flowers solitary on pedicels up to 1.5cm, glandular-pubescent. Calyx c 4mm, reddish, lobes ovate, c 2.5×1.5mm, obtuse. Petals obovate or oblanceolate, 7—8×2—3mm, yellow.

Bhutan: C—Thimphu district (Dochong La) and Bumthang district (Bumthang), N—Upper Mo Chu district (Laya), Upper Pho Chu district (Tranza) and Upper Bumthang Chu district (Pangotang); **Sikkim:** Lingmuthang. On rocks and open grassy slopes, 2900—3950m July—September.

34. S. bergenioides Marquand

Caespitose; stems 4—15cm, sparsely brownish lanate. Basal leaves ovate-elliptic or oblanceolate, 1.5—2×0.3—1(−1.25)cm, obtuse or subacute, base attenuate into petiole up to 2cm, brown-lanate especially on upper surface; stem leaves similar, smaller, usually sessile. Flowers solitary or up to 3, pendulous at least at first. Calyx 6—7mm, purplish, lobes ovate 5×2.5—3mm, sparsely lanate, margin ciliate. Petals obovate 15—18×5—6mm, pink to purple.

Bhutan: N—Upper Pho Chu, Upper Bumthang Chu, Upper Kuru Chu and Upper Kulong Chu districts. On peaty turf and scree, 4000—4400m. July—September.

35. S. lychnitis Hook. f. & Thomson

Similar to *S. bergenioides* but stems smaller, 4—10cm, densely brown-lanate; basal leaves ± rosetted, elliptic, 7—10×3—5mm, acute, base attenuate, subpetiolate, brown-lanate; stem leaves similar, narrower, upper ones glandular-pubescent; flowers solitary; calyx 7—10mm, reddish-brown;

lobes oblong-ovate 6−9×2.5−3mm, glandular-pubescent; petals elliptic, 12−15×2−3mm, yellow.

Bhutan: N—Upper Pho Chu district (Gafoo La and Kangla Karchu La) and Upper Kulong Chu district (Me La); **Sikkim.** On rocks, 4265−4880m. June−August.

36. S. viscidula Hook. f. & Thomson; *S. pseudohirculus* Engler & Irmscher
Similar to *S. bergenioides* and *S. lychnitis* but more densely caespitose; stems 5−15cm, with brownish-lanate and shorter glandular hairs; leaves oblong-ovate, 8−15×2−3mm, acute, base attenuate, sub-petiolate, glandular-pubescent; flowers 1−3 on pedicels 2−5mm, usually solitary; calyx 4−6mm, purplish, lobes oblong-ovate 3−4×1.5mm, glandular-pubescent, becoming reflexed; petals oblanceolate, 12−13×2−3mm, yellow.

Bhutan: N—Upper Mo Cho district (Laya) and Upper Bumthang Chu district (Waitang); **Sikkim:** Tangu; **Chumbi.** Grassy slopes and open sandy screes, 4100m. August−September.

37. S. nigroglandulifera Balakrishnan; *S. nutans* Hook. f. & Thomson non D. Don, nec Adams
Tufted; stems surrounded by foliar remains at base, 4−30cm, glabrous below, glandular-pubescent above. Basal leaves ovate-elliptic, 1.5−4× 0.4−1.5cm, subacute, base attenuate into petiole up to 6cm, margins glandular-ciliate; stem leaves similar, smaller, sessile. Flowers pendulous or nodding, 4−6 in a short raceme, pedicels 5−10mm densely blackish-glandular. Calyx 6−7mm, reddish-brown, glandular, lobes ovate 4−5×2.5−3mm. Petals oblanceolate, c 12×3mm, yellow.

Bhutan: N—Upper Bumthang Chu district (Marlung); **Sikkim:** Thangu. Grassy slopes, 4265−4730m. July−August.

38. S. wardii W. W. Smith; *S. gouldii* Fischer var. *gouldii*, *S. megalantha* Marquand
Caespitose; stems erect, 5−15cm, glandular-pubescent especially in upper parts. Leaves narrowly to broadly ovate, 8−12×4−6mm, acute, base rounded, sessile, margin regularly spinulose-ciliate with gland-tipped hairs. Flowers solitary. Calyx c 8mm, lobes broadly ovate c 7×4−5mm, often reddish-brown tinged, densely glandular and ciliate. Petals obovate, c 10× 7mm yellow, minutely glandular-ciliate with purplish glands.

Bhutan: N—Upper Bumthang Chu district (Ju La); **Chumbi.** On rocks and cliffs, 4265m. July.

39. S. erinacea H. Smith
Similar to *S. wardii* but stems 1−2cm, densely leafy; leaves lanceolate, 6−7×1.5−2mm, acuminate to a fine bristle c 1.5mm, margin

whitish-cartilaginous with 6−8 spinulose cilia on each side; flowers solitary, ± sessile amongst upper leaves; calyx 5−6mm, lobes ovate-triangular 4−5×1.5−2mm, glandular-ciliate; petals oblong-lanceolate c 8×2mm, abruptly tapered into basal claw c 2.5mm.

Bhutan: N—Upper Kulong Chu district (between Me La and Cho La). On open hillsides, 4100−4250m. July−August.

40. S. serrula H. Smith

Similar to *S. wardii* and *S. erinacea* but stems 4−16cm; leaves linear-oblong c 10×1.5mm, mucronate to a fine stiff point, margins glandular-ciliate with short cartilaginous hairs; calyx lobes narrowly ovate c 5×2mm, glandular-ciliate; petals elliptic c 7.5×4mm narrowed at base into a short claw c 1.5mm.

Bhutan: C—Tongsa district (Chendebi). Among grass and on dry soil, 3050m. August.

41. S. latiflora Hook. f. & Thomson

Stems simple, 7−17cm, erect, ± stout, often bearing bulbils at base, glandular-pubescent especially in upper parts. Leaves elliptic-oblanceolate 3−6×1−2cm, acute, lower ones narrowed at base into a broad petiole, upper ones sessile, glabrous or sparsely ciliate. Flowers solitary or sometimes 2−3, pedicels glandular-pubescent. Calyx 8−10mm, lobes broadly ovate 6−9mm, ± obtuse, glandular-ciliate. Petals oblong-ovate 8−12×6−8mm, yellow, spotted with orange within.

Bhutan: C—Tongsa district (Omta Tso), **N**—Upper Mo Chu district (Yale La); **Chumbi:** Chomo Lhari; **Sikkim:** Alukthang. Grassy ravines, 4265−4880m. August−September.

42. S. harry-smithii Wadhwa

Similar to *S. latiflora* but stems slender 15−30cm, pale brown eglandular-pubescent; leaves ovate 10−20×5−10mm, acute, base rounded or cordate, lower ones borne on narrow petioles 2−3cm, upper ones sessile, sparsely pubescent on both surfaces; flowers usually solitary sometimes 2−4; calyx 6−11mm, lobes ± obtuse, glabrous or eglandular-ciliate; petals obovate or suborbicular 8−10×5−10mm.

Bhutan: N—Upper Mo Chu district (Laya), Upper Bumthang Chu district (Pangotang) and Upper Kulong Chu district (Me La). Screes and rocky slopes, 4100−4570m. August−September.

Endemic to Bhutan.

43. S. kingiana Engler & Irmscher

Caespitose; basal leaves absent at flowering time. Stems 30−45cm, leafy throughout, sparsely brownish-lanate. Leaves decreasing in size towards base and apex of stem, ovate, 1.5−4×1−2cm, acute or acuminate, base cordate, sessile, semiamplexicaul, with 5−7 parallel veins, sparsely lanate on both surfaces. Flowers (1−)5−15, ± corymbose, pedicels covered with long brownish glandular hairs. Calyx 7−8mm, lobes ovate, acute 4.5×2.5mm. Petals obovate 8−10×4−5mm, yellow, sometimes spotted with orange within.

Bhutan: C—Tongsa district (Lachu La); **Sikkim:** Changu and Sherabthang. Rocky hill slopes and scree, 3960−4570m. August−September.

44. S. moorcroftiana (Seringe) Sternberg; *S. diversifolia* sensu F.B.I. p.p. non Seringe. Fig. 35m

Similar to *S. kingiana* but stems ± glabrous; leaves elliptic-oblanceolate, 3.5−4(−6)×0.5−1.5(−1.8)cm, acute or subacute, base narrowed or rounded but scarcely cordate, usually widest near middle, glabrous but with a few long brownish hairs at axils, weakly 5-veined at base; flowers usually several sometimes solitary, pedicels glandular-pubescent; calyx 5−6mm, lobes oblong-ovate c 4×2.5mm, ± obtuse, glandular-ciliate; petals obovate, c 8×3mm.

Bhutan: C—Thimphu district (Tarka La) and Tongsa district (Padima Tso), **N**—Upper Mo Chu district (Cheypechey to Chumiten) and Upper Kulong Chu district (Me La); **Sikkim.** Among rocks and boulders, 4100−4400m. August.

45. S. parnassifolia D. Don; *S. sphaeradena* H. Smith subsp. *dhwojii* H. Smith

Stems slender 10−20(−30)cm, usually glabrous at base, glandular-pubescent above. Basal leaves broadly ovate, 1−3.5×1−3.5cm, obtuse or subacute, base cordate, sparsely lanate above, usually glabrous but sometimes villous beneath; petioles 1−5cm; stem leaves similar, decreasing in size above, sessile, cordate at base, usually ciliate, otherwise glabrous, ± glaucous beneath, uppermost ones glandular-ciliate. Flowers 1−6, ± corymbose. Calyx 4−5mm, lobes ovate, c 3×2mm, erect at first, later spreading or reflexed, glandular-ciliate. Petals obovate, 7−8×3mm, yellow.

Bhutan: C—Tongsa district (Rukubji), Bumthang district (Bumthang) and Mongar district (Donga La), **N**—Upper Bumthang Chu district (Pangotang); **Sikkim.** Cliff ledges and open, grassy banks, 3050−3800m. August−September.

46. S. diversifolia Seringe; *S. diversifolia* sensu F.B.I. p.p.

Similar to *S. parnassifolia* but more robust, stems 20−30cm; basal leaves broadly ovate, cordate at base, petioles up to 12cm; stem leaves ovate-oblong, 1−3×0.5−1.5cm, rounded at base and unlike basal leaves, not glaucous beneath; inflorescence up to 30-flowered, more coarsely glandular-pubescent.

Bhutan: C—Thimphu district (Pajoding and Barshong) and Tashigang district (Dib La), **N**—Upper Mo Chu district (Lingshi) and Upper Bumthang Chu district (Tolegang, Chamka, etc); **Sikkim**. In moist scree, 2900–3950m. September–November.

Probably not specifically distinct from *S. parnassifolia*.

47. S. tangutica Engler; *S. hirculus* L. var. *subdioica* Clarke. Fig. 35 p

Caespitose; stems 3–10cm, densely reddish-brown lanate especially in upper parts, surrounded at base by foliar remains. Basal leaves elliptic-oblanceolate 1.5–2.5×0.3–0.7cm, subacute, base attenuate into petiole 2–4cm, glabrous or ciliate with long brownish-lanate hairs; stem leaves similar, decreasing in size and becoming sessile. Flowers 5–10 in ± compact corymbs. Calyx c 4.5mm, lobes oblong-ovate, c 3×1.5mm, brownish-lanate, becoming reflexed. Petals elliptic, c 4×1mm, orange or brownish-red.

Bhutan: N—Upper Mo Chu district (Lingshi) and Upper Kulong Chu district (Me La); **Sikkim**. Rocky hillsides, 4400–4880m. July–September

References to the occurrence in Sikkim of *S. hirculoides* Decaisne, a West Himalayan plant, and to the West Chinese *S. przewalskii* Engler probably refer to this species.

48. S. umbellulata Hook. f. & Thomson

Caespitose; rootstocks slender. Stems 3–5cm, densely glandular-pubescent. Basal leaves spathulate, 7–15×1.5–3mm, numerous in rosettes, obtuse or subacute, attenuate and ciliate at least near base; stem leaves similar, smaller, glandular, alternate but ± whorled at base of inflorescence. Flowers 8–15 in terminal subumbels; pedicels 1–2.5cm, glandular-pubescent. Calyx 3–4mm, lobes ovate-triangular, c 2×1mm, glandular. Petals oblanceolate, 8–10×2mm, yellow.

Bhutan: N—Upper Mo Chu district (Laya); **Sikkim:** Tangu and Lhonak; **Chumbi:** Kungpu. Dry hillsides, 3650–5300m. August–September.

49. S. hookeri Engler & Irmscher; *S. corymbosa* Hook. f. & Thomson non Boissier, ?*S. lamninamensis* Ohba

Stems 7–35cm, sometimes with small scaly bulbils c 5mm at base, brownish lanate in lower parts, glandular-pubescent above. Leaves ovate, 7–15×5–12mm, subacute, base rounded, the lower ones petiolate (petioles up to 2cm), the upper sessile, glandular-pubescent. Flowers (1–)3–10, loosely racemose or corymbose, pedicels glandular-pubescent. Calyx including ovate lobes 5–6mm, usually glandular-pubescent. Petals obovate, 6–7×2–3mm, yellow.

Bhutan: C—Ha, Thimphu and Tongsa districts, **N**—Upper Mo Chu, Upper Pho Chu and Upper Kuru Chu districts; **Sikkim**. Grassy banks and streamsides, 3200–4100m. July–September.

Var. *glabrisepala* Engler & Irmscher was created for plants with glabrous calyces but the presence or absence of glands can vary within a single gathering.

50. S. sikkimensis Engler

Similar to *S. hookeri* but stems eglandular-lanate throughout; leaves lanceolate, 5−13×2−4mm, acute, base attenuate to petiole up to 1cm, upper leaves ± sessile, sparsely lanate-ciliate; flowers (1−) 3−5(−9); calyx c 5mm, lobes oblong-elliptic 3.5×1.5mm, spreading; petals obovate c 7× 4.5mm.

Bhutan: C—Thimphu district, **N**—Upper Kulong Chu district (Me La); **Sikkim:** Changu, Sherabthang, Jongri, etc. Scree slopes, 3650−4400m. August.

51. S. petrophila Franchet

Similar to *S. hookeri* and especially *S. sikkimensis* but stems 10−15cm, brownish-lanate at base, glandular-pubescent above, ± densely leafy throughout their length; leaves ovate 5−8×2.5−3mm, subacute, base rounded, sessile, upper surface and margins glandular-pubescent as least in upper leaves, lower surface lanate, eglandular; flowers 3−5; calyx 4−5mm, lobes ovate-triangular, c 3×1.5mm, glandular-pubescent; petals obovate 5−6×2.5−3mm.

Bhutan: N—Upper Kulong Chu district (Me La). Open hillsides, 3800m. August.

52. S. haematochroa H. Smith

Rhizomes slender; stems 2−4cm, densely leafy. Leaves oblanceolate or obovate, 5−7×2−4mm, subacute, base rounded and narrowed to a broad membranous petiole, ± stiffly brown ciliate at margin, otherwise glabrous. Flowers solitary, ± sessile among uppermost leaves or on peduncles up to 10mm. Calyx c 7mm, crimson, lobes oblong-elliptic c 6×3.5mm, glabrous. Petals obovate, c 8×5mm, deep crimson.

Bhutan: N—Upper Kulong Chu district (Me La). On rocky shady hillsides, 4570m. August.

53. S. parva Hemsley

Similar to *S. haematochroa* but rhizomes stout and densely caespitose; stems 1.5−3.5cm brownish-lanate; basal leaves elliptic 7−10×3mm on petioles up to 1.5cm, glabrous or brownish-ciliate; calyx c 5mm, lobes oblong-ovate c 3.5×1.5mm, becoming reflexed, ciliate; petals oblanceolate 4−5×1.5−2mm, yellow or sometimes red outside.

Bhutan: locality unknown (73).

54. S. cordigera Hook. f. & Thomson; *S. palpebrata* Hook. f. & Thomson sensu F.B.I. p.p.

Caespitose; stems slender, 2–6cm, brownish-lanate. Basal leaves ovate or obovate, 3–5×2.5–3mm, obtuse or subacute, base rounded and narrowed into a broad petiole 3–5mm, sparsely brownish-ciliate; stem leaves ovate-oblong, 5–7×3–5mm, obtuse, base shallowly cordate, sessile, semiamplexicaul, brownish-lanate especially at margin. Flowers solitary. Calyx 4–5mm, lobes oblong-obovate, 3–4×2–2.5mm, subacute, ciliate. Petals obovate, 6–10×4–5mm, yellow.

Sikkim: Dobinda Pass, Thanka La and Yakla. Moist rocks and gravel at streamsides, 4570m. July–September.

The record (80) of *S. palpebrata* from Sikkim refers to *S. cordigera* (73).

55. S. aristulata Hook. f. & Thomson; ?*S. subaristulata* Engler

Densely caespitose; stems slender, 2–3.5cm, brownish-lanate at base, usually glandular-pubescent above. Basal leaves linear-lanceolate, 5–10×0.75–1.25mm, acuminate to bristle-like point 0.5–1mm, base narrowed, subpetiolate, margin narrowly inrolled, glabrous; cauline leaves similar, decreasing in size upwards. Calyx 3–4mm, lobes ovate, c 2.5× 1mm, glabrous, becoming reflexed. Petals obovate 4–5×1.5–2.5mm, yellow.

Sikkim: Lhonak, Alookthang, Dadong and Sittong; **Chumbi.** 4265m. July–September.

56. S. llonakhensis W. W. Smith

Similar to *S. aristulata* but less densely caespitose; basal leaves usually absent, stem leaves oblong-elliptic 4–6×1–1.5mm, acuminate to an apical bristle up to 0.75mm, base attenuate, margin ± revolute, ciliate; calyx c 4mm glandular-pubescent at base, lobes oblong-ovate c 3×1.5mm, obtuse; petals obovate, 4–7×2.5–3.5mm.

Sikkim: Lhonak and near Thango. 4265m.

57. S. lepida H. Smith; *S. lepidostolonosa* H. Smith

Similar to *S. aristulata* but stems up to 8cm, sparsely lanate at leaf axils below, usually glabrous, rarely glandular above; basal leaves oblanceolate c 8× 1.5mm, obtuse or subacute, base attenuate into petiole up to 10mm, ciliate or glabrous; stem leaves similar, smaller, becoming sessile, glabrous; calyx 4–5mm, lobes ovate c 3×2mm; petals obovate 5–7×2.5–5mm.

Bhutan: N—Upper Mo Chu district (Laya), Upper Mangde Chu district (Ju La) and Upper Bumthang Chu district (Maruthang and Pangotang); **Chumbi:** Chomo Lhari. In grassy crevices and on wet moss on cliff faces, 3650–4420m. July–September.

58. S. caveana W. W. Smith; *S. diapensia* H. Smith

Caespitose; acaulescent, rhizomes bearing numerous leaves of previous years. Leaves mostly basal, oblanceolate, 7–10×3–5mm, subacute, base attenuate into petiole up to 10mm, glabrous. Flower solitary, pedicels 1−2(−7)cm, minutely brownish glandular-pubescent in upper parts. Calyx 7−8mm, lobes ovate 5−6×3−4mm, glabrous or glandular-ciliate. Petals ovate or obovate, 8−10×5−6mm, yellow.

Bhutan: N—Upper Mo Chu, Upper Kuru Chu and Upper Kulong Chu districts; **Sikkim**. Screes and rock crevices, 3050−4880m. July−September.

59. S. glabricaulis H. Smith

Caespitose; stems 2−4(−6)cm, glabrous. Basal leaves ± rosetted, ovate-lanceolate, 4−6×1.5−3mm, obtuse or subacute, base attenuate into petiole up to 1.5cm, ciliate and usually pubescent on upper surface; stem leaves similar, decreasing in size and becoming sessile. Calyx 5−6mm, lobes ovate-elliptic, 4−5×1.5−3mm, glabrous. Petals obovate, 7−8(−10)× 3−5mm, yellow.

Bhutan: C—Thimphu district (Tremo La), Tongsa district (Black Mountain), **N**—Upper Mo Chu, Upper Pho Chu, Upper Bumthang Chu and Upper Kulong Chu districts; **Sikkim**. On rocks and cliff ledges, 4200−4600m. June−July.

60. S. sphaeradena H. Smith

Similar to *S. glabricaulis* but stems usually taller, 5−15cm, densely and shortly blackish-glandular in the upper part; leaves all glabrous; calyx lobes ovate c 4×3mm, glandular; petals obovate, 7−10×4.5−7mm.

Sikkim: Lampokri. 4200m. August.

61. S. montana H. Smith; *S. montanella* H. Smith, *S. hirculus* L. var. *indica* Hook. f., *S. indica* (Hook. f.) Wadhwa,? *S. alookthangensis* Wadhwa

Caespitose; stems 3−10(−25)cm, densely brown-lanate especially in the upper part. Basal leaves elliptic-lanceolate, 5−15×2−4.5mm, subacute, base attenuate into petiole 5−15mm, glabrous. Flowers solitary or few. Calyx 5−7mm, lobes oblong-ovate 4−15×1.5−2mm, brown-pubescent and ciliate. Petals elliptic or obovate 6−10×3−4mm, yellow.

Bhutan: N—Upper Mo Chu to Upper Kulong Chu districts; **Sikkim**. In peaty soil and screes, 4100−4900m. July.

If *S. montanella* has any real identity, it can only be regarded as a dwarf form of *S. montana* in which the stems are normally 5cm but may only be as much as 1cm tall. It is always proportionately a larger-flowered plant than the related *S. parva* Hemsley and its leaves are less crowded than in the latter.

62. S. elliptica Engler & Irmscher

Similar to *S. montana* but smaller and more slender, stems up to 7cm tall, less densely brown lanate; basal leaves ovate-elliptic, 5−10×2.5−3.5mm, acute or subacute, base rounded, narrowing to petiole up to 1.5cm, margin

inrolled, purplish and sparsely brown-lanate beneath, upper surface ± glabrous and with finely granular epidermis; calyx c 4mm, brown lanate at base, lobes ovate, c 2.5×1.5mm, glabrous; petals obovate, 5−6×3−4mm.
Sikkim: Jongri; **Chumbi:** Chomo Lhari. 4880m. September.

63. S. tsangchanensis Franchet;? *S. thimpuana* Wadhwa

Caespitose; stems 7−10cm bearing a few brown-lanate hairs at lower leaf axils, glandular above. Basal leaves ovate-lanceolate, 7−10×3−5mm, acute or obtuse, base attenuate into petioles up to 20mm, bearing scattered long white hairs on upper surface, glabrous beneath; stem leaves elliptic-oblanceolate, decreasing in size. Calyx c 5mm, lobes ovate, 4−5×2mm, glabrous or glandular-ciliate. Petals obovate, 8−9×3.5−4mm.
Bhutan: C—Thimphu district (Barshong); **Chumbi:** above Tsethanka. 3800−4600m. July.

64. S. subspathulata Engler & Irmscher

Similar to *S. tsangchanensis* but leaves smaller and glabrous, basal ones lanceolate or subspathulate, c 3×1−1.5mm, long petiolate, stem leaves linear-lanceolate, 2−4×0.5mm, glabrous, sessile; sepals oblong 1.5mm glabrous; petals obovate c 3.5×2mm.
Bhutan: N—Upper Bumthang Chu district (Tolegang); **Sikkim:** locality unknown; **Chumbi:** Chomolhari. Steep grassy banks, 4880m. September.

65. S. chumbiensis Engler & Irmscher

Caespitose; stems 7−15cm, bearing a few brown lanate hairs at leaf axils, otherwise glabrous. Basal leaves lanceolate or spathulate, c 10×4mm, subacute, base attenuate into petiole up to 2cm, sparsely pilose on upper surface with hairs c 1mm; stem leaves similar, decreasing in size and becoming sessile upwards. Calyx lobes ovate, obtuse, 5.5−6.5×3.5mm, ciliate. Petals obovate 7−7.5×4mm.
Sikkim: between Cho La and Dobinda; **Chumbi:** Chakung La. In screes and peaty meadows, 3600−4600m. July.

66. S. kinchingingae Engler; ?*S. nigroglandulosa* Engler & Irmscher

Similar to *S. chumbiensis* but stems glandular-pubescent above; upper surface of leaves pilose with long pale-coloured hairs 1.5−2mm, glabrous beneath; flowers solitary or 2.
Bhutan: N—Upper Kulong Chu district (Me La); **Sikkim:** Kangchenjunga. In scree, 4100m. August.
Possibly not specifically distinct from *S. chumbiensis*.

67. S. inconspicua W. W. Smith; ?*S. exigua* H. Smith

Densely caespitose. Leaves thickish, subterete, oblong-elliptic, c 2.5× 0.7mm, subacute, glabrous. Flower ± sessile among upper leaves. Calyx c

3mm, glabrous, lobes ovate c 2.5×1.5mm. Petals as long as or shorter than sepals, oblong-elliptic, c 1.5−2×0.5mm, greenish.

Sikkim: Yumchho La, Tanka La, etc. On rocks, 4265−5000m. September.

68. S. saginoides Hook. f. & Thomson; *S. matta-viridis* H. Smith, *S. deminuta* H. Smith

Caespitose, forming tight mats, apparently subdioecious. Leaves linear-elliptic c 4−5×1mm, obtuse or subacute, base narrowed to sheathing brown-ciliate petiolar part, lamina glabrous rarely sparsely ciliate. Flowers solitary, sessile or on brown lanate pedicels up to 1cm. Calyx c 4mm, glabrous, lobes oblong-ovate, c 3×1mm. Petals elliptic c 5×1mm, yellow.

Bhutan: N—Upper Mo Chu, Upper Pho Chu, Upper Mangde Chu and Upper Kulong Chu districts; **Sikkim.** On mossy rocks and screes, 4265−5180m. June−August.

In view of the slight differences that are claimed to exist between this species and *S. matta-viridis* (petioles glabrous but membranous and fimbriate at base) and *S. deminuta* (with leaf margin ciliate) it is doubtful if they can be regarded as specifically distinct as on the Me La the three are recorded to occur together.

69. S. perpusilla Hook. f. & Thomson

Similar to *S. saginoides* but forming small, less dense tufts; leaves spathulate, 2−2.5×1mm, obtuse or subacute, white-ciliate without brown lanate hairs, sessile; flowers solitary on glandular-pubescent pedicels 0.5−1.5cm; calyx c 2.5mm, lobes ovate c 1.5×1mm, ciliate; petals oblong-elliptic c 3.5−4×1.5mm, yellow sometimes orange-spotted below middle within.

Bhutan: N—Upper Bumthang Chu district (Chagen La); **Sikkim:** Mt Donkiah, Chapopla, Zemu valley. On bare earth and scree slopes, 4880−5300m. June−September.

70. S. stella-aurea Hook. f. & Thomson; *S. finitima* W. W. Smith, *S. jacquemontiana* Decaisne var. *stella-aurea* (Hook. f. & Thomson) Clarke

Caespitose; stems 1−1.5cm, densely leafy. Leaves ± rosetted, obovate-spathulate, 2−4×1−2mm, obtuse, narrowed at base, usually glandular-ciliate. Flowers solitary on naked pedicels, 1.5−3(−5)cm, glandular-pubescent. Calyx 3−5mm, lobes ovate-oblong, 2.5−4×1.5−2mm, glandular, becoming reflexed. Petals elliptic, 3.5−6×1.5−2.5mm, yellow.

Bhutan: N—Upper Mo Chu, Upper Pho Chu, Upper Kuru Chu and Upper Kulong Chu districts; **Sikkim.** On open sandy slopes, 3950−4570m. June−September.

A variable species which has been divided into several varieties: var.

ciliata Marquand & Shaw with subacute eglandular-ciliate leaves; var. **polyadena** H. Smith with leaves densely glandular on the back. *S. finitima* W. W. Smith can only be regarded as a larger form of this species.

71. S. jacquemontiana Decaisne

Similar to *S. stella-aurea* but flowers subsessile; leaves usually larger, spathulate, c 5×1.5−2.5mm, glandular-pubescent on the back; calyx c 5mm, glandular, lobes oblong-ovate c 4×2mm, sometimes reddish; petals elliptic 5−6×2−3mm, yellow.

Bhutan: N—Upper Bumthang Chu district (Marlung) and Upper Kuru Chu district (Narim Thang); **Sikkim**. On open sandy screes, 4100−4880m. July−September.

72. S. contraria H. Smith

Loosely caespitose; stems thin, c 5cm, glabrous and sparsely leafy below, becoming more densely leafy and shortly eglandular-pubescent on flowering shoots. Leaves opposite, elliptic-obovate 1.5−2.5×1−1.5mm obtuse, base rounded, sessile, glabrous. Pedicels 1.5−2.5cm, shortly pubescent. Calyx c 3mm, lobes ovate c 2.5×2mm, glabrous. Petals elliptic, 3.5−4×1.5mm, yellow sometimes orange-spotted within or entirely orange-red.

Bhutan: N—Upper Pho Chu district (Woji and Gafoo La) and Upper Bumthang Chu district (Waitang and Marlung). Streamsides and screes, 3950−4400m. June−July.

73. S. punctulata Engler

Stems 3−5cm covered with subsessile glands. Basal leaves spathulate c 5×2mm, coarsely ciliate, warted and wrinkled on upper surface at least when dry; stem leaves elliptic, c 5×1.5mm, densely glandular-pubescent. Flowers (1−)2−6 in axils of upper leaves, pedicels 1−2.5cm. Calyx 4−5mm reddish-brown, lobes ovate 2−3×1.5mm. Petals obovate, 6−8×3mm, white or pale yellow with several (12 or more) red spots on each within.

Sikkim: Cholomo and Lhonak; **Chumbi.** On open screes, 4600−4900m. August−September.

74. S. engleriana H. Smith

Loosely caespitose; basal stems up to 5cm, ± prostrate, glabrous. Leaves alternate ± widely spaced, becoming more crowded at branch ends, alternate, elliptic or obovate, 2.5−3.5×1.5−2mm, obtuse or subacute, base narrowed, sessile, glabrous. Flowers solitary on pedicels 1.5−2cm usually with 1−2 leaves in lower part, eglandular-pubescent. Calyx c 3.5mm, glabrous, lobes broadly ovate, c 2.5×2mm, obtuse. Petals obovate, 3.5−4×2.5mm, yellow, sometimes orange-spotted at base within.

Bhutan: N—Upper Mo Chu district (Lingshi La and Yale La) and Upper Kuru Chu district (Narim Thang); **Sikkim:** Chulung. On open sandy hillsides, 4265−4880m. July−September.

75. S. microphylla Hook. f. & Thomson

Similar to *S. engleriana* but smaller and with green petals; basal stems 1−2.5cm; leaves obovate-spathulate 3.5−5×2mm, obtuse or subacute, glabrous; pedicels c 1cm, pubescent; calyx 5−6mm, lobes oblong c 3×1mm, glabrous; petals elliptic 2.5−3×1mm, sometimes absent.

Sikkim: Cho La. Rock crevices and ledges, 4400m. July−September.

Family 68. PARNASSIACEAE

by A. J. C. Grierson

Perennial herbs. Stems 1-flowered, often bearing one leaf near the middle, rarely naked. Leaves mostly radical, entire, palmately veined, petiolate, exstipulate. Calyx tube ± adnate to ovary, lobes 5 narrowly decurrent on peduncle. Petals 5, white or creamy-white, rarely green. Stamens 5 alternating with 5 shorter nectariferous staminodes. Ovary semi-inferior, 1-celled, ovules numerous, placentation parietal, style short with 3−4 lobes or branches. Fruit a 3−4-valved capsule.

1. PARNASSIA L.

Description as for Parnassiaceae.

1. Petals green; staminodes undivided, ± suborbicular at apex .... **7. P. tenella**
+ Petals white or creamy-white; staminodes divided at apex into 3−7 lobes or segments 2

2. Connective elongated into a thickened point, sometimes longer than anther cells .... **5. P. delavayi**
+ Connective not elongated above anther cells 3

3. Petals lanceolate, ciliate all around except at basal claw .... **6. P. cooperi**
+ Petals obovate, ± entire at apex, toothed, fimbriate or entire near base . 4

4. Staminodes divided into 5−7 linear segments at apex .... **4. P. wightiana**
+ Staminodes 3-lobed 5

5. Leaves rounded, cuneate or shallowly cordate at base, attenuate into petiole; petals 12−17mm .... **1. P. nubicola**
+ Leaves distinctly cordate at base; abruptly narrowed into petiole; petals 5−9mm 6

6. Petals dentate or ciliate below middle; cauline leaves usually well developed .... **2. P. chinensis**
+ Petals entire below middle; cauline leaves often poorly developed or absent .... **3. P. pusilla**

1. P. nubicola Royle. Fig. 36 e−h

Stems erect 15−35cm. Radical leaves ovate 3−10×1.5−4.5cm, acute or subacute, base rounded, cuneate or shallowly cordate, attenuate into petiole up to 12cm, glabrous or petiole with a few long brownish hairs; cauline leaves similar, slightly smaller, usually inserted below middle of stem, subsessile, ± sheathing and bearing long brown ciliate hairs at base. Calyx lobes ovate, 7−10×4.5−6mm, subacute, usually ciliate at base with a few thick brown hairs. Petals obovate, 12−17×7−10mm, fimbriate in lower half. Staminodes c 5mm, broadened and 3-lobed above, mid lobe longer and narrower than lateral ones. Style broadly 3-lobed. Capsule obovate-ellipsoid 1−1.5cm.

Bhutan: C—Thimphu district (Chapcha) and Bumthang district (Kyi Kyi La), **N**—Upper Mo Chu district (Chumiten); **Chumbi; Sikkim.** Sandy soil among grass and low shrubs, 3500−4400m. July−September.

2. P. chinensis Franchet; *P. mysorensis* sensu F.B.I. p.p. non Wight & Arnott

Similar to *P. nubicola* but smaller, stems 2−15(−20)cm; leaves broadly ovate 3−20mm long and broad, subacute, base cordate; petioles 1−10cm; cauline leaves similar but smaller, usually with a few long brown hairs on margin at base; calyx lobes ovate, 2−3×1−2mm, usually with a few brown hairs at basal angles; petals obovate 5−9×3−6mm, ciliate or dentate near base; staminodes c 3.5mm ± equally 3-lobed at apex.

Bhutan: C—Thimphu, Tongsa and Bumthang districts, **N**—Upper Mo Chu, Upper Bumthang Chu, Upper Kuru Chu and Upper Kulong Chu districts; **Sikkim.** On open bare or grassy hillsides, 2550−4400m. June−September.

3. P. pusilla Arnott; *P. mysorensis* sensu F.B.I. p.p. non Wight & Arnott, *P. ovata* sensu F.B.I. non Ledebour, *P. affinis* Hook. f. & Thomson

Similar to *P. nubicola* and more especially to *P. chinensis* but cauline leaves often poorly developed or absent; angles between calyx lobes often without hair; petals entire near base.

Bhutan: N—Upper Pho Chu district (Gafoo La); **Sikkim:** Lachung; **Chumbi:** Phari. On grassy rocks, 2650−4250m. July−September.

4. P. wightiana Wight & Arnott; *P. ornata* Arnott. Fig. 36 j & k

Stems 12−25cm. Leaves broadly ovate 1.5−4cm long and broad, acute or subacute, base deeply cordate with rounded lobes; basal lobes of cauline leaves ± surrounding stem usually with a tuft of brown hairs up to 5mm at axil. Calyx lobes oblong-ovate 7−9×3.5−5mm, subacute, ciliate with long brown hairs at base. Petals obovate 8−13×7−10mm, abruptly tapering into a ciliate-fimbriate claw. Filaments c 3mm. Staminodes c 4mm divided at apex into 5−7 narrow segments.

Bhutan: C—Tongsa district (Rinchen Chu), N—Upper Kuru Chu district (Singhi Dzong); **Sikkim.** In Spruce forests and alpine pastures, 2450−4700m. July−August.

5. P. delavayi Franchet. Fig. 36 i

Similar to *P. wightiana* but petals denticulate or shortly ciliate near base; connective elongated into a long thick point above anther cells and sometimes longer than them; staminodes 3-lobed at apex, mid lobe narrow.

Bhutan: C—Thimphu and Tongsa districts, N—Upper Mo Chu, Upper Pho Chu, Upper Mangde Chu, Upper Bumthang Chu and Upper Kulong Chu districts. On open grassy hillsides and in Fir forests, 3800−3950m. June−July.

6. P. cooperi Evans

Similar to *P. wightiana* but petals lanceolate c 13×3mm, acuminate, tapering abruptly in lower third into a narrow claw, margin ciliate all around except on claw; staminodes 3-lobed, mid lobe minute; style c 2 mm bearing 3 branches c 1mm.

Bhutan: C—Ha district (near Ha); **Sikkim:** near Toong. On sandy banks, 3050m. July−September.

7. P. tenella Hook. f. & Thomson

Stems slender 7−12cm. Radical leaves suborbicular 1.5−3×1.5−3.5cm, rounded or emarginate, base deeply cordate, margin entire or crispate, texture thin; petioles 3−7cm; cauline leaves inserted in upper half of stem, similar to radical ones but smaller. Calyx lobes oblong 2−3×1−1.5mm, rounded. Petals obovate 5×4−5mm, dark green, minutely ciliate. Staminodes c 1.5mm, undivided, flattened and suborbicular at apex. Styles 3, c 0.5mm. Capsule obovoid, c 3mm.

Sikkim: Lachen, Lachung, Kanglasa and Laghep. On damp mossy rocks and banks in mixed forests, 3050−3650m. July.

Family 69. HYDRANGEACEAE

by A.J.C. Grierson

Trees or shrubs sometimes climbing. Leaves opposite, simple, exstipulate, pinnately veined. Flowers in terminal cymosely branched corymbs or panicles, bisexual, or outer flowers sterile with 4–5 large petaloid sepals. Calyx tube ± adnate to ovary, 4–5-lobed. Petals 4–5 free or cohering. Stamens 8–12. Disk absent. Ovary inferior or semi-inferior, 2–6-celled, styles as numerous, ovules many on parietal placentae. Fruit a capsule or berry.

1. Sterile flowers usually present at margin of corymb; fruit capsular
   1. Hydrangea
+ Sterile flowers always absent; fruit a berry .... **2. Dichroa**

1. HYDRANGEA L.

Description as for Hydrangeaceae but with the differential characters shown in the key.

1. Sterile flowers many, more numerous than bisexual flowers; teeth at leaf margins obtuse (cultivated plants) .... **5. H. macrophylla**
+ Sterile flowers absent or fewer than bisexual flowers; teeth at leaf margins acute or acuminate (indigenous plants) 2

2. Hairs on lower surface of leaves confined to vein axils; teeth at leaf margin acute; petals cohering and falling together as a cap . **1. H. anomala**
+ Hairs on lower surface of leaves dense or sparse sometimes only along veins; teeth at leaf margin acuminate; petals free, spreading separately . 3

3. Teeth at leaf margins ± regularly alternating large and small; bracts densely appressed-pubescent on the back; styles much broadened at apex
   2. H. aspera
+ Teeth at leaf margins ± uniform in size; bracts sparsely pubescent or absent; styles not or slightly broadened at apex 4

4. Fertile flowers white; bracts linear-lanceolate, sparsely pubescent or ciliate; styles connate at base .... **3. H. heteromalla**
+ Fertile flowers blue; bracts apparently absent; styles free at base
   4. H. stylosa

FIG. 36. **Hydrangeaceae, Parnassiaceae, Grossulariaceae, Philadelphaceae, Iteaceae** and **Pittosporaceae. Hydrangeaceae.** a–d, *Hydrangea anomala:* a, leaves and inflorescence; b, bud; c, unfolding flower with cap of petals; d, mature capsule. **Parnassiaceae.** e–h, *Parnassia nubicola:* e, habit; f, flower with one calyx lobe and two petals removed; g, staminode; h, top of ovary with style. i, *Parnassia delavayi:* upper part of stamen with elongated connective, j & k, *Parnassia wightiana:* j, petal; k, staminode. **Grossulariaceae.** l–o, *Ribes griffithii:* l, portion of shoot with leaves and inflorescence; m, flower; n, flower in section; o, portion of infructescence. **Philadphaceae.** p–t, *Deutzia corymbosa:* p, portion of shoot with leaves and inflorescence; q, magnified portion of under side of leaf; r, flower with two petals removed; s, stamen; t, mature capsule. **Iteaceae.** u–w, *Itea macrophylla:* u, portion of shoot with leaf and inflorescence; v, flower; w, maturing capsule. **Pittosporaceae.** x–z, *Pittosporum napaulense:* x, portion of shoot with leaves and inflorescence; y, flower; z, dehisced capsule. Scale: a × ⅖; p, u, x × ½; e, l, o × ⅔; f × 1½; m, n, r × 2; j, t, w, y, z × 3; b, c, q, s × 4; d, i, v × 5; k × 6; g, h × 7.

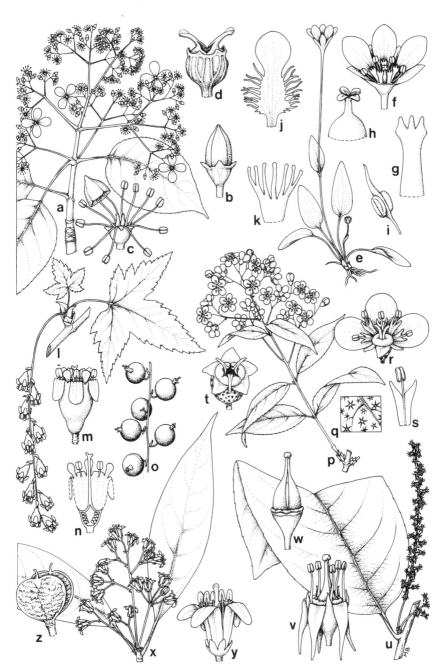

69. HYDRANGEACEAE

1. H. anomala D. Don; *H. altissima* Wall. Fig. 36 a–d
Climbing shrub, young growth finely and sparsely pubescent. Leaves ovate 6–15×4–9cm, acuminate, base rounded or cuneate, margins serrate with acute teeth, pubescent along the midrib above and with a few brown hairs in vein axils beneath, otherwise glabrous; petioles 3.5–8cm. Corymbs loose, spreading, up to 25cm across; bracts elliptic 0.7–1.5×0.1–0.3cm, glabrous. Fertile flowers: calyx including lobes 2.5–3mm, glabrous, lobes acute or subacute; petals c 2.5mm cohering into a conical cap and falling as stamens elongate; stamens c 4mm; styles 2 or 3, spreading, c 2mm, slightly thickened above; capsules subglobose 3–4mm, ± strongly ribbed; seeds compressed, broadly elliptic 1.5–2mm, surrounded by a wing. Sterile flowers: calyx lobes broadly obovate, 1–2×0.8–1.5cm, rounded, entire, whitish.
Bhutan: S—Chukka and Deothang districts, **C**—Punakha and Tashigang districts, **N**—Upper Kuru Chu district; **Sikkim.** Climbing on trees in Warm broad-leaved forests, 1800–2133m. April–May.

2. H. aspera D. Don; *H. robusta* Hook. f. & Thomson. Sha: *Sungmulagu Shing;* Nep: *Bhogote* (34), *Bhokote* (34)
Similar to *H. anomala* but usually an erect shrub 3–5m, young growths densely greyish pubescent; leaves broadly ovate 12–27×6–17cm, margin ± doubly serrate with fine acuminate teeth alternating in size, pilose on both surfaces usually more dense beneath; petioles 5–12cm; bracts lanceolate up to 15×5mm, densely appressed pubescent, deciduous; fertile flowers: calyx including lobes c 1.5mm, lobes ± acute, petals lanceolate c 3mm, spreading and falling separately, purplish, stamens 5–8mm, styles c 2mm abruptly broadened ± reniform at apex, c 1mm broad; capsule broadly campanulate c 3.5mm long and broad, truncate above, seeds ellipsoid c 1mm tapering at either end into a tail; sterile flowers: calyx lobes broadly ovate or elliptic 1.5–3cm long and broad, creamy white, margin ± serrate.
Bhutan: S—Samchi, Chukka, Gaylegphug and Deothang districts, **C**—Thimphu and Tashigang districts, **N**—Upper Mo Chu district; **Sikkim.** Warm broad-leaved forests, 1000–2133m. August.

The above description applies to the broad-leaved subsp. **robusta** (Hook. f. & Thomson) McClintock; subsp. **aspera** with lanceolate or narrowly ovate leaves has also been recorded from Darjeeling and Sikkim.

3. H. heteromalla D. Don; *H. vestita* sensu F.B.I. non Wall. Nep: *Halonre*
Similar to *H. anomala* but always a bush or tree 2–10m; young growths with scattered weak hairs at first, soon glabrous; leaves ovate-elliptic 10–18×4.5–11cm, margin uniformly serrate with acuminate teeth, puberulous along main veins above, sometimes almost glabrous beneath with pilose hairs along veins to densely greyish pilose; petioles 2–5cm; corymbs ± densely flowered, bracts linear-oblanceolate 1–2×0.1–0.4cm,

sparsely pubescent or ciliate; fertile flowers: calyx including lobes c 2mm, lobes acute, petals c 3mm white, spreading, stamens 2−4mm, styles usually 3, c 2mm united in lower half, slightly thickened above, capsules ellipsoid, c 5×2.5mm including styles, bearing persistent calyx lobes slightly above middle; seeds 1.5−2mm ellipsoid, tapering at each end into a fine tail; sterile flowers: calyx lobes ovate 1.5−2×1−1.5cm, rounded or acute, margin entire.

Bhutan: S—Chukka district, C—Thimphu and Bumthang districts, N—Upper Mo Chu district; **Sikkim.** Cloud forest especially Spruce and Hemlock, 2450−3350m. June−August.

4. H. stylosa Hook. f. & Thomson; *H. macrophylla* (Thunberg) Seringe subsp. *stylosa* (Hook. f. & Thomson) McClintock

Erect shrub 1−2m, young growths finely pubescent. Leaves ovate-elliptic 6−10×3−5cm, acuminate, base cuneate, margin sharply serrate in upper half, scattered weakly pilose above, pilose along veins beneath; petioles 1−2cm. Corymbs usually small up to 12cm across, bracts apparently absent. Fertile flowers: calyx including obtuse lobes c 2.5mm; petals lanceolate c 3 mm, bluish, reflexed; stamens 4−4.5mm; styles slender c 2mm spreading; capsules ovoid c 3.5mm. Sterile flowers: calyx lobes broadly elliptic 1−2cm long and broad, white or bluish, usually with a few coarse teeth in upper part.

Bhutan: C—Punakha, Tongsa, Bumthang and Tashigang districts, N—Upper Mo Chu district; **Sikkim.** In mixed forest, 2130−2750m. June−July.

5. H. macrophylla (Thunberg) Seringe. Sha: *Dosem Metog*

Similar to *H. stylosa* but leaves obovate 7−12×5−7cm, ± abruptly acuminate, base cuneate, margin bluntly serrate, glabrous above, usually with a few brownish hairs in vein axils beneath. Fertile flowers: calyx including lobes c 2mm; petals lanceolate c 3.5mm, white or purplish, reflexed; stamens ± as long as petals; styles c 1.5mm; sterile flowers: calyx lobes broadly ovate or elliptic, up to 1.5cm long and broad, blue or white.

Bhutan: cultivated e.g. Tongsa district (Shamgong). 2000m. June−August.

2. DICHROA Loureiro

Description as for Hydrangeaceae but with the differential characters shown in the key.

1. D. febrifuga Loureiro. Dz: *Hindo Nam, Hogena;* Nep: *Basak* (34), *Bashak, Hil Bashak, Basuri*

Shrub 1–3m. Leaves elliptic-oblanceolate 12–15×3.5–7cm, acuminate, base cuneate or attenuate, margin serrate, sparsely pubescent on both surfaces; petioles 1–3.5cm. Calyx including acute lobes 3–4mm. Petals elliptic 5–6×1.5–2.5mm, pale blue, becoming reflexed. Stamens 5–6mm purplish. Styles 3–5, c 3mm, thickened at apex. Berries subglobose 6–8mm diameter, intense metallic blue.

Bhutan: S—Samchi, Phuntsholing and Gaylegphug districts, C—Punakha, Tongsa, Mongar and Tashigang districts; **Sikkim.** Margins of Warm broad-leaved and Evergreen Oak forests, 1000–2300m. May–November.

The shoots and bark of the roots are used to prepare a febrifuge (34).

Family 70. GROSSULARIACEAE

by A.J.C. Grierson

Shrubs, sometimes spiny, often glandular. Leaves alternate, simple, often palmately lobed, exstipulate. Flowers mostly in axillary racemes on short lateral shoots, rarely solitary or few, axillary, bisexual or unisexual (the plants then dioecious). Calyx shallowly cup-shaped or tubular, 5-lobed, often petaloid. Petals 5, adnate to calyx and smaller than calyx lobes. Stamens 5, opposite calyx lobes, usually present but functionless in female flowers. Ovary inferior, 1-celled, style single, bifid above, stigmata usually capitate, ovules numerous on two parietal placentae, styles but not ovaries usually present in male flowers. Fruit a berry, bearing persistent flower remains above.

Some species of *Ribes* are important as alternate hosts of Himalayan coniferous rust fungi (16).

1. RIBES L.

Description as for Grossulariaceae

1. Flowers bisexual, solitary, or few, axillary, or numerous in pendent racemes; calyx above ovary, including lobes, 6–9mm 2
+ Flower unisexual, few–numerous in ± erect racemes; calyx above ovary, including lobes, 2–5mm 3

2. Stems spiny; flowers borne singly .... **1. R. alpestre**
+ Stems unarmed; flowers in racemes .... **Species 2 & 3**

3. Leaves suborbicular, 1−2(−3)cm long and broad, shallowly 3−5-lobed, lobes rounded, crenately toothed .... **4. R. orientale**
+ Leaves ovate or broadly ovate in outline, (2−)3−9cm long and broad, unlobed or shallowly or deeply 3−5-lobed, lobes acute or acuminate, acutely toothed 4

4. Leaves large, commonly 5−9cm long and broad, unlobed or shallowly 3−5-lobed, mid-lobe longer than lateral ones, acuminate; racemes, especially the male ones, elongate (up to 15cm) .... **5. R. acuminatum**
+ Leaves small, rarely more than 4(−6)cm long and broad, shallowly or deeply 3−5-lobed; racemes 2−4cm 5

5. Leaves deeply lobed, mid-lobe gradually acuminate, more than twice as long as lateral ones .... **6. R. laciniatum**
+ Leaves shallowly lobed, mid-lobe acute or subacuminate as long as or up to 1.5 times as long as lateral ones .... **Species 7 & 8**

1. R. alpestre Decaisne; *R. grossularia* sensu F.B.I. non L.

Shrub 2−3m, shoots bearing a 3-fid spine 0.5−1.5 (−2)cm at each leaf axil. Leaves ± suborbicular in outline, 1.5−2(−4)cm across, 3−5-lobed, obtusely toothed, base truncate or shallowly cordate, sparsely pubescent with glandular and eglandular hairs; petioles 0.5−1.5cm, densely pubescent on upper surface. Flowers solitary or 2, greenish or flushed crimson, pendulous, peduncles 5−7mm, glandular-pubescent; bracts ovate, c 2mm. Sepals oblong c 4×1.5mm, reflexed. Petals elliptic, c 3×1mm. Stamens slightly longer than petals. Ovary obovoid c 3mm, covered with long glandular hairs. Fruit ellipsoid, c 1.5×1cm, purplish or yellow, sparsely glandular-pubescent.

Bhutan: N—Upper Mo Chu district (Yabu Thang to Laya) and Upper Bumthang Chu district (Pangotang). Forest margins, 3400−3800m. May−June.

2. R. griffithii Hook. f. & Thomson. Fig. 36 l−o

Shrub 1−4m. Leaves broadly ovate, 3−5-lobed, 3.5−12×4.5−12cm, acute or acuminate, base cordate, margin sharply serrate, often bearing scattered gland-tipped hairs on upper surface at first, whitish pubescent along veins beneath; petioles 4−9cm, sparsely ciliate near base with cilia up to 7mm. Flowers ± precocious, greenish-yellow or purplish in pendulous racemes up to 15cm, bracts oblong-ovate, 3−7×1mm, pubescent, pedicels up to 5mm. Calyx tube campanulate, 5−6mm, lobes oblong 3×1.5mm obtuse, reflexed, glabrous or sparsely pubescent. Petals obovate, c 2× 1.5mm, erect. Stamens ± as long as petals. Berries ovoid, 1−1.5cm long and broad, red.

Bhutan: C—Ha, Thimphu and Tongsa districts, **N**—Upper Mo Chu,

Upper Bumthang Chu and Upper Kulong Chu districts; **Sikkim**. In forests and on hillsides, 2750−3500m. April−May.

3. R. himalense Decaisne; *R. rubrum* sensu F.B.I. non L., *R. emodense* Rehder

Similar to *R. griffithii* but reddish-brown epidermis peeling to reveal greyish-black bark; leaves whitish pubescent beneath, bracts oblong, 1−1.5mm, somewhat reflexed; sepals obovate, c 3×2.5 mm, erect; petals c 1.5×1mm.

Bhutan: C—Ha, Thimphu, Tongsa and Bumthang districts. Beside streams, 2450−3050m. April.

4. R. orientale Poiret

Dioecious shrub 30−130cm. Leaves broadly ovate or suborbicular, 1−2.5×1.5−3cm, shallowly 3−5-lobed, rounded, base cordate, margin shallowly serrate, pubescent with fine eglandular hairs interspersed with coarser glandular bristles on both surfaces but especially beneath; petioles 0.5−1.5cm, pubescent and glandular. Racemes erect, 5−10-flowered (female racemes usually shorter and fewer-flowered than males), rachis pubescent and glandular; bracts elliptic c 3×1mm, glandular-ciliate, pedicels ± as long. Calyx purplish, lobes ovate, c 1.5×1mm glandular-pubescent. Petals half as large as calyx lobes. Male flowers shallowly campanulate; female flowers more deeply so. Berries ± globose, 4−5mm, reddish, pubescent especially near apex.

Bhutan: C—Thimphu district (Shodug), N—Upper Mo Chu district (Lingshi Dzong). 3800m. May.

5. R. acuminatum G. Don; *R. desmocarpum* Hook. f. & Thomson, ?*R. takare* D. Don. Dz: *Chamze Nam*

Dioecious shrub 2−4m, bark often blackish. Leaves broadly ovate, shallowly 3-lobed or sometimes unlobed, 4−9×3−7cm, acuminate, base cordate, margin subacutely serrate, sparsely pubescent with coarse glandular hairs on upper surface, ± densely pubescent with glandular and eglandular hairs beneath; petioles 1−6cm, glandular-pubescent. Racemes ± erect, 4−8(−15)cm, densely flowered; bracts narrowly elliptic, 5−7mm, longer than pedicels. Flowers green, sometimes tinged crimson, 5−6mm diameter, calyx lobes ovate-oblong c 3×1.5mm acute. Petals obovate, c 0.5 mm. Berries broadly ellipsoid, c 7× 5mm, red, glandular-pubescent.

Bhutan: C—Thimphu, Punakha, Tongsa, Bumthang and Mongar districts, N—Upper Mo Chu and Upper Kulong Chu districts; **Sikkim**. Margins of coniferous forests, 2740−3950m. May−June.

Hooker and Thomson are to be followed in regarding *R. takare* D. Don as indeterminable; the original description could apply to several species and as the type specimen appears to be lost, it certainly cannot be accepted as an earlier name for *R. acuminatum*.

6. R. laciniatum Hook. f. & Thomson; *R. tenue* Janczewski

Slender, dioecious shrub 1−4m. Leaves usually deeply 3-lobed, 2−2.5(−7)×1.5−2(−6)cm, mid-lobe longer than lateral ones, gradually acuminate, base rounded, truncate or cordate, margin coarsely and sharply serrate, sparsely glandular-pubescent above, glabrous or with scattered subsessile glands beneath; petioles 1−3cm. Racemes 3−5cm, rather distantly flowered, minutely pubescent with eglandular and glandular hairs; bracts lanceolate, 5−8×0.75−1mm, sparsely ciliate. Flowers crimson or purple. Calyx lobes ovate-lanceolate, 2−2.5× 0.75mm. Berries ellipsoid, c 4.5×3.5mm, red, glabrous.

Bhutan: C—Thimphu, Tongsa, Bumthang and Mongar districts, **N**—Upper Kulong Chu district (Me La); **Sikkim**. In Fir/Rhododendron forests, 2440−3950m. April−June.

7. R. luridum Hook. f. & Thomson

Dioecious shrub 1−1.5m, bark usually blackish. Leaves broadly ovate in outline, 2−3.5cm long and broad, shallowly 3-lobed, mid-lobe ± as long as lateral ones, acute or subacute, base truncate or cordate, sparsely and minutely pubescent on both surfaces interspersed with subsessile glands beneath; petioles 1−2.5cm, finely pubescent and glandular. Racemes erect, 2−3.5cm, finely pubescent. Flowers dark purplish-black, c 5mm diameter, minutely pubescent, pedicels 1−4mm, bracts obovate c 5×2.5mm. Berries globose c 7mm, black.

Bhutan: C—Thimphu district (Barshong), **N**—Upper Mo Chu district (Lingshi); **Sikkim**: Lachen. Open hillsides, 3800−4250m. June.

8. R. glaciale Wall. Nep: *Kimbu* (34)

Similar to *R. luridum* but a taller shrub up to 5m; leaves with subacuminate mid-lobes, more acutely serrate, sparsely glandular pubescent on upper surface, glabrous beneath; racemes puberulous, bracts linear-lanceolate 2−4×1mm, glabrous; flowers purplish-crimson; fruit globose c 7mm diameter, red.

Bhutan: C—Thimphu district (Cheka), Punakha district (Pele La) and Tongsa district (Tunle La and Phobsikha), **N**—Upper Mo Chu district (Lingshi); **Sikkim**: Zemu Valley. Forest margins and hillsides among Rhododendrons, 2740−3050m. April−May.

Family 71. PHILADELPHACEAE

by A. J. C. Grierson

Erect shrubs sometimes with stellate indumentum. Leaves opposite, simple, exstipulate, pinnately or palmately 3−5-veined at base. Flowers bisexual, in terminal racemes or cymose corymbs on short lateral shoots. Calyx tube adnate to ovary, lobes 4−5. Petals 4−5, imbricate or valvate. Stamens 10−numerous, filaments sometimes lobed or toothed above. Ovary inferior, 3−5-celled, styles 3−5, free or united at base. Fruit a capsule.

Most species of *Deutzia* and *Philadelphus* are valued as ornamental shrubs.

1. Leaves pinnately veined, bearing stellate hairs; flowers in cymose corymbs; petals 5; stamens 10 in two series, filaments broadened above and often ending in a tooth on either side of the anther; styles free **1. Deutzia**
+ Leaves palmately 3−5-veined at base, bearing simple hairs; flowers in racemes; petals usually 4; stamens numerous (c 30), filaments slender, unappendaged; styles usually united at base .... **2. Philadelphus**

1. DEUTZIA Thunberg

Description as for Philadelphaceae but with the differential characters shown in the key.

1. D. corymbosa G. Don; *D. hookeriana* (Schneider) Airy Shaw, *D. compacta* auct. non Craib. Fig. 36 p−t

Shrub 1−3m. Leaves ovate, 3.5−7×1.5−2cm, acuminate, base rounded, margin finely serrate, sparsely pubescent on both surfaces with minute stellate hairs; petioles 3−5mm. Calyx including ovate, rounded or subacute lobes 3−4mm, sparsely or densely stellate pubescent. Petals broadly obovate, 4−6×3−5mm, imbricate, white or pink tinged. Stamens 2−4mm, inner series shorter. Styles 3−3.5mm. Capsules broadly campanulate, 3−4mm, opening by 3−5 valves at apex.

Bhutan: S—Chukka district, **C**—Thimphu, Bumthang, Mongar and Sakden districts, **N**—Upper Mangde Chu and Upper Kuru Chu districts; **Sikkim.** Blue Pine, Hemlock and Fir forest margins, 2100−3350m. June−July.

2. D. staminea Wall.; *D. bhutanensis* Zaikonnikova

Similar to *D. corymbosa* but petals oblong-elliptic 7−10×3−5mm, valvate, white or purplish, stamens 6−7.5mm.

Bhutan: S—Chukka district (near Chukka), **C**—Tashigang district (Yonpu La); **Sikkim.** Margins of Warm broad-leaved and Evergreen Oak forests, 1050−2550m. April−June.

3. D. crenata Siebold & Zuccarini

Similar to *D. corymbosa* and, more especially, to *D. staminea* in petal shape; leaves subcoriaceous, broadly ovate 4−5.5×2−3.5cm; inflorescence slender almost racemiform; petals ovate-oblong, 8−10×2−3mm.

Sikkim: the double form (forma **candidissima** (Bonard) Hara) is cultivated around Darjeeling, 2100m. June. (69).

2. PHILADELPHUS L.

Description as for Philadelphaceae but with the differential characters shown in the key.

1. P. tomentosus G. Don; *P. coronarius* L. var. *tomentosus* (G. Don) Clarke.

Shrub 2−6m. Leaves ovate, 4−10×2−5cm, gradually acuminate, base rounded, margin minutely and distantly serrate, glabrous above, sparsely or densely pubescent beneath, rarely completely glabrous; petioles 0.5−1cm. Racemes (3−)5−7-flowered, pedicels 5−10mm. Calyx tube 4−5mm, usually sparsely pubescent with a line of hairs beneath each of the angles between the lobes; lobes ovate 5−6×3mm acuminate, pubescent within. Petals obovate 10−15×7−12mm, white. Stamens c 8mm. Styles c 7mm, sometimes pubescent at base. Capsule c 10×5mm, slightly 4-angled. Seeds ellipsoid c 1mm, tapering at one end into a tail ± as long.

Bhutan: C−Ha to Tashigang districts, N−Upper Mo Chu and Upper Kulong Chu districts; **Sikkim**; **Chumbi.** Margins of Blue Pine and Evergreen Oak forests and in dry scrub, 2250−3050m. May−July.

A variable species in respect to the degree of pubescence on leaves, calyces and styles.

Family 72. ITEACEAE

by A. J. C. Grierson

Trees. Leaves alternate, simple, pinnately veined, exstipulate. Racemes in axillary clusters, many flowered. Calyx campanulate, 5-lobed, adnate to lower half of ovary. Petals 5, larger than calyx lobes, valvate. Stamens 5. Disc present, lobes alternating with stamens. Ovary semi-inferior, carpels 2, connate above into a simple style and capitate stigma. Carpels eventually separating above and dehiscing along ventral suture at maturity.

1. ITEA L.

Description as for Iteaceae.

1. I. macrophylla Roxb. Nep: *Tilki* (34). Fig. 36 u−w

Tree 5−7m. Leaves elliptic or ovate, 15−23×7−13cm, acute or shortly acuminate, base rounded, margin distantly and shallowly serrate, glabrous, veins 7−10 pairs; petioles 2−4cm. Calyx including narrowly triangular lobes c 3mm. Petals triangular c 3×1mm, becoming reflexed. Stamens c 2mm. Carpels 6−8mm, lower half surrounded by calyx cup; perianth segments deciduous to leave a distinct rim.

Bhutan: S—Samchi district (near Samchi), Chukka district (Marichong) and Deothang district (near Samdrup Jongkhar); **Sikkim.** Subtropical forest, 400−600m. May.

Family 73. PITTOSPORACEAE

by D. G. Long

Evergreen trees or shrubs, sometimes climbing; branchlets often whorled. Leaves alternate, often clustered at branch-ends, simple, entire, pinnately veined, exstipulate. Flowers in terminal subumbellate panicles, bisexual, actinomorphic. Sepals 5, free or connate at base, imbricate. Petals 5, free and erect or sometimes connate in lower part, spreading above. Stamens 5, opposite sepals. Ovary superior, 2-celled; style 1, columnar, stigma minute; ovules numerous, axile or parietal. Fruit a 2-or 3-valved capsule.

1. PITTOSPORUM Solander

Description as for Pittosporaceae.

1. Petals c 6mm; capsules c 0.5cm, 2-valved .... **1. P. napaulense**
+ Petals 10−12mm; capsules c 2cm, 3-valved .... **2. P. podocarpum**

1. P. napaulense (DC.) Rehder & Wilson; *P. floribundum* sensu F.B.I. p.p. non Wight & Arnott, *P. napaulense* var. *rawalpindiense* Gowda. Nep: *Phurke* (34), *Khorsane* (34), *Tsaulane*. Fig. 36 x−z

Shrub, sometimes epiphytic or climbing, or small tree to 6m. Leaves crowded near branch-ends, coriaceous, elliptic, 7−20×2.5−5.5cm, acuminate, base attenuate, glabrous; petioles 0.8−2.1cm. Panicles few to many, each 2−5cm, whitish pubescent. Flowers sweet-scented. Sepals ovate, c 2mm, shortly connate at base. Petals yellow, obovate-oblong, c 6× 2mm, free to base. Capsules compressed, broader than long, c 5×7mm, striate within, with 4−6 red, sticky seeds.

Bhutan: **S**—Chukka and Gaylegphug districts, **C**—Punakha, Tongsa and Mongar districts; **Sikkim.** Warm evergreen broad-leaved forests, 1370−2100m. May−August.

Merits cultivation as an ornamental (48).

2. P. podocarpum Gagnepain; *P. glabratum* sensu F.B.I. non Lindley

Similar to *P. napaulense* but leaves oblanceolate, 6−10×2−3cm; flowers larger, petals 10−12×2mm; fruit larger, c 2cm, 3-valved.

Sikkim: Darjeeling. Warm broad-leaved forests. April−May.

This record is based on a single specimen collected by Griffith; according to Hooker & Thomson (80) the specimen may be mis-labelled.

Family 74. ROSACEAE

by A. J. C. Grierson & D. G. Long

Trees, shrubs or herbs; sometimes armed, indumentum of simple hairs (sometimes stellate in *Rubus*). Leaves alternate, simple or pinnately, palmately or pedately compound (triternate in *Aruncus*), pinnately veined; stipules usually present. Flowers solitary or often in fascicles, racemes, cymes, corymbs or panicles, actinomorphic, bisexual, rarely unisexual. Calyx lobes, petals and stamens inserted at margin of calyx cup or tube (hypanthium). Calyx lobes 4−6, sometimes with as many outer epicalyx segments. Petals 4−6 or more, free. Stamens 4−many. Ovary of 1−many free, superior carpels or 4−6-celled, inferior or semi-inferior and united to calyx-tube; styles simple, free, sometimes united; ovules 1 or more per cell. Fruit an achene (often aggregated), follicle, drupe or pome.

1. Leaves pinnately, palmately, pedately or triternately compound 2
+ Leaves simple (sometimes lobed) 16

2. Shrubs, stems prickly or at least stiffly bristly 3
+ Trees, shrubs or herbs; stems not prickly or bristly 4

3. Corolla less than 3cm diameter (up to 4cm in *R. wardii* but then leaves trifoliate); fruits an aggregate of fleshy drupelets on a conical receptacle .... **7. Rubus**
+ Corolla more than 3cm diameter; leaves pinnate; fruit of numerous achenes enclosed within a fleshy enlarged calyx tube (hip) .... **17. Rosa**

4. Herbs with triternate leaves; exstipulate .... **2. Aruncus**
+ Herbs, shrubs or trees with trifoliate, pinnate, palmate or pedate leaves; stipulate 5

5. Flowers in compact globose heads; petals absent .... **16. Sanguisorba**
+ Flowers solitary or in corymbs, racemes or panicles; petals present 6

6. Leaves pinnate, lateral leaflets in 2–many pairs 7
+ Leaves 3-foliate or palmately or pedately 5-foliate (in *Fragaria nubicola*, *Potentilla saundersiana* and *P. fragariodes* sometimes with an additional pair of small leaflets and appearing pinnate) 12

7. Trees, more than 1m tall .... **21. Sorbus**
+ Herbs or shrubs, up to 1m 8

8. Flowers 10–50 in racemes; styles 2; achenes completely enclosed in calyx tube 9
+ Flowers 3–15 in cymes or corymbs, 1–5 in short racemes or 1–2 in axils; styles 5 or more; achenes not enclosed in calyx tube 10

9. Leaflets serrate to base; racemes spike-like, 20–50-flowered; pedicels 1–2mm; calyx bearing hooked spines .... **14. Agrimonia**
+ Leaflets 2–3-fid at apex; racemes lax, 10–20-flowered; pedicels (10–)15–40mm; calyx without hooked spines .... **15. Spenceria**

10. Terminal leaflet of basal leaves much larger than laterals (in *G. elatum* lateral leaflets sessile and broadly attached to rachis); styles usually persistent (in *G. macrosepalum* styles deciduous to leave a beak) **13. Geum**
+ Terminal leaflet of basal leaves not or scarcely larger than laterals; styles completely deciduous 11

11. Petals 3–13mm long; stamens 10 or more .... **8. Potentilla**
+ Petals c 2mm long; stamens 5 .... **9. Sibbaldia** *(S. micropetala)*

12. Epicalyx segments absent; leaves palmately or pinnately 3-foliate, or pedately 5-foliate .... **7. Rubus**
+ Epicalyx segments present; leaves palmately 3 or 5-foliate 13

13. Rosetted herbs with long slender stolons; receptacle enlarged and becoming fleshy at maturity 14
+ Erect, prostrate or sometimes rosetted herbs, or erect shrubs, never stoloniferous; receptacle neither enlarged nor fleshy at maturity 15

14. Epicalyx segments elliptic-lanceolate, acuminate, entire or with a few teeth; petals white .... **10. Fragaria**
+ Epicalyx segments obovate, rounded, coarsely toothed at apex; petals yellow .... **11. Duchesnea**

15. Petals 1−3mm; stamens usually 5, occasionally 10. .... **9. Sibbaldia**
+ Petals 3−10mm; stamens more than 10 .... **8. Potentilla**

16. High alpine tufted herbs; leaves 3-fid at apex or with petioles broadly winged and ciliate 17
+ Shrubs and trees from various altitudes; leaves sometimes lobed but not 3-fid at apex; petioles never broadly winged and ciliate 18

17. Leaves 3-fid at apex; petioles not winged and ciliate; carpels c 10; epicalyx present .... **9. Sibbaldia** (*S. trullifolia*)
+ Leaves entire; petioles broadly winged and ciliate; carpels 2; epicalyx absent .... **12. Brachycaulos**

18. Mature stems (and leaves in *Rubus*) bearing spines or prickles 19
+ Mature stems unarmed 24

19. Prickles superficial, up to 5mm, randomly borne on stems, leaves and petioles .... **7. Rubus**
+ Spines robust, 10−35mm, derived from modified lateral branchlets . 20

20. Flowers 1−3 in clusters; petals 10−30mm; fruit an ovoid or subglobose pome 2.5−7cm 21
+ Flowers numerous in racemes or cymes; petals 3−7mm; fruit a pome or drupe less than 1.5cm 22

21. Stipules leafy, suborbicular 10−15mm; leaves elliptic, obtuse or acute; cultivated .... **24. Choenomeles**
+ Stipules subulate, 2−4mm; leaves ovate, acuminate; native
25. Docynia

22. Leaves deeply pinnatifid .... **20. Crataegus**
+ Leaves unlobed 23

23. Deciduous; leaves elliptic-lanceolate, acuminate; flowers in short axillary racemes; petals 6−7mm; drupes 10−15mm .... **6. Prinsepia**
+ Evergreen; leaves oblong or obovate, obtuse; flowers in corymbose cymes on short lateral shoots; petals 3−5mm; pomes 6−7mm
19. Pyracantha

24. Calyx with epicalyx segments .... **8. Potentilla** (*P. arbuscula* var. *unifoliolata*)
+ Calyx without epicalyx segments 25

25. Calyx tube flat or shallowly cup-shaped, carpels visible in flower 26
+ Calyx tube campanulate or tubular, carpels hidden in flower 27

26. Exstipulate; petals 1.5–3mm, carpels 5 on an almost flat receptacle; fruit a cluster of 5 follicles .... **1. Spiraea**
+ Stipulate; petals 10mm or more, carpels many, on a convex receptacle; fruit an aggregate of fleshy 1-seeded drupes .... **7. Rubus**

27. Leaves 3-lobed (or if unlobed then lower racemes supra-axillary and fruit a single follicle) 28
+ Leaves unlobed (except occasionally in juvenile plants of *Docynia*); fruit always fleshy 29

28. Leaves mostly shallowly 3-lobed, sometimes unlobed; flowers in racemes or panicles .... **3. Neillia**
+ Leaves 3-lobed to middle; flowers sub-fasciculate on lateral branches
21. Sorbus (*S. bhutanica*)

29. Leaf margins towards base bearing numerous gland-tipped hairs; stipules 15–20mm, often persistent; calyx lobes and petals similar
4. Maddenia
+ Leaf margins without gland-tipped hairs (margins or marginal teeth often sessile-glandular in *Prunus*); stipules 2–15mm, mostly caducous, or if persistent then smaller; calyx lobes and petals dissimilar 30

30. Style 1; fruit a 1-seeded drupe, usually fleshy, rarely dehiscent
5. Prunus
+ Styles 2–6; fruit a (1–) 2–15-seeded pome 31

31. Petals showy, 8–30mm; flowers 1–3 in fascicles or 3–12(–20) in corymbs; fruit 25–80mm (7–15mm in *Malus* but then borne on slender pedicels 15–40mm) 32
+ Petals small, 2–8mm; flowers many in corymbose cymes or panicles (sometimes few-flowered in *Cotoneaster*); fruit 5–18mm, on short pedicels up to 12mm 35

32. Flowers 1–3 at branch ends; calyx, including tube and lobes, 16–23mm, white tomentose externally; fruit yellow, usually 15-seeded .
25. Docynia
+ Flowers 3–12(–20) in terminal corymbs; calyx 8–14mm, glabrous or pubescent externally (densely in *Malus pumila*); fruit yellow-green, brown or red, 3–10-seeded 33

33. Leaf undersides, pedicels and calyx externally densely pale pubescent; fruit ellipsoid or subglobose, c 1.5cm or 4–8cm, flesh thick, pithy, not granular .... **26. Malus**
+ Leaf undersides, pedicels and calyx externally subglabrous or sparsely

pubescent; fruit ellipsoid, 0.7−0.8cm, or 2−3cm, or obovoid-subglobose, 4−6cm, flesh thin, pithy, or thick, granular 34

34. Leaves elliptic-lanceolate, rarely ovate, base cuneate or ± rounded; petals usually pinkish-white; fruit ellipsoid, 0.7−0.8cm, flesh thin, pithy; native plant .... **26. Malus** (*M. baccata*)
+ Leaves mostly ovate, base rounded, shallowly cordate, rarely cuneate; petals usually pure white; fruits ellipsoid, 2−3cm or obovoid-subglobose, 4−6cm, flesh thick, granular; cultivated plants **27. Pyrus**

35. Leaves quite entire 36
+ Leaves serrate or at least serrulate near apex 37

36. Leaves acute or rounded (acuminate in *C. acuminatus* but 3−8cm) **18. Cotoneaster**
+ Leaves strongly acuminate, 7−15cm .... **22. Photinia** (*P. integrifolia*)

37. Flowers in conical panicles usually longer than broad, with stout axis; leaves usually narrow, elliptic, oblong, lanceolate or oblanceolate; fruits 10−18mm mostly 1(−2)-seeded .... **23. Eriobotrya**
+ Flowers in rounded corymbose panicles usually broader than long, with slender axis; leaves broad to narrow, ovate, obovate, elliptic or rarely oblanceolate; fruits 4−10mm, or if larger (10−20mm) then 2−5-seeded 38

38. Mature leaves white or brownish tomentose beneath .... **21. Sorbus**
+ Mature leaves becoming glabrous beneath 39

39. Petioles 12−40mm; leaves 9−15cm, mostly with 12−24 pairs of lateral veins 40
+ Petioles 4−10mm; leaves 5−10cm, mostly with 7−13 pairs of lateral veins 41

40. Petioles 12−18mm; lateral veins prominent beneath, 12−14 pairs; leaves acuminate, sharply serrulate; fruit 4-seeded, calyx lobes not persistent .... **21. Sorbus** (*S. rhamnoides*)
+ Petioles 20−40mm; lateral veins not prominent beneath, 14−24 pairs; leaves acute, crenate-serrate; fruit 1−2-seeded, with persistent calyx lobes .... **22. Photinia** (*P. griffithii*)

41. Leaves membranous, deciduous, 6−8cm, serrulate only in upper half; calyx lobes deciduous .... **21. Sorbus** (*S. thomsonii*)
+ Leaves thinly coriaceous, evergreen, 5−12cm, serrate almost to base; calyx lobes persistent .... **22. Photinia**

74. ROSACEAE

1. SPIRAEA L.

by A. J. C. Grierson

Dioecious or bisexual shrubs, stems usually branched, sometimes simple. Leaves simple, unlobed; stipules absent. Flowers in terminal corymbs or corymbose panicles. Calyx cup concave, lobes 5 triangular. Petals 5. Stamens 20−25, reduced to staminodes in female flowers. Carpels usually 5, free, absent or reduced to pistillodes in male flowers; styles short. Follicles 5, ovoid, immersed or exserted from calyx cup, few-seeded.

Many species of the genus are valued as ornamental shrubs.

1. Stems unbranched, 20−40cm tall .... **1. S. hemicryptophyta**
+ Stems branched, 0.5−5m tall (sometimes less than 0.5m in *S. alpina* but then plants very twiggy) 2

2. Leaves 4−15cm; inflorescences (2−)5cm or more broad .... **Species 2 & 3**
+ Leaves (0.6−)1−2cm; inflorescences 1−2cm broad .... **Species 4−6**

1. S. hemicryptophyta Grierson

Rhizomatous shrub, stems 20−40cm, simple, sparsely fine pubescent. Leaves ovate, 2.5−4.5×1.75−3cm, acute, base rounded, margin serrate, glabrous or sparsely pubescent beneath; petioles 2−5mm. Dioecious; flowers in terminal corymbs 3−5cm broad. Calyx cup 1.5−2×2.5−3mm, lobes 1.5−2mm. Petals white or pink, obovate 2.5−3×2mm. Male flowers with stamens c 3mm attached at margin of thickened glandular ring, pistillodes minute. Female flowers with ellipsoid carpels 1.5−2mm, styles 1−1.5mm, staminodes 0.5−1mm. Fruit unknown.

Sikkim: Phedup, Chowbhanjan and Kangling. 3350−4250m. June−August.

2. S. bella Sims. Nep: *Lahare Phul* (34)

Very similar to *S. hemicryptophyta* but stems 1−2.5m, branched; leaves narrowly ovate 3−6×0.7−1.5cm, acuminate, base cuneate, margin serrate or doubly serrate; corymbs (2−)4−7cm broad, terminal on lateral shoots; petals white or pink; follicles c 3mm glabrous or pubescent.

Bhutan: S—Deothang district, C—Ha, Thimphu, Mongar and Sakden districts, N—Upper Mangde Chu, Upper Bumthang Chu and Upper Kulong Chu districts; **Sikkim.** Amongst scrub and by streamsides, 2300−3950m. May−August.

3. S. micrantha Hook. f. Sha: *Khangtsalu*

Similar to *S. hemicryptophyta* and more especially to *S. bella* but a branched shrub 1−1.5m, leaves ovate-lanceolate 4−15×2−5cm, gradually acuminate, base cuneate, margin usually coarsely doubly serrate; petioles up to 10mm; panicles 8−20cm broad, petals white or pink; follicles 1.5−2mm, pubescent.

Bhutan: S—Deothang district, C−Thimphu, Punakha, Tongsa, Bumthang, Mongar, and Sakden districts, N−Upper Mo Chu, Upper Mangde Chu and Upper Kulong Chu districts; **Sikkim.** Margins and clear areas in forests, 2200−3800m. June−August.

4. S. canescens D. Don

Shrub 2−5m with arching branches, twigs slightly ribbed, greyish pubescent. Leaves elliptic to oblanceolate, 0.5−1.5×0.2−0.5cm, obtuse or subacute, base attenuate, subsessile, margin entire or sometimes with a few teeth near apex, glabrous above, appressed pubescent beneath. Flowers 15−20, bisexual, 4−5mm diameter, in corymbs 1.5−2(−3)cm across on short lateral shoots. Calyx cup 1−1.5mm, pubescent, lobes c 1mm. Petals obovate, 2−3×1.5mm, white or pink. Follicles 1.5−2mm, pubescent, ± immersed in calyx cup.

Bhutan: C—Ha district (Ha) and Bumthang district (Kuktang, Shabjetang and Ura La); **Sikkim.** In open Blue Pine forest, 2775−3350m. May−June.

5. S. arcuata Hook. f. Fig. 38 f−h

Similar to *S. canescens* but twigs more strongly ribbed, reddish-brown, ± glabrous; leaves obovate 0.7−2.5×0.3−1.3cm, margins usually with a few teeth near apex, finely pubescent on both surfaces at first; flowers somewhat larger, 7−8mm diameter; follicles 2−3mm glabrous, exserted from calyx cup.

Bhutan: C—Ha, Thimphu, Tongsa and Bumthang districts, N—Upper Mo Chu, Upper Pho Chu, Upper Bumthang Chu and Upper Kulong Chu districts; **Sikkim.** Amongst Juniper/Rhododendron scrub, 3350−4250m. June−July.

The record (109) of *S.* x *vanhouttei* (Briot) Zabel from Kalimpong requires confirmation. This hybrid is only known in cultivation but resembles *S. arcuata*.

6. S. alpina Pallas; *S. ulicina* Prain

Similar to *S. canescens* but a bush 60cm or less; twigs slightly ridged, reddish brown, glabrous, bark exfoliating from older shoots; leaves greyish green, becoming fascicled on older shoots, oblanceolate, 6−10× 1.5−2.5mm, acute, entire, sparsely and shortly pubescent above, glabrous and glaucous beneath; corymbs c 1cm across, 10−15-flowered; calyx purplish c 2mm including lobes; petals obovate 1.5−2mm; follicles becoming exserted from calyx cup, puberulous.

Sikkim: Lhonakh and Jonsong La Valley. On alpine banks, 4600−4900m. July−August.

2. ARUNCUS L.

by A. J. C. Grierson

Perennial herbs. Leaves triternate; exstipulate. Flowers in elongate loose terminal panicles, branches racemose, spike-like. Calyx cup concave, glandular at rim, lobes 5. Petals 5. Dioecious; male flowers with c 20 stamens and rudimentary ovary; female flowers with vestigial stamens and 3−5(−8) free ovaries. Follicles glabrous, each with 3−5 seeds.

1. A. dioicus (Walter) Fernald subsp. **triternatus** (Maximowicz) Hara; *Spiraea aruncus* L.

Stems 0.5−1m, sparsely pubescent. Leaves 20−30cm, leaflets ovate-lanceolate 3−7(−10)× 1.25−3.5(−7)cm, acuminate, base rounded, glabrous above, appressed pubescent beneath, margins coarsely and sharply doubly serrate or sometimes lobed. Calyx cup 0.5−1mm, lobes triangular c 0.75mm. Petals white or purplish, obovate, c 1.5×0.75mm. Stamens c 2mm. Ovaries ovoid-ellipsoid, c 1×0.75mm; style c 0.75mm. Follicles 2.5−3mm, seeds linear 1.5−2mm.

Bhutan: C—Ha, Thimphu, Tongsa, Mongar and Tashigang districts, **N**—Upper Mo Chu, Upper Mangde Chu and Upper Kuru Chu districts; **Sikkim.** Conifer/Rhododendron forests and marshy meadows, 2300−3650m. July−August.

3. NEILLIA D. Don

by A. J. C. Grierson

Shrubs. Leaves simple, mostly 3-lobed, serrate; stipules ovate or lanceolate, deciduous or ± persistent. Flowers in racemes or panicles. Calyx cup campanulate, lobes 5. Petals 5. Stamens 8−25 borne in 2 or 3 series on calyx cup. Carpel 1, style terminal. Follicle dehiscent along inner suture, seeds 2−10.

1. N. thyrsiflora D. Don. Dz: *Khenzem;* Sha: *Totom;* Nep: *Jhikre* (34)

Shrub 1−3m. Leaves ovate, 5−12×2.5−6cm, acuminate, base rounded or shallowly cordate, margins irregularly serrate, shallowly 3-lobed, occasionally unlobed; petioles 5−10mm; stipules ovate c 7mm, margins serrate. Flowers in large leafy panicles, branches naked at base, borne up to 1cm above leaf axils usually with 2−3 small dormant buds between them. Calyx cup c 2.5mm appressed pubescent, lobes triangular 2.5−3.5mm, erect. Petals elliptic 2−5mm white. Stamens 8−10, filaments c 1mm. Carpel ovoid, c 2mm, style 2mm. Fruiting calyx cup bearing stiffly stalked capitate glands c 1.5mm, ± concealing follicle c 5mm; seeds 8−10.

Bhutan: S—Samchi, Phuntsholing, Chukka, Gaylegphug and Deothang districts, **C**—Punakha, Tongsa and Bumthang districts, **N**—Upper Mo Chu district; **Sikkim**. In moist broad-leaved forests, 1200−2550m. July−August.

2. N. rubiflora D. Don. Dz: *Totuma*

Similar to *N. thyrsiflora* but leaves 3−10×2−5cm, more deeply 3-lobed, base usually deeply cordate; petioles up to 2cm; stipules 0.7−1cm, usually entire; inflorescence a narrow raceme or compact panicle, flowering shoots scaly at base and without dormant buds; calyx cup 4−5mm broadly campanulate, densely velutinous, often reddish, lobes c 2.5mm; petals white or pink c 3mm; stamens 20−25, filaments 1.5−2mm; fruiting calyx cups with stalked capitate glands at least at base.

Bhutan: S—Deothang district, **C**—Thimphu, Punakha, Tongsa, Bumthang, Mongar and Tashigang districts, **N**—Upper Mo Chu and Upper Kulong Chu districts; **Sikkim**. In Cool wet broad-leaved forests, 2100−3300m. May−July.

4. MADDENIA Hook. f. & Thomson

by A. J. C. Grierson

Deciduous shrubs or trees, inner bud scales accrescent, shortly persistent. Leaves simple, unlobed; stipules paired, ± persistent. Flowers in terminal racemes, functionally unisexual. Calyx lobes 5−10, similar to 5−8 petals, borne on margin of receptacular cup. Male flowers with 20−30 stamens and 1 carpel with short style. Female flowers with 2 carpels, one or both developing, stigma sessile. Drupe thinly fleshy, stone woody.

1. M. himalaica Hook. f. & Thomson; *M. hypoleuca* Koehne

Shrub or tree 1−10m, twigs thinly or densely brown tomentose, inner bud scales ovate-elliptic up to 1.5×0.7cm. Leaves ovate or elliptic, 4−12×2−5cm, acuminate, base rounded or shallowly cordate, margins pectinately serrate, teeth capitate-glandular in lower half, veins 12−15 pairs, glabrous or densely brown pubescent on veins beneath, subsessile or petioles up to 5mm; stipules oblong-lanceolate 1.5−2×0.2−0.5cm. Racemes 10−20-flowered, receptacular cup 4−5mm, green tinged crimson. Calyx lobes and petals narrowly lanceolate, 3−4×0.75−1mm, pubescent. Stamens cream-coloured 5−8mm. Ovary and style glabrous, 7−10mm. Drupes ellipsoid, 8−10×6−8mm, crimson.

Bhutan: C—Thimphu, Punakha, Tongsa, Bumthang and Mongar districts, **N**—Upper Mo Chu and Upper Kulong Chu districts; **Sikkim**. Margins of Spruce, Hemlock and Fir forests, 2500—3200m. April—May.

Var. **glabrifolia** Hara described from W Central Bhutan has thinly pubescent twigs and glabrous leaves.

5. PRUNUS L.

by A. J. C. Grierson

Trees or shrubs, evergreen or deciduous. Leaves simple, unlobed, usually with one or several sessile glands near apex of petiole, or on the adjoining leaf margin, or scattered on lamina beneath; stipules usually deciduous. Flowers in axillary racemes or fascicles, sometimes solitary. Calyx deciduous, tubular or obconic, lobes 5. Petals 5 alternating with calyx lobes. Stamens 10—35. Ovary a single carpel, style simple; calyx cup circumscissile to leave a circular scar or rim at base of developing fruit. Fruit an ovoid or globose drupe containing a hard one-seeded stone, indehiscent or rarely dehiscent.

1. Flowers 15 or more in racemes 2
+ Flowers 1—3 in clusters or subfasciculate racemes 4

2. Evergreen trees; leaves with disc-like or oval, flat or sunken glands on undersides, or glandless; racemes always leafless at base 3
+ Deciduous trees; leaves with small raised glands at apex of petiole or on leaf margin near base; racemes leafy at base (except *P. venosa*)
   Species 5—7

3. Racemes and calyx cups glabrous or sparsely pubescent; drupes ovoid
   Species 1 & 2
+ Racemes and calyx cups pubescent; drupes transversely ovoid or bilobed
   Species 3 & 4

4. Flowers and fruits on pedicels 1—2cm; drupes fleshy, indehiscent 5
+ Flowers and fruits ± sessile; drupes velutinous or lanate, either fleshy and indehiscent or (in *P. dulcis*) dry and dehiscent .... **Species 13—15**

5. Stipules pinnately divided into narrow lobes; petals obovate
   Species 8 & 9
+ Stipules linear, entire, ciliate; petals suborbicular .... **Species 10—12**

1. P. undulata D. Don, non sensu F.B.I.; *P. acuminata* (Wall.) Dietrich, non *Pygeum acuminatum* Colebrooke, *Prunus wallichii* Steudel, *Laurocerasus undulata* Roemer, *L. acuminata* (Wall.) Roemer. Nep: *Lekh Arupate* (34), *Lali* (34)

Evergreen shrub or tree 2−12m. Leaves elliptic or oblong, 8−15×2.5−5(−6)cm, long acuminate, base cuneate, margin entire or shallowly serrate, veins 6−10 pairs slightly prominent beneath, glands small, 2 at base and usually with others on the lower surface ± parallel to midrib; petioles 2.5−10mm; stipules c 5×1mm entire, caducous. Racemes solitary or in fascicles of 2−3, rachis 6−10cm, glabrous or sparsely pubescent, pedicels 3−7mm. Calyx lobes broadly triangular c 1mm, minutely pubescent at apex. Petals elliptic 2−4mm white or creamy, glabrous. Filaments c 3mm, glabrous. Drupes fleshy, ovoid, stone 15−20×8−12mm.

Bhutan: S—Chukka and Gaylegphug districts, **C**—Punakha and Tongsa districts; **Sikkim.** Moist Warm broad-leaved forests, 900−1900m. September−November.

Fruit edible (34).

2. P. jenkinsii Hook. f. & Thomson

Similar to *P. undulata* but leaves always serrate, veins 10−13 pairs; basal glands only 2; racemes solitary; pedicels 2−3mm; drupes ellipsoid 17−22×12−15mm.

Bhutan: S—Gaylegphug district (Rani Camp, 38). Warm broad-leaved forests, 1650m. November.

3. P. ceylanica (Wight) Miquel; non *Pygeum ceylanicum* Gaertner sensu F.B.I.; *Pygeum acuminatum* Colebrooke, *Pygeum glaberrimum* Hook. f. Nep: *Dharani* (34)

Evergreen tree up to 24m. Leaves elliptic or ovate 5−18×3−8cm, acuminate, base rounded or cuneate, margin entire, sometimes with a few circular glands (c 1mm diameter) near base, glabrous; petioles 10−18mm; stipules lanceolate, 9−12mm, caducous. Racemes simple, axillary c 10cm, rachis and calyx cup pubescent. Calyx lobes triangular, 0.5−1mm. Petals elliptic, 1−1.5mm, white. Stamens c 4mm. Fruit transversely ellipsoid or bilobed 1.5−2×2.5−3.5cm, glabrous.

Bhutan: S—Samchi district (near Gokti) and Gaylegphug district (Taklai Khola); **Sikkim.** Subtropical forest slopes, 350−500m. August−November.

4. P. arborea (Blume) Kalkman; *Pygeum montanum* Hook. f.

Similar to *P. ceylanica* but twigs pubescent at first; leaves elliptic to ovate 7−20×2−7cm, brownish pubescent beneath, basal glands 2 sunk at base on either side of midrib; racemes 7−9cm in clusters of 2−6; rachis and calyx cups densely pubescent; stamens c 7mm; ovary glabrous or sparsely hairy, drupes globose to transversely ellipsoid, 5−12×7−17mm.

Sikkim: Darjeeling district. Warm broad-leaved forests. September–October.

The above description refers to var. **montana** (Hook. f.) Kalkman.

5. P. cornuta (Royle) Steudel; *P. padus* sensu F.B.I. non L., *P. racemosa* Lamarck, *P. glauciphylla* Ghora & Panigrahi, *P. wattii* Ghora & Panigrahi. Eng: *Bird Cherry*

Tree 3–15m. Leaves elliptic, 5–15×2–6cm, acuminate, base cuneate, rounded or subcordate, margin finely serrate, glabrous or pubescent beneath but always with tufts of hairs in lateral vein axils; petioles 1–2.5cm usually bearing 1 or 2 glands near apex; stipules linear-lanceolate 8–10(–15)mm. Racemes 15–25cm, terminal on leafy lateral shoots, flowers fragrant, pedicels 3–5mm. Calyx cup 2–3mm, ± glabrous, lobes broadly triangular c 1mm, shortly ciliate with thick glandular hairs. Petals suborbicular, 2.5–3.5mm, white. Stamens 2–3mm. Drupes ellipsoid to subglobose, stone 7–8mm.

Bhutan: C—Ha, Thimphu, Tongsa and Bumthang districts, **N**—Upper Mo Chu, Upper Bumthang Chu and Upper Kulong Chu districts; **Sikkim.** Steep scrub-covered slopes, 2300–3500m. May–June.

6. P. venosa Koehne; *P. undulata* sensu F.B.I. non D. Don

Similar to *P. cornuta* but glands borne on leaf margin near base, usually with tufts of hair only in lower axils of lateral veins; racemes 8–12cm, leafless.

Bhutan: C—Tongsa district (near Chendebi) and Tashigang district (Shapang), **N**—Upper Mo Chu district (Tamji to Gasa); **Sikkim.** Scrub-covered slopes, 1800–2450m. April–May.

Fruit edible (34).

7. P. napaulensis (Seringe) Steudel. Nep: *Arupate* (34)

Similar to *P. cornuta* but a larger tree up to 20m; leaves narrowly elliptic-lanceolate 6–18×1.5–5cm, lateral vein axils without tufts of hair beneath; petioles 0.5–1cm bearing a pair of glands at apex or on margin of lamina at its base; racemes leafy at base, 15–25cm, rachis and calyx cups pubescent; calyx lobes eglandular; petals obovate, 4–5mm; drupes ovoid 1.3–1.5×1–1.3cm.

Bhutan: C—Thimphu district (near Thimphu), **N**—Upper Kuru Chu district(Dengchung); **Sikkim.** Evergreen Oak forests, 2100m. April.

8. P. cerasoides D. Don; *P. puddum* (Seringe) Brandis, including *Maddenia pedicellata* Hook. f. Nep: *Paiyun* (34), *Paiyu*. Fig. 38 c–e

Tree 5–15m. Leaves ovate or oblong-elliptic, 5–12×3.5–4cm, shortly acuminate, base rounded or cuneate, margin finely serrate, glabrous; petioles 1–2cm, bearing near apex 2–5 raised disc-like glands 0.75–1mm diameter; stipules linear-lanceolate 7−10mm, pinnately divided into linear segments, deciduous. Flowers 1−3 in fascicles, unfolding with young leaves, produced from lateral buds and surrounded by rounded bud scales at base; pedicels 1−2cm. Calyx cup tubular-campanulate c 1cm, glabrous, green flushed pink, lobes c 5mm ovate, acute. Petals obovate, 1.5−1.8×0.8cm, pink, spreading. Filaments of outer stamens ± as long as petals. Drupes ellipsoid 1−1.3×0.8cm.

Bhutan: S—Chukka and Deothang districts, **C**—Ha, Tongsa and Tashigang districts, **N**—Upper Mo Chu, Upper Bumthang Chu and Upper Kulong Chu districts; **Sikkim.** Warm broad-leaved forests, 1050−2000m. October−November.

Wood used in furniture-making (34); often cultivated for its ornamental flowers (16); fruit edible (16).

9. P. carmesina Hara; *P. cerasoides* D. Don var. *rubea* Ingram. Shamgong: *Sishi*

Similar to *P. cerasoides* but a larger tree up to 30m; flowers precocious; bud scales and bracts early deciduous; calyx tube crimson, lobes obtuse; petals deep pink, not spreading.

Bhutan: S—Chukka and Gaylegphug districts, **C**—Tongsa, Mongar and Tashigang districts; **Sikkim.** Mixed evergreen broad-leaved forests, 1900−2600m. March−April.

10. P. rufa Hook. f. Nep: *Lekh paiyun* (34)

Shrub or tree 2−10m. Leaves ovate-elliptic 3−10×1.5−3.5cm, acute or acuminate; base rounded or cuneate, margin serrate, sometimes doubly serrate, usually with 1 or 2 glands on leaf margin on either side at base; petioles usually eglandular; stipules linear-lanceolate, 5−15mm, glandular ciliate. Flowers unfolding with young leaves, solitary or in pairs, bud scales caducous. Calyx tube 7−10mm, glabrous or pubescent, lobes ovate 3−4mm. Petals suborbicular, 6−8mm, pink or white. Fruit ellipsoid 1−1.5×0.7−1cm.

Bhutan: C—Ha, Thimphu, Tongsa, Bumthang, Mongar and Tashigang districts, **N**—Upper Mo Chu, Upper Bumthang Chu, Upper Kuru Chu and Upper Kulong Chu districts; **Sikkim; Chumbi.** Clearings in mixed forests, 2750−3950m. April−May.

A variable species; plants which display to varying degrees more elongate leaves, pubescent pedicels and calyx tubes, and petals appressed pubescent on the back have been segregated as var. **trichantha** (Koehne) Hara (*P. trichantha* Koehne).

Young leaves poisonous (34).

11. P. cerasus L. Eng: *Cherry*

Tree 2−7m. Leaves obovate, 5−8×2.5−3.5cm, acute, base cuneate, margins bluntly serrate, 2 lowermost teeth gland-tipped, white villous beneath at first, later glabrous. Flowers in clusters of 2 or 3, pedicels 3—4cm, precocious or unfolding with young leaves. Calyx cup broadly campanulate c 5mm, lobes ovate c 5mm, reflexed. Petals broadly obovate c 12×8—10mm, white. Fruit globose c 2cm diameter, stone broadly ellipsoid 8×7×6mm.

Sikkim: Darjeeling district, sometimes cultivated (34). April—May.

Probably native of W. Asia; cultivated for its edible fruits (17).

12. P. domestica L.; *P. communis* Hudson. Eng: *Plum*

Similar to *P. cerasus* but leaves elliptic 6—10×3.5—5cm, pubescent beneath; pedicels c 1cm; calyx cup c 2.5mm, lobes ovate c 3mm, obtuse; petals obovate 8—9×5—6mm; drupes ovoid c 6×3.5cm, red or purple, glaucous, longitudinally grooved down one side, stone ellipsoid c 2.5× 1.5cm.

Sikkim: Mungsong, cultivated (34). April—May.

Native of S. Europe and SW Asia; cultivated for its edible fruit and useful timber (17, 48).

13. P. armeniaca L.; *Armeniaca vulgaris* Lamarck. Eng: *Apricot*

Tree 3—10m. Leaves broadly ovate 4—9×3.5—6cm, acute or shortly acuminate, base rounded or subcordate, pubescent at least in vein axils beneath, margin crenate-serrate and with a pair of glands at base of lamina; petioles 1.5—3.5cm sometimes bearing glands at apex. Flowers precocious, usually solitary from lateral buds, ± sessile. Calyx tube c 5mm, lobes ovate 4×3mm, glabrous. Petals obovate 10—12×6—8mm, pink. Fruit subglobose 3—4cm, orange-yellow, shortly velutinous; stone broadly ellipsoid c 1.8× 1.5cm, smooth.

Sikkim: occasionally cultivated (34). April—May.

Probably native to SW Asia; cultivated for its edible fruit (17). A useful firewood crop (48); seeds yield a useful oil (126).

14. P. persica (L.) Batsch; *Persica vulgaris* Miller. Sha: *Lengsey;* Nep: *Aru;* Eng: *Peach, Nectarine*

Similar to *P. armeniaca* but leaves lanceolate 7—12×2—4cm, acuminate, base rounded, margins serrate, pubescent beneath at first, later usually with a few hairs persisting along midrib at base; petioles 7—12mm with 2—4 glands near apex; flowers usually solitary; calyx tube 5—6mm, ± concealed by bud scales, calyx lobes oblong, rounded 3.5—5×3.5mm, pale lanate; petals obovate c 1.5—2×1cm, pink or red; style lanate; fruit globose 5—7cm, velutinous, pink or purplish; stone ovoid-ellipsoid 3—4×2—3cm, irregularly furrowed and pitted.

Bhutan: S—Phuntsholing district (Phuntsholing), **C**—Ha, Thimphu, Punakha and Tongsa districts; **Sikkim.** Cultivated, especially in the Terai, 250—2800m. February—April.

Cultivated for its edible fruit; possibly native to China (17).

15. P. dulcis (Miller) Webb; *P. amygdalus* Batsch, *Amygdalus communis* L.
Eng: *Almond*

Similar to *P. persica* but leaves narrowly ovate 4−6×1.5−2.5cm, acute or acuminate, base rounded, margins finely serrate, lowermost teeth gland-tipped, usually glabrous; petioles 1−2cm sometimes bearing small glands; flowers often in pairs; calyx lobes finely ciliate; petals rose pink to white; fruit ovoid, somewhat flattened, 3.5−4.5×2.5−3cm, greyish lanate, leathery, dehiscent to release 1−2 woody, pitted stones.

Bhutan: C—Thimphu district (Yosepang); **Sikkim.** Scrub amongst Blue Pine, 2700m. April.

Sometimes naturalised; cultivated for its edible kernels, which also yield an aromatic oil used as a substitute for Tragacanth (126).

6. PRINSEPIA Royle

by A. J. C. Grierson

Deciduous spiny shrubs. Leaves simple, unlobed, stipules minute deciduous. Flowers in short axillary racemes borne at base of spines. Calyx persistent cup-shaped, lobes 5, concave. Petals 5. Stamens c 30. Ovary monocarpellate, style subterminal. Drupes ellipsoid, thinly fleshy.

1. P. utilis Royle. Dz: *Dushi Tsang*

Shrub 1−4m, branches bearing ascending spines (modified lateral shoots) 0.7−3.5cm. Leaves elliptic-lanceolate 2.5−7×0.5−2(−2.5)cm, acuminate, base attenuate or rounded, minutely serrate, subsessile or on petioles up to 10mm, glabrous. Racemes 3.5−7cm, 2−5-flowered. Calyx lobes suborbicular, 4−6mm. Petals elliptic or obovate 6−7×5mm, white. Stamens 2−3mm, filaments crimson at base. Drupes 10−15mm, borne on persistent calyx cup, purplish, style near base due to unequal growth, stone c 10×7−8mm.

Bhutan: S—Chukka and Gaylegphug districts, **C**—Ha, Thimphu, Tongsa and Bumthang districts; **Sikkim; Chumbi.** River banks, 2200−2750m, March−April.

Useful as a hedge-plant (17); seeds contain an oil used in lamps (16).

7. RUBUS L.

by D. G. Long

Erect or spreading shrubs or creeping herbs, often prickly or bristly. Leave alternate, simple, often lobed, or 3-foliate, or pinnately, palmately or pedately compound; stipulate. Flowers in terminal and axillary panicles or corymbs, sometimes few or solitary. Calyx with a broad cup-shaped tube and 5 persistent lobes. Petals 5. Stamens many. Carpels many, on a convex receptacle, each with subterminal style. ovules 2. Fruit a cluster of fleshy 1-seeded drupes on a conical receptacle.

1. Stems creeping, herbaceous (± woody in *R. sengorensis*), regularly rooting at nodes; leaves simple or 3−5-foliate 2
+ Stems erect, arching, scrambling or climbing, woody, not regularly rooting at nodes; leaves simple or 3−11-foliate 7

2. Leaves simple (sometimes shallowly lobed) 3
+ Leaves pinnately 3-foliate or pedately 5-foliate 4

3. Stems and petioles hirsute and with slender prickles; petals white; spring flowering .... **1. R. calycinus**
+ Stems and petioles softly bristly and minutely pubescent, not prickly; petals pink; summer and autumn flowering .... **2. R. pectinaroides**

4. Stems and petioles with slender prickles; stipules divided into 3 linear lobes to base; petals pink, 4−5mm .... **6. R. sengorensis**
+ Stems and petioles glabrous or bristly, never prickly; stipules ovate, unlobed; petals white, 10−15mm 5

5. At least some leaves pedately 5-foliate; stipules 3−4mm; calyx puberulous, without bristles; petals 7−10mm long .... **5. R. fragarioides**
+ All leaves pinnately 3-foliate; stipules 5−7mm; calyx with a mixture of fine pubescence and bristles; petals 9−14mm long 6

6. Stems and petioles pubescent, without bristles; calyx lobes entire; petals elliptic, 10−14×3−4mm .... **3. R. föckeanus**
+ Stems and petioles finely pubescent and bristly; calyx lobes toothed at apex; petals obovate, 9−11×6−9mm .... **4. R. nepalensis**

7. Leaves simple (sometimes shallowly or deeply lobed) 8
+ Leaves compound (3-foliate, palmate, pedate or pinnate) 21

8. Leaves lanceolate, elliptic or ovate, unlobed, base cuneate, rounded or truncate (shallowly cordate in *R. griffithii*) 9
+ Leaves ovate, broadly ovate or suborbicular, shallowly to deeply lobed (occasionally almost unlobed in *R. efferatus, R. insignis* and *R. paniculatus*), base deeply cordate (sometimes becoming truncate in uppermost leaves) 13

9. Lateral veins 20−25 pairs; petals deep red .... **11. R. calophyllus**
+ Lateral veins 7−12 pairs; petals white 10

10. Branchlets and leaf undersides densely white tomentose
 10. R. preptanthus
+ Branchlets and leaf undersides glabrous or sparsely pubescent, never white 11

11. Leaves thinly coriaceous, brown when dry, shortly and abruptly acuminate; bracts deeply laciniate; pedicels 3−8mm ... **9. R. hamiltonii**
+ Leaves membranous, green when dry, caudate or long acuminate; bracts subulate, entire or linear, serrate; flowers sessile or on pedicels 8−12mm 12

12. Leaves caudate-acuminate; petioles 12−16mm; panicles shortly branched; bracts subulate, entire; flowers on pedicels 8−12mm
 7. R. acuminatus
+ Leaves gradually acuminate; petioles 7−9mm; panicles long-branched; bracts linear, serrate; flowers sessile .... **8. R. griffithii**

13. Flowers in terminal panicles; leaves distinctly longer than broad (except in *R. tiliaceus*), shallowly lobed (lobes 2−5 (−10)mm deep) 14
+ Flowers in axillary clusters or in racemes on short lateral shoots (in *R. kumaonensis* flowers in axillary racemes and terminal panicles); leaves often as broad as or broader than long, with deep lobes 10−40mm ... 17

14. Lateral veins 10−15 pairs; leaf margins crenulate-denticulate
 15. R. insignis
+ Lateral veins 6−8 pairs; leaf margins sharply serrate 15

15. Leaves pubescent but not white tomentose beneath; calyx lobes pectinately toothed on margin .... **14. R. efferatus**
+ Leaves white tomentose beneath; calyx lobes entire 16

16. Leaves distinctly longer than broad; panicles with long spreading branches often 6−10cm or more long; bracts with lanceolate teeth
 12. R. paniculatus
+ Leaves scarcely longer than broad; panicles narrow, branches rarely over 4cm; bracts with linear teeth .... **13. R. tiliaceus**

17. Stems and petioles glandular-bristly .... **19. R. treutleri**
+ Stems and petioles eglandular 18

18. Leaf lobes round; petioles ± equalling leaf laminae .... **20. R. cooperi**
+ Leaf lobes acute or subacute (occasionally obtuse in *R. kumaonensis*); petioles mostly much shorter than leaf laminae 19

19. Leaves soft to touch beneath with thick white tomentum; stems and petioles with few, thin, straight or weakly hooked prickles
    18. R. kumaonensis
+ Leaves rough to touch beneath with stiff hairs (sometimes mixed with minute white tomentum); stems and petioles with few to many strongly hooked prickles 20

20. Leaves 10–20cm long and broad, lobes acute; stipules 1.5–2cm, subpersistent; pedicels 8–15mm; calyx lobes 10–12mm, often toothed
    16. R. calycinoides
+ Leaves 8–12×6–10cm, lobes rounded or subacute; stipules 1–1.5cm, early caducous; pedicels 2–7mm; calyx lobes 8–9mm, entire or minutely toothed at apex .... **17. R. rugosus**

21. Leaves digitately or pedately 3–7-foliate, if 3-foliate then petiolules subequal 22
+ Leaves pinnate or pinnately 3-foliate with terminal petiolule distinctly longer than laterals 27

22. Leaflets with 15–30 pairs of lateral veins; leaflets white sericeous beneath at least when young (in *R. senchalensis* densely pubescent at first becoming glabrescent) 23
+ Leaflets with 8–12 pairs of lateral veins; leaflets glabrous or sparsely pubescent beneath 26

23. Leaves usually 3-foliate; terminal leaflet 3.5–7.5cm broad; stems, petioles and inflorescences glandular-bristly 24
+ Leaves mostly 5-foliate, occasionally 3 or 7-foliate; terminal leaflet 1.5–4cm broad; plants eglandular 25

24. Stems unarmed; gland-tipped bristles long, 1.5–4mm; flowers in broad panicles .... **21. R. splendidissimus**
+ Stems with short prickles; gland-tipped bristles short, c 0.5mm; flowers in fascicles or short condensed panicles .... **22. R. phengodes**

25. Leaflets white sericeous beneath at least when young; lateral veins 25–50 pairs; stipules 4–8mm broad; calyx lobes 8–11mm
    23. R. lineatus
+ Leaflets whitish pubescent or glabrescent beneath; lateral veins 14–22 pairs; stipules 1.5–2mm broad; calyx lobes 6–8mm
    24. R. senchalensis

26. Leaflets sessile, terminal not much larger than laterals; stipules glandular; flowers 1−3 in terminal fascicles; calyx lobes 10−15mm, spiny; petals white .... **25. R. pentagonus**
+ Leaflets on petiolules 2−4mm, terminal leaflet much larger than laterals; stipules eglandular; flowers 1−6 in axillary panicles; calyx lobes 3−5mm, glabrous; petals red .... **26. R. thomsonii**

27. Leaves pinnately 3-foliate 28
+ Leaves pinnate, leaflets 5−11 39

28. Leaflets broadly elliptic or obovate, rounded; branchlets densely eglandular-bristly and prickly .... **27. R. ellipticus**
+ Leaflets mostly ovate, acute or acuminate; branches glabrous or prickly and/or glandular-bristly 29

29. Branchlets, petioles and often calyces with spreading gland-tipped hairs or bristles 30
+ Branchlets, petioles and calyces eglandular 35

30. Leaves white-tomentose beneath 31
+ Leaves green beneath, glabrous or pubescent 34

31. Branches with slender, straight often dense prickles; calyx lobes spreading 32
+ Branches with scattered, stout, curved prickles; calyx lobes erect 33

32. Low shrub to 1m; petals white .... **30. R. irritans**
+ Tall shrub to 2.5m; petals ?reddish-purple or pink
   31. R. sikkimensis var. **canescens**

33. Stems, petioles, pedicels and calyx glandular-hairy; calyx cup densely prickly, lobes pubescent, 10−13mm .... **28. R. alexeterius**
+ Petioles glandular-hairy, other parts rarely so; calyx cup and lobes glabrous 6−9mm .... **29. R. biflorus**

34. Terminal leaflet ovate, 4−6×2.5−4cm; stipules linear, entire; flowers 1.5−2cm diameter .... **31. R. sikkimensis** var. **sikkimensis**
+ Terminal leaflet ovate-rhombic, 8−11×5−9cm; stipules ovate, laciniate; flowers 5−6cm diameter .... **32. R. wardii**

35. Leaves white tomentose beneath, glabrous or pubescent and green above 36
+ Leaves green beneath, glabrous (except veins) when mature 38

36. Stems with white bloom; calyx lobes glabrous outside, petals white
 29. R. biflorus
+ Stems without white bloom; calyx lobes pubescent or sericeous outside; petals pink or purple 37

37. Flowers solitary or 2–5 in axillary racemes; calyx lobes lanceolate 8–12mm, with long subulate apex; drupelets pubescent throughout
 33. R. hypargyrus var. **niveus**
+ Flowers 5–15 in terminal and axillary corymbs; calyx lobes triangular c 7mm, shortly pointed; drupelets with a tuft of hair at apex
 34. R. mesogaeus

38. Prickles slender, 2–3mm; terminal leaflet ovate, coarsely serrate, pubescent on veins beneath; calyx lobes 8–12mm, tomentose; petals red .... **33. R. hypargyrus** var. **hypargyrus**
+ Prickles stout, 4–10mm; terminal leaflet ovate-lanceolate, finely serrate, glabrous or sparsely hairy on veins; calyx lobes 6–7mm, thinly pubescent on outside; petals white, sometimes tinged pink
 35. R. macilentus

39. Leaves white tomentose beneath 40
+ Leaves green beneath, glabrous or pubescent 41

40. At least some leaves 3-foliate; flowers solitary or 2–3 in fascicles; calyx lobes 6–9mm, glabrous; petals white .... **29. R. biflorus**
+ All leaves 5–9-foliate; flowers numerous in terminal corymbs; calyx lobes 4–6mm, whitish pubescent; petals pink .... **36. R. niveus**

41. Calyx cup with dense slender prickles and shorter gland-tipped bristles
 38. R. pungens
+ Calyx cup not prickly, pubescent or softly glandular-bristly 42

42. Branchlets, petioles and inflorescences with numerous short or long gland-tipped bristles 43
+ Branchlets, petioles and inflorescences eglandular or with minute sessile glands 44

43. Gland-tipped bristles short, straight, 0.5–1mm; calyx lobes 13–17mm; petals 10–13mm .... **39. R. indotibetanus**
+ Gland-tipped bristles long, flexuose, 2–4mm; calyx lobes 6–10mm; petals c5mm .... **40. R. sumatranus**

44. Leaves and inflorescences without minute sessile glands, leaflets mostly 7 or 9; calyx lobes 4–5mm; petals pink, 4–5mm .... **37. R. inopertus**

\+   Leaves and inflorescences with minute sessile glands, leaflets mostly 5; calyx lobes 7−15mm; petals white, 8−17mm .... **41. R. rosifolius**

1. R. calycinus D. Don;? *R. subherbaceus* Kuntze. Nep: *Bin Aselu* (34)

Herb with long creeping slightly woody stems, softly hirsute and with scattered slender prickles. Leaves simple, cordate-orbicular to reniform, 3−6cm diameter, apex rounded, base deeply cordate, unlobed or shallowly lobed, margins denticulate, hirsute especially on veins, and with scattered slender prickles on veins beneath; petioles 3−8cm, hirsute and prickly; stipules broadly ovate, 5−11×4−10mm, serrulate. Flowering branches suberect, 4−10cm bearing 1−4 leaves and 1−2 erect flowers. Pedicels 1−3cm, prickly. Calyx lobes green, ovate, 8−10×6−7mm, obtuse, serrate. Petals white, obovate, 10−15×8−10mm. Fruit red, globose, c 1.5cm, drupelets 25−30.

Bhutan: C—Punakha district (above Lometsawa), Tongsa district (Changkha), Mongar district (Saleng) and Tashigang district (Tashi Yangtsi); **Sikkim**. Mossy slopes in Cool broad-leaved forests, 1800−2450m. April−May.

Fruit edible (34).

2. R. pectinaroides Hara

Similar to *R. calycinus* but stems and petioles with spreading soft slender flexuose bristles and minute pubescence; stipules ovate-oblong, 6−10×3−5mm, serrate only at apex; flowers nodding; calyx lobes purplish, entire or 3−5-toothed at apex; petals deep pink.

Bhutan: C−Tongsa district (Longte Chu), Mongar district (between Sengor and Sheridrang) and Tashigang district (Sana and Tashi Yangtsi),**N**—Upper Kulong Chu district (Lao); **Sikkim:** Karponang and Kanglasa. Rocky river banks and ravines in Hemlock forests, 2560−3650m. July−September.

3. R. fockeanus Kurz; *R. nutans* G. Don var. *fockeanus* (Kurz) Kuntze

Creeping herb, stems pubescent, not prickly or bristly. Leaves pinnately 3-foliate, leaflets broadly ovate or suborbicular, 1.5−3×1−2cm, apex round, base cuneate, oblique on lateral leaflets, margins serrate, veins prominent beneath, pubescent; petiolules 2−6mm; petioles 1−4.5cm, pubescent; stipules ovate, 5−7mm, obtuse. Flowers solitary on pedicels 1.5−5mm, axillary on short erect leafy shoots 1−3cm. Calyx cup 5−6mm diameter, puberulous and sparsely bristly, lobes lanceolate 9−10mm, entire. Petals white, elliptic, 10−14×3−4mm. Carpels c 10−15. Fruit red, c 7mm, concealed within calyx.

Bhutan: C—Ha, Thimphu, Punakha, Tongsa and Bumthang districts, **N**—Upper Mo Chu district (Gasa); **Sikkim:** Kalapokri, Sandakphu, etc. Mossy ground in Oak, Spruce and Hemlock forests, 2700−3500m. May−July.

4. R. nepalensis (Hook.f.) Kuntze; *R. nutans* G. Don var. *nepalensis* Hook. f., *R. nutantiflorus* Hara var. *nepalensis* (Hook. f.) Balakrishnan

Very similar to *R. fockeanus* but stems and petioles finely pubescent and thinly bristly; calyx cup densely bristly, lobes triangular, toothed at apex; petals obovate, 9−11×6−9mm.

Sikkim: locality unknown (69); **E. Nepal:** Tambur River. 2750−3960m.July.

The occurrence of this species in Sikkim awaits confirmation. It has sometimes been wrongly united with the NW Himalayan *R. nutantiflorus* Hara (*R. nutans* G. Don non Vest, *R. barbatus* Rehder non Fritsch) which differs in its larger size, densely bristly stems and petioles, stipules 8−14mm and lanceolate calyx lobes 11−15mm.

5. R. fragarioides Bertoloni; *R. arcticus* L. var. *fragarioides* (Bertoloni) Focke

Similar to *R. fockeanus* but stems and petioles minutely puberulous; leaves mostly pedately 5-foliate, some occasionally 3-foliate (but pedately 5-veined); leaflets obovate, 1−3.5×0.7−2.5cm; stipules small, 3−4mm; calyx cup puberulous, without bristles, lobes entire or shortly 1−2-toothed near apex; petals obovate, 7−10×6mm; carpels 3−6.

Bhutan: C—Tongsa district (Yuto La) and Bumthang district (Shabjethang), **N**— Upper Mo Chu district (Kangla Karchu La) and Upper Kulong Chu district (Me La); **Chumbi; Sikkim:** Jongri, Yumthang etc. Mossy ground in Fir and Juniper forests, 3050−4260m. May−August.

6. R. sengorensis Grierson & Long

Creeping stems elongate, woody, bearing hairs and slender straight prickles, with prostrate or suberect woody leafy branches 15−30cm. Leaves pinnately 3-foliate, terminal leaflet obovate 4−7×1.5−5.5cm, acute or obtuse, margins crenate-serrate, thinly hairy on both surfaces, prickly on midrib beneath, lateral leaflets smaller; petioles 2−7cm, prickly; stipules divided almost to base into 3 simple or branched linear lobes 4−8mm. Flowers 1−2, axillary; pedicels 1.5−2cm. Calyx cup pilose and with a few prickles, lobes ovate, 5−8mm, acuminate, entire or 1−2-toothed. Petals pink, obovate, 4−5×2−3mm. Carpels 20−25.

Bhutan: C—Mongar district (below Sengor and near Zimgang). Hemlock and Fir forests, creeping on mossy banks and streamsides, 2700−3175m. June−July.

Endemic to Bhutan.

7. R. acuminatus Smith; *R. latifolius* Kuntze. *R. pilocalyx* Kuntze. Dz: *Komatsang*; Nep: *Biraley Kara* (117), *Sanu Aselu* (34)

Large climbing shrub; stems glabrous, with few small recurved prickles. Leaves green when dry, ovate, 6−13×3−6cm, caudate-acuminate, base rounded or truncate, margins and apex serrate, reticulate beneath, sparsely pubescent only on veins beneath; petioles slender, 1.2−1.6cm; stipules subulate, c 5mm, deciduous. Flowers in short little branched terminal panicles or racemes 6−10cm. Pedicels 8−12mm; bracts subulate, 6mm, entire. Calyx lobes triangular, 4−5mm, entire, glabrous with pubescent margins. Petals white, equalling calyx lobes. Fruit red, enclosed by calyx, drupelets 5−8.

Bhutan: S—Chukka district (Chasilakha), **C**—Punakha district (Sewla Gompa), Tongsa district (Mangde Chu) and Mongar district (Khinay Lhakang), **N**—Upper Mo Chu district (Gasa); **Sikkim**. Climbing on rocks and shrubs in Warm broad-leaved forests, 1700−2250m. July−September.

Sterile specimens from Deothang, Bhutan and Lama Hills, Sikkim resemble plants from W. China labelled *R. acuminatus,* but these may represent an undescribed taxon differing from typical *R. acuminatus* and *R. griffithii* in the densely prickly stems, gradually acuminate more irregularly serrate leaves which are subcordate at the base, and larger broader panicles.

8. R. griffithii Hook.f.; *R. excurvatus* Kuntze

Similar to *R. acuminatus* but leaves gradually acuminate, margins doubly serrate; petioles 7−9mm; panicles wide branched; bracts linear, serrate; flowers sessile; calyx lobes tomentose.

Sikkim: Darjeeling district; **Assam Duars:** Dewangiri Hills.

Described from a Griffith specimen labelled 'Darjiling' but according to Hooker (80) the locality is doubtful. Another sterile specimen from Deothang, Bhutan may be conspecific but bears deeply laciniate stipules.

9. R. hamiltonii Hook. f.

Similar to *R. acuminatus* but leaves usually brown when dry, shortly and abruptly acuminate with entire apex; petioles 10−12mm; stipules pectinate, 7−8mm; flowers on widely branched terminal panicles; bracts large, deeply pectinate, 7−9mm.

Bhutan: S—Samchi district (Chepuwa Khola), Sarbhang district (near Phipsoo) and Gaylegphug district (Gaylegphug, 117); **Sikkim:** Darjeeling district; **Assam Duars:** Durunga. In ravines in subtropical forests, 150−550m. September−October.

10. R. preptanthus Focke

Large climbing shrub; stems with few small recurved prickles, branchlets white tomentose. Leaves ovate 6−10×3−4cm, acuminate, base truncate, margins doubly serrate, green and subglabrous above, densely white tomentose beneath, lateral veins 8−11 pairs; petioles 5−7mm, tomentose; stipules lanceolate 10−15mm, entire. Flowers 3−8 in short terminal racemes 4−7cm, bracts similar to stipules. Pedicels stout, 10−20mm, tomentose. Calyx lobes triangular, c10mm, with subulate apex and 2−4 marginal teeth, densely white tomentose. Petals white, c10mm. Carpels 25−30.

Bhutan: C—Tashigang district (above Khaling). Cool broad-leaved forest slopes, 2550m. June−July.

A disjunct species known elsewhere only from Yunnan, W China.

11. R. calophyllus Clarke

Suberect shrub 1−3m, stems with minute scattered recurved prickles, branchlets whitish pubescent. Leaves ovate 13−18×5−8cm, finely acuminate, base rounded, margins coarsely serrate with finely pointed teeth; white sericeous beneath, lateral veins 20−25 pairs, closely parallel, prominent beneath; petioles 5−8mm; stipules linear, 12−15mm, entire. Flowers crowded in dense rounded 2−7-flowered axillary panicles. Calyx lobes triangular, 10−12mm. Petals deep wine red, obovate, c 10×8mm, stamens deep wine red; carpels c 40.

Bhutan: C—Mongar district (Latu La and Sheridrang) and Tashigang district (Kori La). Margins of wet Hemlock and Cool broad-leaved forests, 2200−2900m. June−July.

Related to *R. lineatus, R. splendidissmus* and their allies but differing in its simple leaves.

12. R. paniculatus Smith; *R. paniculatus* sensu F.B.I. p.p., *R. poliophyllus* Kuntze, *R. moluccanus* L. var. *paniculatus* (Smith) Kuntze. Dz: *Domaytsalu*

Climbing shrub, stems with scattered small recurved prickles; branchlets creamy white tomentose, eglandular. Leaves ovate, shallowly 2−4-lobed on each side, 8−14×5−9cm, acuminate, base deeply cordate on lower leaves, becoming almost truncate on uppermost leaves, margins finely serrate, lateral veins 6−8 pairs, pubescent above, whitish tomentose beneath; petioles 1.5−4cm, prickly or not; stipules oblong-lanceolate, c 12mm, laciniate at apex. Flowers in broad tomentose panicles up to 30×25cm. Bracts lanceolate, 8−10mm, laciniate with lanceolate teeth. Calyx lobes triangular, c 8mm, entire, sericeous. Petals white, oblong, 3−4mm. Drupelets c 25, black.

Bhutan: S—Deothang district, **C**—Thimphu, Punakha, Tongsa and Mongar districts, **N**—Upper Kulong Chu district (Tobrang); **Sikkim:** Darjeeling, Karponang etc. Cool broad-leaved forest margins, 610−2800m. June−November.

13. R. tiliaceus Smith; *R. paniculatus* sensu F.B.I. p.p. non Smith, *R. paniculatus* Smith forma *tiliaceus* (Smith) Hara, *R. moluccanus* L. var. *tiliaceus* (Smith) Kuntze, *R. cordifolius* D. Don non Noronha

Closely allied to *R. paniculatus* but branchlets, leaves and panicles more thickly white tomentose; leaves broadly ovate or suborbicular

6−14×6−10cm; petioles 2−5cm; panicles narrow, 9−15×4−9cm, with short branches; bracts laciniate with linear-subulate teeth; calyx lobes ovate-lanceolate.

Sikkim: Darjeeling district (101). 2000−2400m.

A closely allied species, *R. glandulifer* Balakrishnan *(R. lanatus* Hook. f. non Focke) was reported by Focke (Bibliotheca Botanica 72: 111, 1910) to have been collected in Sikkim by J. D. Hooker, but no specimen has been located. It is a W Himalayan taxon differing from *R. tiliaceus* in having gland-tipped hairs mixed with the white tomentum on the branchlets, petioles and panicles.

14. R. efferatus Craib; *R. ferox* Focke non Trattinick,? *R. sterilis* Kuntze, *R. kurzii* Balakrishnan *nom. superfl., R. moluccanus* L. var. *ferox* Kuntze. Nep: *Pan-kara* (117)

Similar to *R. paniculatus* but leaves pubescent on both surfaces (not white tomentose); flowers in smaller terminal panicles and on short axillary branches or occasionally solitary; calyx lobes triangular 6−7mm, pectinately toothed on margins.

Bhutan: S—Gaylegphug district (Sureylakha, 117); **Sikkim:** Leebong. Warm broad-leaved forests, 1200−1600m.

15. R. insignis Hook. f.; *R. insignis* var. *ochraceus* Focke, *R. moluccanus* L. var. *insignis* (Hook. f.) Kuntze

Similar to *R. paniculatus* but leaves oblong-ovate, more deeply cordate, shortly and abruptly acuminate, margins irregularly crenulate-denticulate, lateral veins 10−15 pairs; petioles stout, more prickly; stipules deeply pectinate; panicles very broad, with flowers aggregated near ends of branches; calyx lobes 4−5mm, entire; petals orbicular, equalling sepals.

Bhutan: S—Deothang district (Deothang), **C**—Tashigang district (Phalang). Amongst shrubs in Subtropical and Warm broad-leaved forests, 300−2000m. November−December.

16. R. calycinoides Kuntze; *R. moluccanus* auct. p.p. non L., *R. moluccanus* L. var. *calycinoides* (Kuntze) Kuntze, *R. rugosus* auct. p.p. non Smith, *R. bhotanensis* Kuntze, *R. darschilingensis* Kuntze, *R. difissus* Focke, *R. fallax* Kuntze, *R. himalaicus* Kuntze. Nep: *Aselu* (34)

Large scrambling shrub, stems brownish pubescent and with scattered short recurved prickles, eglandular. Leaves broadly ovate or suborbicular, 10−20cm long and broad, base cordate, 3−5-lobed or with a few additional minor lobes, lobes acute, sharply serrate, stiffly pilose to subglabrous and often rugose above, tomentose, pilose or subglabrous beneath, main veins often prickly beneath; petioles 4−9cm, with recurved prickles; stipules ovate, 1.5−2cm, deeply toothed or laciniate, often persistent. Flowers in axillary clusters or short racemes up to 6cm. Pedicels 8−15mm. Bracts elliptic, 1–1.5cm, toothed at apex. Calyx lobes triangular, 10–12mm, brownish or yellowish sericeous, often toothed on margins. Petals white, equalling calyx lobes. Fruits red, drupelets 40–50.

Bhutan: S—Phuntsholing, Chukka and Sarbhan, districts, **C**—Punakha district; **Sikkim:** common in Darjeeling district. Subtropical and Warm broad-leaved forest margins, 780–2100m. June–August.

A very variable species, particularly in leaf shape, depth of lobes and indumentum.

17. R. rugosus Smith; *R. moluccanus* auct. p.p. non L.

Closely allied to *R. calycinoides* but leaves smaller, 8–12×6–10cm, lobes rounded or subacute, margins crenate-serrate; stipules smaller, 1–1.5cm, early caducous; flower clusters dense; pedicels short, 2–7mm; calyx lobes 8–9mm, entire or minutely toothed at apex.

Sikkim: locality unknown.

Records of *R. moluccanus* L. from Sikkim (e.g.80) could refer to this species, which occurs in Nepal, Assam and Khasia. A single specimen from the Sikkim Terai (Siliguri) may belong to a third species; it differs from *R. calycinoides* and *R. rugosus* in its large suborbicular leaves 15×16cm with round lobes.

18. R. kumaonensis Balakrishnan; *R. reticulatus* Hook. f. non. Kerner. Nep: *Aselu* (34)

Similar to *R. calycinoides* but stems and petioles whitish tomentose and with few short (1–2mm) straight or weakly curved prickles; leaves suborbicular, deeply 5–7-lobed, lobes acute or subacute, occasionally obtuse, upper surface reticulate, rugose, pilose, lower surface reticulate, softly and densely white tomentose; stipules lanceolate, c 10mm, deeply laciniate; flowers in tomentose axillary and terminal racemes, often forming dense terminal panicles; calyx lobes whitish tomentose outside, enlarging in fruit; petals white.

Sikkim: Lachen; **Arunachal Pradesh:** Nyam Jang Chu. Ravines in Cool broad-leaved forests, 2300–2750m. June–July.

19. R. treutleri Hook. f.; *R. arcuatus* Kuntze, *R. rosulans* Kuntze, *R. tonglooensis* Kuntze, *R. moluccanus* L. var. *treutleri* (Hook. f.) Kuntze. Nep: *Thalumbo* (131)

Scrambling or climbing shrub with long slender shoots; branchlets with dense gland-tipped bristles and scattered slender prickles. Leaves suborbicular 6–14×6–14cm, 3–5-lobed, base cordate, lobes broadly acute, serrate, pubescent above, softly pilose beneath; petioles 3–6cm, densely covered with soft hairs, gland-tipped bristles and scattered short prickles; stipules c 2cm, divided to base into 4–5 linear segments. Flowers 3–5 in short axillary racemes; bracts pectinate. Calyx lobes ovate,

12−15mm, 3−5-toothed at apex, tomentose and with gland-tipped bristles and prickles. Petals c 10mm, pink. Fruit 1.5−2cm with many drupelets.

Bhutan: C—Punakha, Tongsa and Mongar districts, **N**—Upper Kulong Chu district; **Sikkim; Chumbi.** Cool broad-leaved and Hemlock forests, 2700−3400m. June−August.

Fruit edible (131).

20. R. cooperi Long

Similar to *R. treutleri* but eglandular; stems whitish pilose and with scattered short prickles; leaves 5−8cm diameter, with rounded lobes; petioles 4−7cm; stipules 10−13mm, laciniate into c 10 narrow lobes; calyx lobes 8−10mm, deeply pectinate toothed on margins and at apex; petals white 5−6mm.

Bhutan: C—Mongar district (Sawang). Moist forest, 2740m. June−July.

21. R. splendidissimus Hara; *R. andersonii* Hook. f. non Lefèvre, *R. lineatus* Blume var. *andersonii* Kuntze. Nep: *Phusre Aselu* (34)

Scrambling shrub, stems unarmed, leafy shoots whitish hairy and with long dark gland-tipped bristles 1.5−4mm. Leaves palmately 3-foliate, obovate-elliptic, terminal 9−15×3.5−7cm, abruptly acuminate, base cuneate, margins sharply doubly serrate, lateral veins 15−25 pairs, densely white sericeous beneath, midrib with a few glandular bristles and occasionally several short prickles near base; petioles 3−8cm; stipules elliptic-lanceolate 15−20×6−7mm, entire. Flowers in broad, glandular-bristly corymbose, axillary and terminal panicles; bracts stipule-like but smaller. Calyx lobes triangular, 10−15mm, finely acuminate, entire, whitish tomentose and glandular-bristly. Petals (?white) c 10mm. Fruit red, c 1cm.

Bhutan: C—Thimphu district (Dochong La), Tongsa district (between Tashiling and Charikhachor) and Mongar district (Sawang, Unjar, Rudo La and Sheedrang), **N**—Upper Kulong Chu district (Lao); **Sikkim:** Senchal, Tonglu etc. Cool broad-leaved forests, 2400−2850m. May−August.

22. R. phengodes Focke; *R. lineatus* Blume var. *pulcherrimus* Kuntze

Similar to *R. splendidissimus* but stems with scattered short prickles, glandular bristles fewer and shorter, c 0.5mm; lateral veins 20−35 pairs; stipules ovate, 7−15mm broad; flowers in condensed axillary panicles or fascicles; calyx lobes shortly acuminate, white sericeous but with few bristles only at base; fruit pink.

Bhutan: S—Chukka district (near Takhti Chu); **Sikkim:** Senchal, Karponang and Rungbe. Evergreen oak and Cool broad-leaved forests, 1800−3000m. July−September.

23. R. lineatus Blume; *R. lineatus* var. *angustifolius* Hook. f., var. *intermedius* Kuntze. Nep: *Gempe Aselu* (34)

Scrambling shrub to 3m, eglandular; stems unarmed or with few short prickles; branchlets whitish sericeous. Leaves mostly pedately 5-foliate, sometimes 3- or 7-foliate; leaflets elliptic-oblanceolate, terminal 9−15×1.5−3(−4)cm, finely acuminate, base cuneate, margins sharply serrate, softly white sericeous beneath, at least on veins, lateral veins 25−50 pairs, midrib sometimes with a few prickles near base; petioles 3−4.5cm, unarmed or prickly at apex; stipules ovate, 10−15mm, entire. Flowers few in short axillary clusters or cymes. Calyx lobes triangular, c 10mm, accuminate, entire, sericeous. Petals greenish-white, 4−5mm. Fruit red c1.5cm.

Bhutan: S—Gaylegphug district (Shamkhara and Chabley Khola) and Deothang district (Wamrong, 117), C—Mongar district (Namning); **Sikkim:** Tonglu etc. Marginal scrub and ravines in moist broad-leaved forests, 1800−2700m. March−June.

Dried roots used in treatment of food poisoning (131); stems used to make fences (48).

24. R. senchalensis Hara

Similar to *R. lineatus* but branches unarmed; leaflets whitish pubescent or glabrescent beneath; lateral veins 14−22 pairs; stipules lanceolate, 8−15×1.5−2mm, toothed on margin; calyx lobes 6−8mm, green; petals white, c 3mm.

Sikkim: between Senchal and Tiger Hill (71). 2840m. July−August.

Only known from this locality. Hara (71) suggested that the colony may be of hybrid origin involving *R. lineatus*.

25. R. pentagonus Focke; *R. alpestris* sensu F.B.I. p.p. non Blume, *R. tridactylus* Focke. Dz: *Bjake Tshalu*

Scrambling shrub, stems with broad-based flattened prickles 3−5mm, leafy shoots almost glabrous. Leaves palmately 3-foliate, lateral leaflets sessile, elliptic, terminal one 5−10×2.5−4cm, caudate acuminate, base attenuate, margins doubly serrate, thinly hairy and sometimes with a few prickles on veins beneath, lateral leaflets not much smaller than terminal; petioles 2−4cm, with a few prickles and scattered gland-tipped bristles; stipules narrow linear, 8−12mm, glandular-hairy. Flowers solitary or 2−3 in terminal fascicles, pedicels 1−3cm. Calyx lobes ovate, 10−15mm, subulate tipped, spiny and glandular-hairy, entire or apex 3-fid. Petals white, c 7mm. Fruit red or yellow, 2−3cm diameter, enclosed in calyx, drupelets glabrous.

Bhutan: C—Thimphu, Punakha, Tongsa, Mongar and Tashigang districts, N—Upper Mo Chu and Upper Kulong Chu districts; **Sikkim.** Cool broad-leaved and Hemlock forests, 2400−2900m. May−July.

26. R. thomsonii Focke

Similar to *R. pentagonus* but prickles slender, 1−3mm, terminal leaflet ovate, 5−12×3−5.5cm, much larger than lateral leaflets, margins doubly serrate, petiolules 2−4mm; stipules deeply 2−4-fid into linear segments 4−6mm, eglandular; flowers 1−6 in axillary panicles; calyx lobes lanceolate, 3−5mm, entire, eglandular; petals red; fruit c 1cm, drupelets pubescent.

Bhutan: C—Mongar district (E side of Rudo La), **N**—Upper Kuru Chu district (Julu); **Sikkim:** Rongbe, Karponang etc; **Chumbi:** Lingmatam. Moist Cool broad-leaved and coniferous forests, 2500−3300m. August−October.

27. R. ellipticus Smith; *R. ellipticus* var. *denudatus* Hook. f., *R. flavus* D. Don. Dz: *Tshema Tshelu;* Sha: *Gongsey, Sergong;* Tongsa: *Tshema;* Nep: *Aselu* (34), *Ainselu*

Robust scrambling shrub 2−3m, stems angled, pubescent and with scattered deflexed prickles and dense spreading mostly eglandular bristles 3−6mm. Leaves pinnately trifoliate, leaflets elliptic-obovate to suborbicular, terminal 4−10×3−7cm, obtuse or subacute, base rounded, margins shallowly serrate, finely and densely pale puberulous beneath, midrib prickly; lateral leaflets slightly smaller, terminal petiolule 2−4cm, laterals 2−5mm; petioles 3−8cm; stipules linear, 6−8mm, caducous. Panicles axillary and terminal, few to many-flowered. Calyx cup bristly, lobes ovate, 5−6mm, acute, entire, softly pale pubescent. Petals white, obovate, c 6mm. Fruit yellow, drupelets many.

Bhutan: S—Samchi, Phuntsholing, Chukka and Gaylegphug districts, **C**—Tashigang district; **Sikkim:** Darjeeling district, common. Roadsides, scrub and abandoned cultivation in Warm broad-leaved and Evergreen oak forests, 1200−1900m. February−April.

Fruit commonly eaten and made into preserve (48).

28. R. alexeterius Focke; *R. acaenocalyx* Hara

Shrub 1−2m, stems smooth, with white bloom and stout recurved prickles 5−9mm. Leafy branchlets with shorter slender prickles and a mixture of soft white hairs and gland-tipped bristles. Leaves 3-foliate, leaflets ovate, terminal 3−4×2−3cm, acute, base rounded, margin doubly serrate or shallowly lobed, pubescent above, white tomentose beneath with a few prickles on midrib, terminal petiolule 5−14mm, lateral petiolules 1−2mm; petioles 1−4cm, prickly, white-pilose and glandular bristly; stipules linear 6−8mm, entire. Flowers axillary and terminal, solitary or in fascicles of 2−3; pedicels 1.5−2cm, densely prickly and glandular. Calyx cup with straight prickles, lobes ovate, erect, 10−13mm, caudate acuminate, tomentose and glandular, in fruit enlarging to 15−20mm. Petals white, obovate, c 6mm. Fruit glabrous.

Bhutan: C—Ha district (between Ha and Puduna), Thimphu district (Thimphu Chu valley), Bumthang district (Badar La) and Mongar district (Sengor). Blue Pine and Hemlock forests, in clearings and amongst shrubs, 2200−3350m. May−June.

A specimen from Dobchu valley, Paro may be a hybird with *R. biflorus,* having the vegetative features of *R. alexeterius* but with smaller, unarmed spreading calyx and long petals c 8mm.

29. R. biflorus Smith. Med: *Kentakare* (stem), *Taktse Metog;* Sha: *Thulu Gongsey.* Fig. 38 a&b

Similar to *R. alexeterius* but with few gland-tipped hairs restricted mostly to petioles; some leaves occasionally pinnately 5-foliate; terminal leaflet often deeply lobed; flowers and pedicels subglabrous and eglandular; calyx cup without prickles, lobes ovate, 6−9mm, short-pointed, glabrous; petals white, 7−10mm, early caducous; fruit orange.

Bhutan: C—Ha, Thimphu, Tongsa and Bumthang districts, **N**—Upper Mo Chu and Upper Kulong Chu districts; **Sikkim:** Lachen. In clearings and scrub in coniferous, especially Blue Pine, forests, 2300−3500m. May−July.

Fruits edible. Forms with 5-foliate leaves are rare, known only from Paro valley. Possibly hybridises with *R. alexeterius* (see above) and *R. niveus* (71).

30. R. irritans Focke; *R. purpureus* sensu F.B.I. p.p. non Bunge

Similar to *R. alexeterius* and *R. biflorus* but a low erect shrub 20−60(−100)cm; stems without white bloom, often densely clothed with slender straight prickles, on branchlets and petioles mixed with gland-tipped bristles and minute pubescence; terminal leaflet ovate, 4−7×3−6cm; stipules longer, 10−12mm; flowers similar to those of *R. alexeterius,* calyx cup prickly, lobes 8−12mm, white tomentose; petals white, c 4mm; fruit red, drupelets pubescent.

Bhutan: N—Upper Mo Chu district (Laya and Lingshi La) and Upper Bumthang Chu district (Pangotang). Alpine grassy cliff-ledges, 3960−5200m. June−July.

31. R. sikkimensis Hook. f.

Similar to *R. alexeterius* and *R. irritans* but stems often densely covered with slender straight prickles and gland-tipped bristles; leaflets green and pubescent or white tomentose beneath, glandular above; flowers solitary or paired; calyx cup glandular-bristly and prickly, lobes ovate, pubescent, caudate acuminate; petals red, dark purple or pink.

Two varieties occur:

var. **sikkimensis.** Leaflets green and pubescent beneath.

Bhutan: C—Tashigang district (Praunzong Gompa), **N**—Upper Kulong Chu district (Me La); **Sikkim:** Lachen. Fir forests, 3500−3960m. July−August.

var. **canescens** Long. Leaflets white tomentose beneath.

Bhutan: C—Thimphu district (Barshong) and Tongsa district (Yuto La),**N**—Upper Bumthang Chu district (Pangotang). Moist coniferous forests, 3200−3660m. July.

Var. *canescens* is endemic to Bhutan but flower colour is unknown; if it proves to have white petals it would be better treated as a variety of *R. irritans*, or possibly a distinct species.

32. R. wardii Merrill; *R. hookeri* Focke non Koch, *R. gigantiflorus* Hara, *R. macrocarpus* Clarke non Bentham

Extensive trailing or scrambling shrub, stems and petioles with scattered prickles, often dense gland-tipped bristles, and white pubescence. Leaves pinnately 3-foliate, terminal leaflet rhombic-obovate, 7−11×5−9cm, abruptly caudate acuminate, base rounded, margins serrate and shallowly lobed, thinly glandular or pubescent on both surfaces; terminal petiolule 6−15mm, laterals 1−3mm; petioles 3−7cm; stipules ovate, 10−20×6−12mm, laciniate, hairy. Flowers solitary, axillary; pedicels 4−6cm, bracts stipule-like. Calyx cup 1.5−2cm diameter, prickly, lobes ovate-caudate or laciniate 2−3cm, prickly, pubescent and glandular. Petals white, orbicular, 10−12mm. Fruit red, globose, c 2mm, drupelets many, lower ones dry, pubescent, upper ones fleshy.

Bhutan: S—Deothang district (Tshilingor), C—Tongsa district (near Chendebi), Mongar district (near Sengor) and Tashigang district (Monborong); **Sikkim:** Senchal, Rongbe, Tonglo, etc. Amongst scrub and in ravines in Cool broad-leaved and Evergreen oak forests, 2450−3050m. May−July.

Fruit edible.

A single collection from Zimgang, Mongar district shows affinities with *R. sikkimensis* and *R. wardii* but may be a new species; it differs in being almost eglandular, leaflets broadly elliptic or rhombic, 3−7×2.5−6.5cm with crenate-dentate margins, softly hairy (but not white) on both surfaces, stipules pectinately divided into linear segments c 8mm, sparsely glandular; flowers solitary with ovate calyx lobes 6−8mm, and few, pubescent drupelets.

33. R. hypargyrus Edgeworth; *R. niveus* G. Don non Thunberg, *R. pedunculosus* auct. non D. Don, *R. gracilis* DC. non Presl

Arching or scrambling shrub 2−3m, leafy shoots with scattered to dense weak prickles, pubescent, eglandular. Leaves 3-foliate; leaflets ovate, terminal 3−10×2.5−6cm, acuminate, base rounded or cuneate, margins sharply, shallowly to deeply, often doubly, serrate, thinly pubescent above, white tomentose or rarely almost glabrous and green beneath; terminal petiolule 10−20mm, laterals 2−3mm; petioles 4−8cm; stipules linear, 12−15mm, entire. Flowers solitary or 2−5 in short axillary racemes; pedicels 1−2cm, slender. Calyx cup often with a few prickles, lobes lanceolate 8−12mm with long subulate apex, finely tomentose. Petals pink or red, obovate, 5−6mm. Fruit red or orange, c 1cm, drupelets 30−60 pubescent throughout.

Bhutan: C—Thimphu, Bumthang, Mongar and Tashigang districts, **N**—Upper Mo Chu, Upper Bumthang Chu, Upper Kuru Chu and Upper Kulong Chu districts; **Sikkim.** On banks and in scrub in Fir/Rhododendron forests, 2750–3660m. June–August.

The common E. Himalayan plant belongs to var. **niveus** Hara *(R. niveus* G. Don non Thunberg var. *niveus* sensu F.B.I., *R. pedunculosus* auct. non D. Don) with leaves white tomentose beneath. The typical var. **hypargyrus** *(R. niveus* var. *hypargyrus* (Edgeworth)Hook. f.), a West Himalayan plant, has been recorded from Sikkim (73) but no specimen has been traced; it differs in having leaves green beneath and pubescent only on veins.

A specimen from Lometsawa, Punakha district, differs in its stouter prickles, suborbicular terminal leaflets with minute regular serrations and more numerous but smaller flowers.

34. R. mesogaeus Focke; *R. niveus* var. *microcarpus* Hook. f.

Similar to *R. hypargyrus* var. *niveus* but flowers more numerous (5–15) in terminal and axillary corymbs; calyx cup without prickles; calyx lobes triangular c 7mm, shortly pointed; drupelets with a tuft of hair at apex.

Bhutan: S—Chukka district (Chukka and Chimakothi), **C**—Thimphu district (Chapcha), **N**—Upper Mo Chu district (between Tamji and Gasa); **Sikkim:** Lachen. Evergreen oak forests, 2300–2700m. May–June.

35. R. macilentus Cambessedes

Extensive scrambling shrub, stems angular, glabrous with stout straight prickles 4–10mm; branchlets sparsely pubescent, eglandular. Leaves 3-foliate, terminal leaflet ovate-lanceolate, 2.5–8(–11)×1.5–3(–5)cm, acute or shortly acuminate, base rounded or truncate, margins serrulate, subglabrous, with hooked prickles on midrib beneath; terminal petiolule 1–2(–4)cm, lateral leaflets ovate, smaller, on petiolules 1–2mm; petioles 1.5–4(–8)cm with hooked prickles; stipules linear-lanceolate, 8–15mm, entire. Flowers solitary or in 2–3-flowered axillary or subterminal racemes. Calyx lobes lanceolate, 6–7mm with subulate entire apex, thinly pubescent on outside. Petals white or tinged pink, obovate, c 8mm. Fruit orange, druplets 50 or more, pubescent at apex.

Bhutan: S—Chukka district, **C**—Thimphu, Punakha, Tongsa, Mongar and Sakden districts, **N**—Upper Mo Chu district; **Sikkim:** Lachen. Warm broad-leaved and Evergreen oak forests, 1800–2740m. March–May.

36. R. niveus Thunberg; *R. lasiocarpus* Smith, *R. distans* D. Don. Sha: *Thulu Gong;* Tongsa: *Tsang Guma;* Nep: *Kalo Aselu* (34)

Shrub with long arching reddish shoots with white bloom and scattered stout recurved prickles, eglandular. Leaves pinnate, 5, 7 or 9-foliate, leaflets ovate, sometimes lanceolate or elliptic, 3–6×1–3.5cm, acute or subacute, base narrowly rounded, margins sharply serrate, subglabrous and with impressed veins above, white tomentose and with prominent veins beneath; petioles 2−5cm, prickly; stipules linear-lanceolate 4−8mm. Flowers up to 25 in terminal and sometimes axillary corymbs. Calyx white tomentose, not prickly, lobes ovate 4−6mm, shortly acuminate. Petals pink, 4−6mm. Fruit red, becoming blackish when ripe, drupelets 30 or more, pubescent.

Bhutan: S—Chukka, Gaylegphug and Deothang districts, **C**—Thimphu, Punakha, Tongsa, Mongar and Tashigang districts, **N**—Upper Mo Chu district; **Sikkim**. On cliffs and in scrub in Warm and Cool broad-leaved forests, 1050−2500m. April−June.

The above description and records apply to the widespread var. **niveus**. The more local var. **micranthus** (D. Don) Hara *(R. micranthus* D. Don, *R. lasiocarpus* Smith var. *micranthus* (D. Don) Hook. f.) has been reported from Sikkim; it differs in its smaller ovate, subacute or sometimes obtuse leaflets and smaller flowers with sepals and petals 2.5−3mm.

Closely allied to *R. biflorus* which occasionally has 5-foliate leaves; such plants differ in their sparsely glandular petioles, glabrous calyx lobes and white petals.

Fruit edible (48).

37. R. inopertus (Focke) Focke; *R. lasiocarpus* Smith var. *rosifolius* Hook. f., *R. niveus* Thunberg var. *rosifolius* (Hook. f.) Hara, *R. niveus* Thunberg subsp. *inopertus* Focke. Nep: *Phusre Asaelu*

Similar to *R. niveus* var. *niveus* but leaflets green beneath, pubescent only on veins; flowers in short 3−10-flowered axillary racemes or cymes; calyx lobes finely acuminate, glabrous on outside, white tomentose within and on margins.

Bhutan: S—Chukka district (Chimakothi, 117), **C**—Tongsa district (Tama, 117), **N**—Upper Kulong Chu district (Lao); **Sikkim:** Darjeeling district. Warm and Cool broad-leaved forests, 1200−2740m. May−July.

38. R. pungens Cambessedes; *R. horridulus* Hook. f. non Mueller, *R. parapungens* Hara *nom. superfl.*

Scrambling shrub 1−2m with long slender branches with numerous straight or curved prickles; leafy branchlets short, thinly pilose and with slender hooked prickles, eglandular. Leaves pinnately 7(−9)-foliate, leaflets ovate, 1−5×0.7−4cm, acute or shortly acuminate, base rounded, margins shallowly lobed and sharply serrate, sparsely pubescent on both surfaces with prickles on veins beneath; petioles 2−4cm, grooved, prickly; stipules linear, 5−7mm. Flowers 1−2, terminal on leafy shoots. Calyx cup with dense slender prickles and short gland-tipped hairs, lobes ovate, 6−10mm, caudate-acuminate, appressed pubescent. Petals white or pink 6−8mm. Fruit orange, drupelets 6−20, pubescent.

Bhutan: C—Ha, Thimphu, and Bumthang districts. Amongst *Rosa sericea, Berberis* and other shrubs in dry valleys with Blue Pine and Spruce, 2450−3000m. April−May.

The above description and records apply to var. **horridulus** Hara, which is restricted to Bhutan and SE Tibet.

39. R. indotibetanus Koidzumi; *R. sikkimensis* Kuntze non Hook. f., *R. rosifolius* sensu F.B.I. p.p. non Smith, *R. asper* D. Don non Presl. Nep: *Gempe Aselu* (34)

Similar to *R. pungens* and *R. inopertus* but stems and petioles with sparse to dense short, straight gland-tipped bristles 0.5–1mm; leaflets mostly 5, with short-stalked glands especially beneath; calyx glandular-bristly, not prickly, lobes 13–17mm with triangular base and long filiform point; petals white, 10–13mm; fruit red, large, oblong-globose 2cm diameter.

Bhutan: S—Chukka, Gaylegphug and Deothang districts, **C**—Thimphu and Punakha districts, **N**—Upper Kulong Chu district; **Sikkim:** Kurseong, Lopchu, Darjeeling. Cool broad-leaved and Evergreen oak forests, 1400–3000m. March–May.

Fruits large but not very tasty.

40. R. sumatranus Miquel; *R. rosifolius* sensu F.B.I. p.p. non Smith, *R. sorbifolius* Maximowicz

Similar to *R. pungens* and *R. indotibetanus* but stems and branches with dense long flexuose gland-tipped bristles 2–4mm; leaflets 5 or 7, lanceolate; flowers small, calyx lobes lanceolate, 6–10mm, softly pilose and glandular hairy, not prickly; petals white, c 5mm; fruit cylindric, c $2 \times 1 – 1.5$cm.

Bhutan: S—Phuntsholing district (Phuntsholing) and Gaylegphug district (Tatapani); **Sikkim:** Rishap, Tista etc.. Subtropical forest slopes, 700–900m. February–May.

41. R. rosifolius Smith

Similar to *R. pungens* and its allies but completely without gland-tipped hairs, prickles scattered, straight; leaflets mostly 5, on both surfaces with sessile glistening glands; calyx pubescent and sessile-glandular, lobes lanceolate 10–11mm; petals white, 10–15mm.

Sikkim: Gangtok and Darjeeling. Cultivated, 1600–2100m.

The cultivated plant, forma **coronarius** (Sims) Kuntze *(R. coronarius* (Sims) Sweet) has double flowers with numerous petals. The true *R. rosifolius* is unknown in the Himalaya in the wild; literature reports of it from Sikkim and Bhutan refer mostly to *R. indotibetanus* and *R. sumatranus*.

8. POTENTILLA L.

by A. J. C. Grierson & D. G. Long

Dz: *Choga Sey Sey;* Bumthang: *Yuguli.*

Perennial herbs or shrubs. Leaves 3-foliate, palmate or pinnate, rarely

1-foliate, stipules adnate to petiole. Flowers solitary or few in cymes or corymbs. Calyx tube concave, lobes 5 with 5 epicalyx segments. Petals 5 usually yellow, sometimes white or reddish. Stamens numerous. Carpels numerous, free; style subterminal, deciduous. Achenes numerous, borne on a flat or conical dry receptacle.

1. Erect shrub 60–150cm, sometimes 20–30cm but then stems branching and leafy throughout, never cushion- or mat-forming; stipule lobes united almost to apex .... **1. P. arbuscula**
+ Annual or perennial herbs mostly 5–30cm, sometimes woody at base but then usually cushion- or mat-forming, with little-branched stems leafy only at tips; stipule lobes free 2

2. Leaves 3-foliate or palmately 5-foliate, leaflets sometimes lobed but never pinnatisect (in *P. fragarioides* and *P. saundersiana* sometimes becoming pinnate with a pair of much smaller leaflets below palmately arranged ones) 3
+ Leaves pinnate, lateral leaflets 2–21 pairs (rarely 3-foliate but then leaflets pinnatisect) 17

3. Leaves mostly 3-foliate 4
+ Leaves mostly palmately 5-foliate 15

4. Plants forming dense rounded cushions 5
+ Plants forming rosettes, loose clumps or mats 8

5. Leaflets pinnatifid with 5–13 teeth in upper half .... **6. P. bhutanica**
+ Leaflets simple, linear, entire or divided in upper part into 3 or 5 linear teeth, or divided almost to base into linear lobes 6

6. Leaflets simple, linear .... **2. P. latipetiolata**
+ Leaflets (at least terminal one) 3- or 5-toothed or lobed 7

7. Leaflets 3- or 5-toothed at apex, teeth up to ½ of leaflet length
   4. P. eriocarpa var. **dissecta**
+ Leaflets (at least terminal one) divided to ⅔ or to base into linear lobes
   5. P. eriocarpoides

8. Plants mat-forming, with long spreading woody rhizomes; leaflets 3- or 5-toothed at apex 9
+ Plants mostly rosette- or clump-forming, rhizomes not spreading, herbaceous or woody; leaflets 5–many-toothed 10

9. Rhizomes naked; leaflets 6−12×4−10mm, shallowly 3-toothed; petals slightly emarginate .... **3. P. cuneata**
\+ Rhizomes clothed with persistent petiole bases; leaflets 12−30×5−15mm, deeply 3- or 5-toothed; petals bilobed **4. P. eriocarpa**

10. Leaflets green beneath, subglabrous or thinly pubescent 11
\+ Leaflets white-sericeous beneath 14

11. Leaflets deeply pinnatisect; petals white, with crimson base
 25. P. bryoides
\+ Leaflets serrate or crenate; petals yellow 12

12. Leaflets 15−30×10−25mm, rounded at base .... **9. P. fragarioides**
\+ Leaflets 4−20×3−12mm, attenuate at base 13

13. Leaflets serrate almost to base; petioles 10−70mm; flowers 2−5 in cymes .... **7. P. sundaica**
\+ Leaflets crenate only at apex; petioles 1−5mm; flowers solitary
 8. P. monanthes

14. Robust herb 15−40cm; leaflets 25−40mm; petals 9−11mm
 12. P. atrosanguinea
\+ Dwarf to robust herbs 3−30cm; leaflets 5−25mm; petals 4−15mm .. 15

15. Leaves green and thinly pubescent beneath .... **7. P. sundaica**
\+ Leaves white sericeous beneath 16

16. Large herb 8−30cm; leaflets narrowly obovate, 7−25mm; flowers mostly 3−10 in loose cymes .... **10. P. saundersiana**
\+ Dwarf herb 3−10cm; leaflets broadly obovate, 3−12mm; shoots 1−3-flowered .... **11. P. forrestii**

17. Lateral leaflets entire or sometimes 2-fid at apex .... **13. P. bifurca**
\+ Lateral leaflets crenate, serrate or pinnatisect 18

18. Leaflets crenate or serrate (sometimes pinnatisect in *P. anserina* but then plants stoloniferous) 19
\+ Leaflets pinnatisect (rarely shallowly serrate in *P. microphylla* but then leaflets up to 6mm); plants not stoloniferous 27

19. Lateral leaflets 2−4 pairs 20
\+ Lateral leaflets 4−27 pairs 23

20. Robust herbs, stems 15−55cm; leaflets 0.5−3cm; petals 7−12mm ... 21
\+ Small herbs with prostrate stems 7−20cm; leaflets 0.5−1.5cm; petals 3−5mm 22

21. Leaflets white sericeous beneath; petals 7−10mm .... **14. P. griffithii**
+ Leaflets green beneath; petals 9−12mm .... **16. P. spodiochlora**

22. Leaflets white sericeous beneath; petals 4−5mm .... **15. P. caliginosa**
+ Leaflets green beneath; petals c 3mm .... **17. P. supina**

23. Leaves interruptedly pinnate, with large and small leaflets 24
+ Leaves regularly pinnate, without small leaflets 26

24. Leaflets broadly obovate, green beneath, pubescent but not silvery white sericeous .... **20. P. polyphylla**
+ Leaflets linear, elliptic or narrowly obovate, appressed silvery whitish sericeous beneath 25

25. Plants not stoloniferous; flowers in corymbose cymes ..... **18. P. lineata**
+ Plants stoloniferous; flowers solitary, borne on stolons .. **19. P. anserina**

26. Flowers 2−5 in corymbs; petals 8−13mm .... **21. P. peduncularis**
+ Flowers 5−12, subumbellate; petals 3−5mm .... **22. P. leuconota**

27. Cushion-forming, densely matted, or rosette herbs; flowers yellow
   23. P. microphylla
+ Rosette herbs; flowers white with purple centre 28

28. Leaves 3−10cm, pinnate; leaflets 4−5(−9) pairs .. **24. P. coriandrifolia**
+ Leaves 1−2cm, trifoliate or pinnate with 2 pairs of leaflets
   25. P. bryoides

1. P. arbuscula D. Don; *P. fruticosa* sensu F.B.I. p.p. non L., *P. ribui* Gandoger. Med: *Penma*; Nep: *Chiriya Phal* (34)

Shrub 0.6−1.5m. Leaves (1−)3- or pinnately 5-foliate, leaflets elliptic 6−15×3.5−6mm, acute or obtuse, apiculate, base rounded, margins entire, weakly revolute, sparsely or densely silky pubescent on both surfaces; stipules ovate-lanceolate 8−10mm, brown scarious. Flowers solitary, terminal on short lateral shoots, peduncles up to 5cm usually with a bract consisting of a single leaflet and stipules, near middle. Calyx cup c 5mm diameter often reddish, lobes ovate 5−8mm, epicalyx segments 6−8mm, elliptic. Petals obovate 10−16×8−13mm, rounded, yellow. Achenes 1.5−2mm, ovoid, sparsely white pilose, concealed by long straight white hairs on receptacle.

Bhutan: **C**—Ha and Thimphu districts, **N**—Upper Mo Chu, Upper Pho Chu, Upper Bumthang Chu and Upper Kulong Chu districts; **Sikkim; Chumbi.** Rocky hillsides and amongst Juniper scrub, 3050−4100m. May−August.

Leaves burned as incense.

A polymorphic species varying in habit, indumentum, leaflet number, calyx and epicalyx size; the above records refer to the common var. **arbuscula**. Extremes of variation have attracted varietal distinction; the two following may be significant:

var. **pumila** (Hook. f.) Handel-Mazzetti; *P. fruticosa* L. var. *pumila* Hook. f.

Dwarf shrub 20−30(−60)cm; leaflets smaller 5−7×1−3mm, densely silvery pubescent, margins strongly revolute.

Bhutan: N—Upper Mo Chu, Upper Bumthang Chu and Upper Kulong Chu districts; **Sikkim:** Chaerlung; **Chumbi.** Exposed stony slopes. 3900−4200m. June−September.

var. **unifoliolata** Ludlow

Low shrub clm; leaflets 1(−2), sparsely hairy or subglabrous.

Bhutan: C—Tongsa district (Black Mountain), **N**—Upper Mangde Chu district (Saga La). Rocky hillsides and screes, 3960−4260m. June−July.

This variety is endemic to Bhutan.

2. P. latipetiolata Fischer; *P. armerioides* (Hook. f.) Grierson & Long, *P. fruticosa* L. var. *armerioides* Hook. f.

Dwarf cushion-forming perennial 5−7cm tall; stems becoming woody and densely covered with persistent leaf remains. Leaves 3-foliate, leaflets linear, 5−10×1−1.5mm, apex acute, ± glabrous above, densely hairy beneath, margins strongly revolute; petioles 8−12mm; stipules adnate to petiole along most of their length, c 1mm broad. Flowers solitary, sessile or on peduncles up to 12mm. Calyx lobes ovate c 4.5mm; epicalyx segments lanceolate, c 3.5mm. Petals obovate 6−8mm yellow, emarginate. Achenes obovoid c 1.25mm, glabrous, concealed by long straight white hairs on receptacle.

Sikkim: Kimhin and Chortenima La; **Chumbi:** Naku La. On open hillsides, 5000−5180m. July−August.

3. P. cuneata Lehmann; *P. ambigua* Cambessedes

Rosette or mat-forming perennial herb, stems woody at base; rhizomes slender, naked. Leaves 3-foliate, leaflets obovate, 6−12×4−10mm, apex shallowly 3-toothed, base cuneate, pubescent or glabrous above, appressed hirsute beneath; petioles 0.5−3cm; stipules lanceolate 8−10mm, leafy. Flowers solitary, c 2cm diameter, on peduncles 2−6cm. Calyx lobes ovate, 4−7mm, appressed pubescent; epicalyx segments elliptic, slightly shorter. Petals yellow, broadly obovate, c 10×7−8mm, slightly emarginate. Achenes ovoid, c 1mm, covered with straight pale brown hairs, and hidden by hairs of receptacle.

Bhutan: C—Ha, Thimphu, Tongsa, Tashigang and Sakden districts, **N**—Upper Mo Chu, Upper Pho Chu, Upper Bumthang Chu and Upper Kulong Chu districts; **Sikkim; Chumbi.** On rocky streamsides, cliffs and boulders, 2500−3650m. May−July.

4. P. eriocarpa Lehmann

Similar to *P. cuneata* but lower stems longer and stouter, covered with persistent pubescent petiole bases; leaflets obtriangular, 12−30×5−15mm, deeply 3- or 5-toothed to middle; flowers larger, c 2.5cm diameter; petals bilobed, achenes c 1.5mm.

Bhutan: C—Thimphu and Tongsa districts, **N**—Upper Mo Chu, Upper Pho Chu, Upper Bumthang Chu, Upper Kuru Chu and Upper Kulong Chu districts; **Sikkim**; **Chumbi**. On wet exposed cliff ledges, 4000−4570m. June−August.

Two varieties occur: the widespread var. **eriocarpa** which is a straggling, often mat-forming plant, and var. **dissecta** Marquand which is a small, compact cushion-forming plant, know from alpine Sikkim (Tenchungkar and Dongkong).

5. P. eriocarpoides Krause; *P. eriocarpa* Lehmann var. *tsarongensis* W. E. Evans

Resembles *P. eriocarpa* var. *dissecta* in its compact, cushion-forming habit; differs from *P. cuneata* and *P. eriocarpa* in its leaflets which are divided to c ⅔ or commonly to the base into 1−5 linear lobes; flowers as in *P. eriocarpa*.

Bhutan: C—Ha district (Ha) and Thimphu district (Tremo La); **Sikkim**: Chaerlung; **Chumbi**: Chomo Lhari. Exposed mountain slopes, 2800−4570m. July−August.

6. P. bhutanica Ludlow

Similar to *P. cuneata* and its allies but forming dense cushions with stems thickly covered with persistent leaf remains; leaflets obovate, 10−15×5−9mm, pinnatifid with 5−13 sharp teeth in upper half, densely white pilose beneath; petioles 2−5cm; flowering stems 1−6cm; calyx lobes lanceolate c 8mm; epicalyx segments oblong-elliptic c 7mm; petals 10−12mm long and broad.

Bhutan: N—Upper Pho Chu district (W branch of Pho Chu). On rocks and cliffs, 4265m. June.

7. P. sundaica (Blume) Kuntze; *P. kleiniana* Wight, *P. wallichiana* Lehmann non Seringe

Rosetted herb with spreading prostrate stems up to 45cm. Leaves palmately (3−)5-foliate; leaflets obovate 0.5−2.0×0.3−1.2cm, apex rounded, base cuneate, margins serrate, ± glabrous above, appressed pubescent beneath; petioles up to 7cm; stipules lanceolate 1−1.5cm. Flowers 2−5 in small terminal cymes. Calyx lobes 3−4mm. Petals obovate 3−5×2mm, yellow. Achenes ellipsoid c 1mm, glabrous.

Bhutan: S—Phuntsholing, Chukka and Gaylegphug districts, **C**—Ha, Thimphu, Punakha and Tongsa districts; **Sikkim**. Roadsides and margins of cultivation, 800−2600m. March−July.

8. **P. monanthes** Lehmann

Similar to *P. sundaica* but stems up to 10cm; leaflets always 3, broadly obovate, 4−10mm long and broad, apex rounded and with 7−10 crenate teeth, base cuneate, sparsely pubescent; petioles 1−5mm; stipules ovate or elliptic 3−6mm, entire or 3-toothed; flowers solitary, axillary on pedicels c 1.5cm; petals 4−6×3−4mm.

Bhutan: N—Upper Mo Chu district (Laya) and Upper Kulong Chu district (Me La); **Sikkim; Chumbi.** Alpine grassland, 3650−3900m. May.

The above description relates to the prostrate, small leaved var. **sibthorpioides** Hook. f. which appears to be more common in E Himalaya although the larger more erect var. **monanthes** occurs in Sikkim (Tungu and Lachen).

9. **P. fragarioides** L. agg.

Rosette herb, whitish pilose throughout. Leaves trifoliate, but often becoming pinnate with 2 additional small leaflets on petiole; upper leaflets obovate-elliptic, 1.5−2.5×1−1.7cm, obtuse, coarsely serrate to base; minor leaflets 4−8mm; petioles 3−15cm, densely pilose. Flowers 3−6 in loose cymes on peduncles 4−7cm. Calyx lobes ovate c 4mm, pilose. Petals obovate, 5−6mm, yellow. Achenes 1.4mm, glabrous.

Bhutan: C—Tashigang district (banks of Kulong Chu, Tashi Yangtsi). On moist banks. February.

This description applies to a Griffith collection, which differs in several respects from typical Siberian *P. fragarioides,* which has leaves more clearly pinnate with 1−2 pairs of slightly reduced leaflets below the larger uppermost 3.

10. **P. saundersiana** Royle; *P. multifida* L. var. *saundersiana* (Royle) Hook. f., *P. nivea* sensu F.B.I. p.p. non L.

Rosette herb with stout rootstock; stems spreading or suberect, 8−30cm. Basal leaves palmately (3−)5-foliate, narrowly obovate 7−25×4−10mm, apex rounded, base cuneate, margins serrate, sparsely pubescent above, densely white tomentose beneath; petioles up to 8cm; stipules lanceolate c 10mm; cauline leaves usually 3-foliate and with smaller leaflets. Flowers 3−10 in cymes. Calyx lobes 2.5−3.5mm. Petals obovate 4−5×4−6mm, yellow, emarginate. Achenes c 1.5mm, receptacle sparsely long hairy.

Bhutan: C—Ha and Thimphu districts, N—Upper Mo Chu and Upper Bumthang Chu districts; **Sikkim; Chumbi.** On grassy banks, 3500−4250m. May−September.

This description refers to the common var. **saundersiana.** The var. **potaninii** (Wolf) Handel-Mazzetti (*P. potaninii* Wolf) occurs at Lingshi, 3960m, and differs only in that some leaves bear an additional pair of minor leaflets on the petiole.

11. P. forrestii W. W. Smith; *P. nivea* sensu F.B.I. p.p. non L.

Similar to *P. saundersiana* but a smaller plant 3−10cm; leaves 3- or 5-foliate; leaflets broadly obovate 3−12×2−7mm, margins shallowly to deeply serrate; stems 1−3-flowered.

Bhutan: N—Upper Bumthang Chu and Upper Kulong Chu districts; **Sikkim; Chumbi.** Mountain rocks and meadows, 4000−4900m. May−September.

Three varieties have been recognised by Sojak: the typical var. **forrestii** with 5-foliate leaves with margins shallowly serrate, the common plant in Bhutan, Sikkim and Chumbi; var. **segmentata** Sojak also with 5-foliate leaves but with the leaflets deeply toothed almost to the midrib, from Sikkim (Lhonak and Lachen); and var. **caespitosa** (Wolf) Sojak (*P. saundersiana* Royle var. *caespitosa* Wolf) with consistently 3-foliate leaves, from Bhutan (Pangothang, Upper Bumthang Chu, and Me La, Upper Kulong Chu), Sikkim (Guicha La and Kimhin) and Chumbi.

These Himalayan plants referred by Sojak to *P. forrestii* may merely represent varieties or subspecies of *P. saundersiana*, as intermediates between these variable species seem to be frequent. *P. forrestii* itself is based on a robust cultivated type quite unlike the Himalayan populations in stature. Further research on this complex is desirable.

12. P. atrosanguinea Loddiges; *P. argyrophylla* Lehmann

Similar to *P. saundersiana* but more robust, stems up to 40cm; leaflets always 3, ovate or obovate, 25−40×10−30mm; petioles up to 20cm; calyx lobes 5−6mm; petals yellow, 9−11mm; achenes concealed amongst dense receptacular hairs.

Sikkim: Gamothang and Nepal border. Alpine slopes, 3650−3800m. May.

The above description refers to var. **argyrophylla** (Lehmann) Grierson & Long; the red-flowered var. *atrosanguinea* has not so far been recorded from E Himalaya.

13. P. bifurca L.

Low growing herb with wiry prostrate rhizomes. Leaves pinnate, 2−6cm, lateral leaflets 2−4 pairs, elliptic, 4−7×1.5−3.5mm, acute, base cuneate, margins entire or 2−3-toothed at apex, hirsute especially beneath; petioles 0.5−3cm; stipules lanceolate 5−6mm. Flowers 1−5 in loose cymes on peduncles c 5cm. Calyx lobes ovate 4−5mm. Petals yellow or rarely white, obovate c 6×4mm, rounded at apex. Achenes ovoid 1.5−2mm, glabrous, receptacle sparsely white hairy.

Sikkim: Lhonak and Kungna Lama. Open stony and grassy slopes, 3500−4500m. June−September.

The Himalayan plants belong to var. **moorcroftii** (Lehmann) Wolf (*P. moorcroftii* Lehmann).

14. P. griffithii Hook. f.; *P. sikkimensis* Wolf non Prain

Perennial herb with stout rootstock and prostrate or decumbent shoots 15−55cm, whitish sericeous. Basal leaves pinnate, 5−20cm; lateral leaflets 2−4 pairs, elliptic-obovate 0.5−2.5×0.5−2cm, obtuse, base rounded or cuneate, margins coarsely serrate, green and thinly pubescent above, white sericeous beneath; stipules 1.5−2cm. Cauline leaves similar but smaller and with 1−2 pairs of leaflets; stipules leaflet-like, toothed or lobed. Flowers in loose terminal cymes. Calyx lobes ovate, 5−6cm, white tomentose. Petals broadly obovate 7−10mm, emarginate. Achenes ellipsoid, c 0.75mm.

Bhutan: S—Chukka district, **C**—Ha, Thimphu, Punakha and Bumthang districts, **N**—Upper Mo Chu district; **Sikkim; Chumbi**. Roadsides, margins of cultivation and disturbed ground in Blue Pine forests, 2300−3350m. June−September.

15. P. caliginosa Sojak

Similar to *P. griffithii* but a much smaller plant with prostrate shoots 7−10cm; leaflets 2−3 pairs, obovate-truncate, 5−8mm; calyx lobes 3−4mm; petals 4−5mm.

Sikkim: Yume Chu; **Chumbi**. High mountain slopes, 5000m. June−August.

Possibly only a high-altitude form of *P. griffithii*.

16. P. spodiochlora Sojak

Similar to *P. griffithii* but often more robust, and never white sericeous or tomentose; stems 15−45cm; leaves green beneath, leaflets often larger, up to 3cm, uppermost 3 often united at base; flowers larger, petals 9−12mm.

Bhutan: N—Upper Mo Chu district (Gasa, Laya, Gangyuel and Lingshi); **Sikkim:** Temu La. Open grassy slopes and margins of cultivation. 3000−4800m. April−September.

E Himalayan specimens named as *P. leschenaultiana* Seringe, a peninsular Indian taxon with brownish indumentum, probably belong to this species.

17. P. supina L.

Annual with radiating stems c 20cm. Basal leaves pinnate, 5−9cm, lateral leaflets 2−4 pairs, obovate-elliptic, 5−15×2−10mm, obtuse, base cuneate, margins coarsely crenate-serrate, green beneath, sparsely pubescent, lamina of upper leaflets decurrent on rachis; stipules lanceolate 6−10mm; cauline leaves similar but smaller and with fewer leaflets. Flowers in loose terminal cymes. Calyx lobes ovate c4mm. Petals obovate c 3×2mm, yellow. Achenes ovoid c 0.7mm, glabrous, receptacle shortly hairy.

Bhutan: C—Thimphu district (Thimphu Chu valley) and Punakha district (Mo Chu and Tang Chu valleys); **Sikkim:** terai. Margins of cultivated ground, 300−2350m. April−July.

18. P. lineata Treviranus; *P. fulgens* Hooker, *P. siemersiana* Lehmann
Erect or spreading perennial herb, stems 20−45cm, whitish hairy. Leaves interruptedly pinnate 6−20cm; larger lateral leaflets 4−8 pairs, elliptic or narrowly obovate, 0.7−4×0.5−2cm, obtuse, base rounded or cuneate, margins sharply serrate, green and sparsely pubescent and with impressed veins above, densely silvery white sericeous beneath; minor alternating leaflets similar, up to 1×0.7cm; stipules of basal leaves ovate-lanceolate c 3×0.7cm, brown; stipules of cauline leaves leaflet-like up to 2.5×1.5cm. Flowers in corymbose cymes. Calyx lobes ovate c 5mm, sericeous. Petals obovate 5−9×3−7mm, rounded, yellow. Achenes ovoid 1−1.5mm glabrous.

Bhutan: C—Thimphu, Punakha, Tongsa, Bumthang and Mongar districts; **Sikkim**. On open grassy clearings and meadows, 2250−3600m. June−August.

Var. **intermedia** (Hook. f.) Dixit & Panigrahi is a minor variant found in alpine Sikkim with leaves and calyx lobes more thinly sericeous.

19. P. anserina L.
Similar to *P. lineata* but a much smaller prostrate plant with numerous long slender stolons; stems 5−15cm; leaves 5−10cm; major lateral leaflets 0.5−2×0.3−1cm, deeply toothed or almost pinnatisect; flowers solitary, borne between pairs of leaves on stolons; petals 8−10mm.

Bhutan: C—Thimphu district (between Shodu and Barshong); N—Upper Mo Chu district (Laya and Shingche La); **Sikkim:** Chamgong. Moist meadows, river banks and by cultivation, 3700−4260m. May−July.

20. P. polyphylla Lehmann; *P. mooniana* Wight
Similar to *P. lineata* but stems, leaves and inflorescences spreading pilose-hairy, slightly silky only when young, leaflets green beneath, more broadly elliptic-obovate, margins more bluntly toothed.

Bhutan: C—Thimphu district (Pajoding), N—Upper Kulong Chu district (Me La); **Sikkim:** Senchal, Tonglo, Lachen etc.; **Chumbi**. Forest clearings and meadows, 2400−3650m. June−August.

21. P. peduncularis D. Don
Perennial rosetted herb with thick woody rootstock covered with leaf remains. Leaves mostly regularly pinnate (rarely with minute tooth-like lobes between leaflets) oblong or oblanceolate in outline, 5−20cm; lateral leaflets 9−21 pairs, oblong-elliptic 0.5−3×0.25−0.75cm, ± obtuse, base rounded or cuneate, margin serrate, sparsely pubescent above, usually white sericeous beneath; stipules ovate-lanceolate 1−3cm, brown. Flowers 2−5 ± corymbose, borne on almost leafless peduncles (6−)10−25cm. Calyx lobes ovate 5−6mm. Petals obovate, 8−13×8−10mm, rounded, yellow. Achenes obovoid 1.5−2mm, glabrous.

Bhutan: C—Thimphu, Punakha, Tongsa and Bumthang districts, **N**—Upper Pho Chu, Upper Bumthang Chu and Upper Kulong Chu districts; **Sikkim; Chumbi.** Open hillsides and in Fir forests, 3650–4400m. June–July.

Two poorly-defined varieties occur in Bhutan and Sikkim: the typical and widespread var. **peduncularis** with leaves thickly whitish sericeous beneath, and the more local var. **clarkei** Hook. f. with sparser indumentum, sometimes restricted to the veins beneath.

22. P. leuconota D. Don

Similar to *P. peduncularis* but smaller and prostrate; leaves 4–15cm, lateral leaflets 15–27 pairs, elliptic, 5–15×3–7mm; flowers 5–12 subumbellate on peduncles 6–25cm; calyx lobes c 3mm; petals 3–5mm.

Bhutan: C—Ha, Thimphu and Tongsa districts, **N**—Upper Mo Chu and Upper Kulong Chu districts; **Sikkim; Chumbi.** Alpine pastures and clearings in Juniper, Rhododendron and Fir forests, 2900–3650m. May–July.

23. P. microphylla D. Don

Dense cushion plant or mat- or rosette-forming herb, with stout tap-root or long slender roots. Leaves pinnate, leaflets 3–15 pairs, shallowly to deeply toothed or pinnatifid, often silky hairy beneath. Flowers usually solitary (–2), subsessile or on peduncles 1–5cm. Calyx lobes 4–5mm. Petals yellow, obovate, 7–8mm. Achenes obovoid, c 2.5mm, densely hairy.

var. **microphylla**; *P. microphylla* var. *depressa* Lehmann

Dense cushion-forming herb 4–8cm with stout tap-root and dense old leaf remains. Leaves 6–15mm; leaflets 3–7 pairs, close and often overlapping, obovate in outline, deeply divided into 3–5 narrow upturned acute lobes. Flowers subsessile or on peduncles up to 2cm.

Bhutan: C—Thimphu and Tongsa districts, **N**—Upper Mo Chu, Upper Pho Chu, Upper Mangde Chu, Upper Bumthang Chu and Upper Kulong Chu districts; **Sikkim:** widespread; **Chumbi.** Rocky river banks and screes, 3900–4880m. June–September.

var. **achilleifolia** Hook. f.

Mat-forming herb 4–7cm, roots long and slender, old leaf-remains few. Leaves feathery, 2–4cm; leaflets dense, overlapping, 10–15 pairs, ovate, deeply 8–11-fid into linear strongly upturned segments. Peduncles 2–5cm.

Bhutan: C—Ha and Tongsa districts, **N**—Upper Mo Chu and Upper Bumthang Chu districts; **Sikkim; Chumbi.** Alpine meadows and rocky slopes, 3050–4880m. June–September.

Probably specifically distinct from var. *microphylla* in its very different habit, rootstock and foliage.

var. **latifolia** Lehmann; *P. commutata* Lehmann, *P. microphylla* var. *commutata* (Lehmann) Hook. f.

Rosette herb 3–10cm with stout tap-root and persistent petioles. Leaves flat, 3–6cm; leaflets not overlapping, 7–11 pairs, oblong-elliptic, sharply but not very deeply serrate into 9–15 teeth. Peduncles sometimes 2-flowered, 2–9cm.

Sikkim: Temu La, Tosa, Thanka La etc. Rock crevices and screes, 4420–4880m. July–September.

var. **latiloba** Lehmann

Small rosette-herb 2–3cm, with single tap-root; leaf remains not persistent. Leaves flat, 1–2cm; leaflets not overlapping, 5–7 pairs, obovate, 2–3-fid at apex. Peduncles 1–1.5cm, with leaf-like bract.

Bhutan: N—Upper Kuru Chu district (Narim Thang); **Sikkim:** Lhonak Valley. Alpine meadows, 4000–4880m. July–August.

24. P. coriandrifolia D. Don

Rosette herb with thick woody rootstock and persistent leaf remains. Leaves pinnate, 3–10cm, lateral leaflets 4–5(–9) pairs, 3–12mm, deeply pinnatisect, segments linear, sparsely pubescent. Flowers 2–5 in corymbs on peduncles 7–15cm. Calyx lobes 2–3mm. Petals obovate 6–7mm, emarginate, white shading to deep crimson at base. Achenes oblong-ellipsoid 1.5–2mm.

Bhutan: C—Ha, Thimphu, Tongsa and Bumthang districts, N—Upper Mo Chu and Upper Bumthang Chu districts; **Sikkim; Chumbi.** On open grassy hillsides, 2900–4570m. July–August.

25. P. bryoides Sojak

Similar to *P. coriandrifolia* but smaller and with slender rootstocks; leaves 1–2cm, 3-foliate or pinnately 5-foliate; leaflets 3–4mm deeply pinnatisect; flowers solitary on peduncles 3–5cm; petals 4.5–5.5mm.

Bhutan: C—Tashigang district (Pang La) and Sakden district (Orka La), N—Upper Kulong Chu district (Me La). On mossy rocks, cliff ledges and alpine pastures, 3350–4400m. June–August.

9. SIBBALDIA L.

by A. J. C. Grierson & D. G. Long

Small, often prostrate or mat-forming perennial herbs. Leaves 3–11-foliate, leaflets palmately or pinnately arranged, rarely unifoliate; stipules adnate to petioles. Flowers solitary or in cymes, sometimes

74. ROSACEAE

subumbellate, bisexual, more rarely unisexual. Calyx 4−5-lobed and bearing as many epicalyx segments. Petals 4−6 yellow, white, red or purple. Stamens mostly 4 or 5, rarely 6 or 10. Carpels and achenes 4−35, style subterminal to almost basal.

1. Basal leaves 1- or 3-foliate 2
+ Basal leaves 5−11-foliate 7

2. Leaves 1-foliate .... **6. S. trullifolia**
+ Leaves 3-foliate 3

3. Calyx lobes and petals 5(−6) 4
+ Calyx lobes and petals 4 6

4. Leaflets 3−5mm, 3−5-toothed at apex; flowers solitary, subsessile; petals white or creamy .... **2. S. perpusilloides**
+ Leaflets 5−25mm, 3-fid at apex or toothed to base; flowers 2−10, subumbellate, pedunculate; petals yellow or crimson 5

5. Leaflets 3-fid at apex; petals yellow .... **1. S. parviflora**
+ Leaflets 3−6-toothed in upper part; petals crimson .... **3. S. sikkimensis**

6. Robust densely matted herb, with many persistent dead leaves
   4. S. tetrandra
+ Dwarf slender herb without persistent dead leaves .... **5. S. perpusilla**

7. Basal leaves pinnately 5−11-foliate 8
+ Basal leaves palmately 5-foliate 9

8. Leaves white tomentose beneath, green and pubescent above
   7. S. micropetala
+ Leaves densely white woolly tomentose on both surfaces
   8. S. byssitecta

9. Petals yellow; leaflets deeply pinnatifid with 2−11 lobes
   11. S. compacta
+ Petals purple or red; leaflets entire or 3-fid at apex 10

10. Leaflets 1.5−7×0.5−3.5mm; petioles 0.6−1.5cm; petals c 2mm
    9. S. purpurea
+ Leaflets 5−15(−20)×3−8(−10)mm; petioles 2−7cm; petals 2.5−3mm
    10. S. macropetala

1. S. parviflora Willdenow; *S. cuneata* Kuntze, *Potentilla sibbaldii* sensu F.B.I. non Haller f.

Herb rarely up to 10cm, branches ± prostrate covered with leaf remains.

Leaves 3-foliate; leaflets oblong-obovate 0.5−2×0.25−1.5cm, apex truncately 3-fid, base cuneate or rounded, sparsely pilose especially beneath; petioles 0.5−5cm; stipules linear-lanceolate 5−10mm. Flowers 5-merous, 2−10 ± subumbellate, peduncles 1.5−5cm, lengthening in fruit. Calyx lobes ovate 3−4mm, epicalyx segments linear-lanceolate. Petals yellow, narrowly obovate 3−4×1.5mm. Stamens 5, rarely 10. Achenes 15−20(−30), ovoid c 1.25mm, glabrous.

Bhutan: C—Ha, Thimphu and Punakha districts, **N**—Upper Mo Chu, Upper Bumthang Chu and Upper Kulong Chu districts; **Sikkim; Chumbi.** On rocks and grassy slopes, 3050−4250m. May−June.

In Sikkim the smaller var. **micrantha** (Hook. f.) Dixit & Panigrahi occurs which is characterised by its size and more compact habit, plants usually 2−5cm tall, leaflets 3−6mm; flowers c 4mm diameter. It is sometimes found as low as 2400m.

2. S. perpusilloides (W. W. Smith) Handel-Mazzetti; *Potentilla perpusilloides* W. W. Smith

Similar to *S. parviflora* but much smaller, leaflets 3−5mm, 3−5-toothed at apex, ± glabrous; petioles 4−10mm; flowers solitary subsessile 6−8mm diameter; petals white or creamy up to 4mm; stamens 10; achenes 12−30, glabrous.

Bhutan: C—Tongsa district (Black Mountain), **N**—Upper Pho Chu, Upper Bumthang Chu and Upper Kulong Chu districts; **Sikkim:** Zemu Valley, Cho La etc. On mountain rocks and screes, 3960−4600m. June−August.

3. S. sikkimensis (Prain) Chatterjee; *S. melinotricha* Handel-Mazzetti

Similar to *S. parviflora* but smaller and densely pale brown pubescent; leaflets up to 2.5×1.5cm usually 5-toothed at apex or coarsely serrate to base; petioles up to 6cm; flowers 3−6 subumbellate on peduncles up to 15cm, 5−6-merous, limb c 7mm diameter; petals crimson, obovate c 3× 2mm.

Sikkim: Changu, Gnatong, Jongri, etc. Stony slopes and rocky river banks, 3650−4200m. June−July.

The specimen from Changu is a small shortly stoloniferous form with leaflets c 5mm and flowers 4−5mm diameter.

4. S. tetrandra Bunge; *Potentilla tetrandra* (Bunge) Hook. f.

Robust densely matted herb with many persistent dead leaves, stems 2.5−5cm. Leaves 3-foliate, leaflets obovate-cuneate 4−6mm, 2−3-fid at apex, densely silky pubescent; petioles c 5mm. Dioecious; flowers solitary or in pairs ± sessile, functionally unisexual, 4-merous, limb c 4mm diameter, petals pale yellow. Stamens 4. Achenes 3−4(−6).

Sikkim: Tibetan border (80). Alpine scree slopes and moraines, 4570—4880m.

5. S. perpusilla (Hook. f.) Chatterjee; *Potentilla perpusilla* Hook. f.

Similar to *S. tetrandra* but a more slender plant; leaflets 4—8mm; flowers in short cymes, apparently bisexual, limb c 3mm diameter; achenes 10—15.

Sikkim: Kinchin-jhow (80). 4880m. September.

6. S. trullifolia (Hook. f.) Chatterjee; *Potentilla trullifolia* Hook. f.

Dwarf, mat-forming herb. Leaves unifoliate, obovate 4—5mm long and broad, 3—5-toothed at apex, base rounded or cuneate, ± densely silky pubescent; petioles 2—3mm; stipules triangular ± as long. Flowers solitary, subsessile, c 3mm diameter, 5-merous. Petals not seen. Stamens 5. Achenes c 10, glabrous, borne on a villous receptacle.

Sikkim: Tibetan border. 4880—5180m.

7. S. micropetala (D. Don) Handel-Mazzetti; *Potentilla albifolia* Hook. f.

Rosetted perennial herb with trailing prostrate stems. Leaves pinnate, basal ones 1.5—5cm, lateral leaflets 2—5 pairs, broadly elliptic or suborbicular, 3—15×2—8mm, rounded at apex, base ± cuneate, margins finely serrate, green and appressed pubescent above, white tomentose beneath; stipules brown, membranous, c 6×4mm; cauline leaves often 3-foliate. Flowers 1—2 axillary on peduncles 1—2cm, 5-merous. Calyx c 3mm. Petals yellow, elliptic c 2×1mm. Stamens 5. Achenes 20—35, ovoid, c 1.5mm glabrous.

Bhutan: C—Thimphu and Bumthang districts, N—Upper Mo Chu and Upper Kulong Chu districts; **Sikkim; Chumbi.** Grassy banks in Blue Pine, Hemlock and Fir forests, 3450—3500m. May—July.

8. S. byssitecta Sojak

Similar to *S. micropetala* but leaves densely white tomentose on both surfaces; achenes bearing a few hairs at the apex.

Bhutan: C—Mongar district (E side of Ura La). Grassy slopes at margin of Pine/Spruce forest, 3500—3680m. June.

Endemic to Bhutan.

9. S. purpurea Royle; *Potentilla purpurea* (Royle) Hook. f.

Dwarf mat-forming perennial herb 2—3cm tall. Leaves palmately 5-foliate, leaflets oblanceolate 1.5—7×0.5—3.5mm, apex entire or shallowly 3-fid, base cuneate, pale sericeous on both surfaces; petioles 0.6—1.5cm; stipules ± as long, brown. Dioecious; flowers solitary or 2, unisexual 5-(sometimes 4-) merous, 5—6mm diameter on peduncles up to 1cm. Petals red or purplish c 2mm. Achenes 15—20, ovoid, c 1mm.

Bhutan: C—Tongsa district (Black Mountain), N—Upper Pho Chu district (Kangla Karchu La); **Sikkim.** On rocky mountain slopes, 4570–4900m. June.

Records of *S. pentaphylla* Krause from Sikkim probably refer to 4-merous plants of *S. purpurea;* the red or purple petals of the latter readily distinguish it from the yellow or creamy-flowered *S. pentaphylla* from E Tibet.

10. S. macropetala Muravjeva

Similar to *S. purpurea* but larger and not dioecious; leaflets elliptic or oblanceolate 5–15(–20)×3–8(–10)mm, apex 3-fid, green above, silvery sericeous beneath; petioles 2–7cm; flowers 2–3 on peduncles up to 7cm, 5-merous, bisexual, c 7mm diameter; petals 2.5–3mm, crimson; achenes 12–20.

Bhutan: N—Upper Mangde Chu, Upper Bumthang Chu and Upper Kulong Chu districts; **Sikkim; Chumbi.** On open grassy slopes, 4100–4800m. June–July.

11. S. compacta (Smith & Cave) Dixit & Panigrahi; *Potentilla sericea* L. var. *compacta* Smith & Cave

Cushion-forming herb 1.5–4cm tall, branches densely leafy. Leaves palmately 5-foliate, leaflets 5–10mm, deeply pinnatifid into 2–11 lobes, sericeous; petioles 2–5mm; stipules ± as long. Flowers solitary, 5-merous, 3–6mm diameter, peduncles 2–4mm. Petals yellow. Stamens 5 or 10. Achenes 4–8, receptacle long-hairy.

Sikkim: Lhonak valley. Open mountainsides, 4575m. August.

Endemic to Sikkim.

10. FRAGARIA L.

by D. G. Long

Perennial herbs with slender prostrate stolons. Leaves palmately 3-foliate (sometimes becoming pinnate with 2 additional minor leaflets); stipules adnate to petioles. Scapes 1–3-flowered. Calyx cup 5-lobed and bearing 5 acute or acuminate epicalyx segments. Petals 5 white. Stamens numerous. Carpels numerous, free, borne in pits on a conical receptacle, style short, sub-basal. Achenes ± sunk in surface of enlarged fleshy receptacle and persistent upon it.

1. Leaflets with 4–6 teeth on each side; epicalyx segments spathulate, with 1–3 large teeth on each side; mature fruiting receptacle conical
 2. F. daltoniana
+ Leaflets with 6–14 teeth on each side; epicalyx segments narrow

lanceolate, entire, bifid or 1−2-toothed at apex; mature fruiting receptacle subglobose 2

2. Leaflets with 7−14 teeth on each side, whitish appressed silky beneath; petals 7−10mm .... **1. F. nubicola**
+ Leaflets with 6−8 teeth on each side, densely spreading brownish-pilose beneath; petals 4−6mm .... **3. F. nilgerrensis**

1. F. nubicola (Hook. f.) Lacaita; *F. vesca* L. var. *nubicola* Hook. f.

Prostrate herb with stout rootstock. Leaflets obovate or elliptic, 1.5−5×1−3cm, obtuse, base cuneate, sessile or on very short petiolules c1mm, margins sharply serrate with 7−14 teeth on each side, appressed silky whitish pubescent beneath, at least on veins; petioles 2−10cm, appressed or erect-spreading pale pubescent, sometimes bearing 2 additional minor leaflets. Scapes 2−10cm, with 2 median bracts 3−5mm, 1−3-flowered. Calyx cup 3−5mm diameter, lobes triangular 4−6×1.5−2mm, entire; epicalyx segments elliptic-lanceolate, 3−4mm, acuminate, entire or bifid at apex. Petals broadly obovate, 7−10×6−8mm, white. Achenes 1−1.5mm, borne on succulent red globose receptacle 10−15mm diameter, with pleasant sharp flavour.

Bhutan: S—Chukka district, **C**—Thimphu, Punakha and Tongsa districts, **N**—Upper Mo Chu district; **Sikkim; Chumbi.** Open grassy banks at margins of Cool broad-leaved, Blue Pine and Hemlock forests, 2000−3600m. April−June.

2. F. daltoniana Gay; *F. sikkimensis* Kurz, *F. rubiginosa* Lacaita

Similar to *F. nubicola* but leaflets obovate, 0.8−2.5×0.7−2cm, with 4−6 broad teeth on each side, not silky beneath but with fewer appressed or spreading hairs, on distinct petiolules 1−3mm; petioles 1−4cm, with ± appressed or spreading pale brown or whitish hairs; calyx lobes ovate-triangular, entire or with 1−2 small teeth near apex; epicalyx segments coarsely toothed with 1−3 large teeth on each side; petals smaller, 5−8mm, sometimes blotched reddish; fruiting receptacle globose when young, becoming conical and pink or whitish when ripe, 2−2.2×1−1.3cm.

Bhutan: C—Tongsa district (W side of Yuto La) and Bumthang district (Gyetsa); **Sikkim:** Senchal, Sandakphu, Tonglo, Karponang etc. Mountain meadows and clearings in Hemlock and Fir forests, 2750−3800m. June−July.

A variable species in size and density of indumentum; plants from more open sunny habitats are more hairy and sometimes have petals with reddish blotches (*F. rubiginosa* Lacaita).

3. F. nilgerrensis Gay

Similar to *F. nubicola* but a more robust plant, densely spreading brownish-pilose throughout; leaflets obovate, coarsely 6−8-toothed on each side; petiolules 0.5−2mm; calyx lobes ovate-acuminate, entire; epicalyx segments lanceolate, acuminate, entire or 1−2-toothed; petals white, 4−6mm; fruiting receptacle globose.

Sikkim: Darjeeling, Tonglu and between Sandakphu and Garibans. 2750−3500m. May−June.

11. DUCHESNEA Smith

by A. J. C. Grierson

Similar to *Fragaria* but leaves always palmately 3-foliate; flowers solitary; epicalyx segments obovate; petals yellow; achenes minute borne on surface of enlarged fleshy persistent receptacle and falling from it individually.

1. D. indica (Andrews) Focke; *Fragaria indica* Andrews, *Potentilla indica* (Andrews) Wolf. Fig. 37h−k

Rootstock stout with several prostrate stolons. Leaflets elliptic or obovate, 0.75−3×0.75−2cm, obtuse, margins crenately-serrate, appressed pubescent; petioles up to 10cm. Peduncles 2.5−5cm. Calyx cup c 4mm, lobes ovate 3−5mm, entire; epicalyx segments usually shorter than calyx lobes, obovate, bluntly 3−5-toothed at apex. Petals yellow, obovate, 4−6×2−4mm. Fruiting receptacle conical or subglobose 0.75−1.5cm, red, flesh insipid.

Bhutan: S—Chukka and Gaylegphug districts, **C**—Ha, Thimphu, Bumthang and Mongar districts, **N**—Upper Bumthang Chu and Upper Kulong Chu districts; **Sikkim.** Grassy banks, forest clearings and margins of cultivation from Warm broad-leaved forests to above tree-line, 1060−4420m. February−June.

12. BRACHYCAULOS Dixit & Panigrahi

by A. J. C. Grierson

Dwarf tufted perennial herb. Leaves simple, entire, exstipulate; petioles broadly winged and ciliate. Flowers bisexual, solitary, terminal. Calyx lobes 5 free, epicalyx absent. Petals 5. Stamens 5. Carpels 2 free; styles simple terminal. Achenes 2, smooth.

1. B. simplicifolius Dixit & Panigrahi

Stems 2−5cm covered with remains of leaves and winged petioles. Leaves oblong or lanceolate, 3−6×1−1.5mm, acute or subobtuse, glabrous. Flowers small, 8—10mm across on peduncles 4—5mm. Calyx lobes oblong, 3.5—4×1—1.5mm, obtuse. Petals elliptic-oblanceolate 4—4.5×1.5—2mm obtuse, slightly longer than calyx lobes. Stamens c 3mm. Carpels ovoid c 1mm, style ± as long. Achenes c 1mm black, smooth.

Sikkim: Gaoring. 4575m. July.

Knowledge of this genus is based on a single collection of three plants, clearly indicating the necessity for further material and study before it can be adequately understood.

13. GEUM L.

by A. J. C. Grierson

Perennial rhizomatous herbs. Radical leaves pinnate, cauline leaves lobed or pinnatisect, stipulate. Flowers solitary or few, ± corymbose, peduncles erect. Calyx cup-shaped, 5-lobed and bearing 5 smaller epicalyx segments. Petals 5. Stamens numerous. Carpels numerous, free, borne on a conical receptacle, styles persistent or in part deciduous. Achenes ellipsoid or obovoid, bearing terminal style or stylar remains, sometimes hooked.

1. Terminal leaflet of radical leaves broadly elliptic, similar to lateral ones, not abruptly larger than them .... **1. G. elatum**
+ Terminal leaflet of radical leaves broadly ovate or suborbicular, unlike lateral ones and abruptly larger than them .... **Species 2—4**

1. G. elatum G. Don; *Acomastylis elata* (G. Don) Bolle

Radical leaves 10—30cm, leaflets 10—15 pairs (often with other minor lobes and leaflets between them), broadly elliptic, up to 2.5×2cm, acute or obtuse, base sessile attached to rachis along most of width, margin crenate dentate, ciliate or sparsely pubescent; cauline leaves narrowly elliptic in outline, 2—3×0.3—1cm; stipules narrow, adnate to petiole. Peduncles 8—40cm bearing 1—3 flowers. Calyx cup 6—8mm diameter, pubescent, often reddish purple, lobes broadly ovate 4×3—4mm; epicalyx segments elliptic c 2mm. Petals yellow, suborbicular or obovate, 7—15×10mm. Achenes ellipsoid 2—3mm, silky pubescent, whole style persistent 4—5mm, capitate.

Bhutan: C—Tongsa district, **N**—Upper Mo Chu and Upper Bumthang Chu districts; **Sikkim.** Open grassy hillsides, 3900—4570m. May—August.

Two varieties have been recorded in E Himalaya: var. **elatum** in which the leaves have relatively distant leaflets, peduncles longer than the leaves (30—40cm) and bearing 2—3 flowers; var. **humile** (Royle) Hook.f. in which the leaflets are densely crowded, peduncles shorter than or slightly longer than the leaves (8–15cm) and bearing 1(–3) flowers. However, the characters of the two overlap and there are grounds for regarding var. *humile* as little more than a dwarf alpine state of *G. elatum*.

Some specimens of this species have been misidentified as the Chinese *Coluria longifolia* Maximowicz; the only reliable difference between the two being the completely deciduous style of the latter in fruit.

2. G. macrosepalum Ludlow

Radical leaves 5–15cm, terminal leaflet broadly ovate or suborbicular 3–5×3–5cm, rounded, base deeply cordate, margins crenate or shallowly lobed, lateral leaflets 5–10 pairs, broadly elliptic up to 1×1cm; cauline leaves obovate or broadly oblanceolate 1–4×0.5–2cm, shallowly lobed; stipules ovate, 1–1.5×1cm. Peduncles up to 50cm bearing 1–2 flowers. Calyx cup c 8mm diameter, yellowish, lobes broadly ovate, 1–1.5×0.8–1cm, pubescent. Petals yellow, obovate, 1–1.2×0.6–0.7cm, pubescent on the back. Styles 6–10mm pubescent, constricted near base, upper part deciduous above this point. Achenes ellipsoid 5–6mm, tapering above into a beak 1–2mm.

Bhutan: C—Thimphu and Tongsa districts, N—Upper Mangde Chu, Upper Bumthang Chu and Upper Kulong Chu districts; **Sikkim.** In peat and gravel, 4100–4570m. June–July.

3. G. sikkimense Prain; *Acomastylis sikkimensis* (Prain) Bolle

Very similar to *G. macrosepalum* but calyx lobes smaller 6–10×4–7mm, green or brownish; petals white or pinkish, 8–10×5–6mm, glabrous on the back; achenes c 3mm, styles c 5mm persistent, becoming somewhat hooked apically.

Bhutan: C—Punakha district (Tang Chu) and Tongsa district (Rinchen Chu, Thita Tso, etc.); **Sikkim.** Open grassy hillsides, 3600–4000m. June–August.

4. G. aleppicum Jacquin

Similar to *G. macrosepalum* but basal leaves with 3–4 pairs of lateral leaflets 1–3×0.7–3cm, terminal leaflet broadly ovate 5–6×4–7cm, pale pubescent; cauline leaves pinnatisect 5–10cm; stipules broadly elliptic 1–3×1–1.5cm; flowers 1–5 ± corymbose; calyx cup 6–7mm diameter, lobes ovate-triangular 5×3mm; petals yellow, suborbicular c 5mm; achenes 3.5–4mm, pubescent, style c 5mm hooked near apex, terminal stigmatic portion c 1.5mm deciduous.

Bhutan: S—Chukka district, C—Thimphu, Bumthang and Tashigang districts, N—Upper Mo Chu district. Open hillsides and amongst scrub, 1525–2225m. June–July.

The record (117) of *G. urbanum* L. from Bhutan probably refers to this species

14. AGRIMONIA L.

by A. J. C. Grierson

Erect perennial herbs. Leaves interruptedly pinnate; stipules adnate to petioles. Flowers many in terminal spike-like racemes. Calyx tube turbinate, accrescent, lobes 5, surrounded by numerous hooks at base of teeth; epicalyx absent. Petals 5. Stamens 5–10. Carpels 2 free, included in calyx tube, style slender. Achenes 1 or 2 enclosed in persistent hook-covered calyx tube.

1. **A. pilosa** Ledebour. Tongsa: *Brumzey*. Fig. 37 a–c

Stems 30–100cm, pilose with ± spreading hairs. Leaves 10–15cm, larger leaflets 7–11, elliptic, obovate or suborbicular 1–6.5×0.7–5cm, acute or obtuse, base rounded or cuneate, margins serrate, sparsely pilose and minutely glandular on both surfaces, smaller intermediate leaflets 0.5–1.5×0.2–1cm; stipules obliquely ovate, 1.5–3×1–2cm, leafy. Racemes 20–50-flowered. Calyx tube c 2mm, lobes ovate ± as long. Petals yellow, narrowly obovate 3–4mm. Stamens ± as long as calyx lobes. Fruiting calyx tube c 3mm, deeply 10-grooved on the sides, bearing hooked spines 2–3mm long at apex.

Bhutan: S—Chukka and Deothang districts, C—Thimphu, Tongsa and Bumthang districts, N—Upper Mo Chu district; **Sikkim.** Roadsides and river banks, 1250–3000m. May–July.

The specimens from Bhutan and the majority of those from Sikkim, with ± persistent basal leaves with elliptic or obovate leaflets and 5–10 stamens, belong to var. **nepalensis** (D. Don) Nakai. Var. **zeylanica** (Hook. f.) Purohit & Panigrahi with deciduous basal leaves, suborbicular leaflets and invariably 5 stamens has also been recorded from Sikkim.

FIG. 37. **Rosaceae.** a–c, *Agrimonia pilosa:* a, habit; b, flower; c, fruit. d–g, *Cotoneaster racemiflorus:* d, flowering shoot; e, flower in section; f, fruit; g, fruit in section. h–k, *Duchesnea indica:* h, habit; i, flower; j, fruiting receptacle; k, achene. l & m, *Rosa macrophylla:* l, flowering and fruiting shoot; m, fruiting calyx cup in section. n–q, *Sanguisorba diandra:* n, habit; o, flower; p, carpel and style; q, fruiting calyx tube. Scale: a, h × ½; l, n × ¾; d, m × 1; j × 1½; i × 2; f, g × 2½; b × 3; c, e, q × 4; o × 6; k × 7; p × 8.

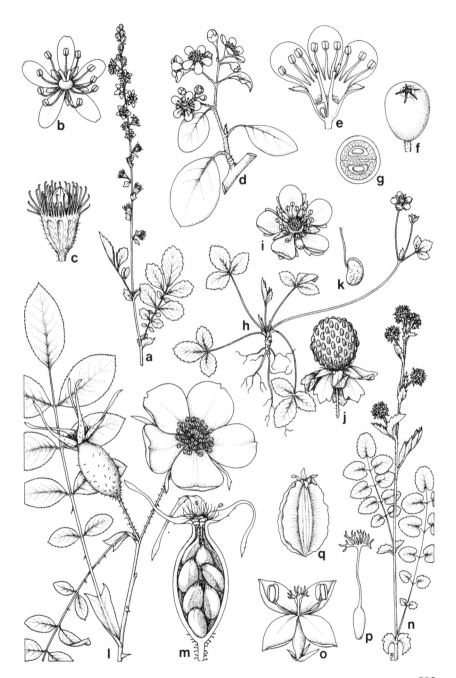

74. ROSACEAE

15. SPENCERIA Trimen

by A. J. C. Grierson

Erect perennial rhizomatous herbs. Leaves pinnate, mostly basal and rosetted, stem leaves few; stipules adnate to petiole. Flowers in simple racemes each surrounded at base by a cup-like involucre of 2 connate irregularly toothed bracts. Calyx tube turbinate, lobes 5, epicalyx absent. Petals 5. Stamens 15—40. Carpels 2, free, borne at base of calyx tube, styles 2, filiform. Fruit a single achene enclosed within slightly enlarged calyx tube.

1. S. parviflora Stapf; *S. ramalana* Trimen var. *parviflora* (Stapf) Kitamura

Rootstock thick, woody, covered in the upper parts with leaf remains. Leaves 5—10cm; leaflets 11—17, obovate-elliptic, 0.3—1.25×0.2—0.7cm, apex shallowly bifid (terminal leaflet usually 3-fid), base rounded, sparsely covered with appressed whitish hairs. Racemes 10—20-flowered on peduncles (1—)1.5—4cm bearing ± spreading long hairs and shorter pubescence, bracts simple ovate-elliptic 5—10×2—4mm, pedicels 1—2cm. Bracts of involucre 3—4mm finely glandular-pubescent. Calyx tube c 2mm, lobes lanceolate c 3mm. Petals obovate c 6×3.5mm, yellow. Stamens c3mm. Styles c 4mm. Fruiting calyx tube 3—4mm.

Bhutan: C—Mongar district (Donga La) (111).

This is possibly not distinct from *S. ramalana* Trimen from which it differs principally in the smaller size of its parts.

16. SANGUISORBA L.

by A. J. C. Grierson

Erect perennial herbs. Leaves pinnate; stipules adnate to petiole. Flowers subsessile in globose heads, bisexual or unisexual. Calyx accrescent, tube ellipsoid, lobes 4 or absent; epicalyx absent. Petals absent. Stamens 1—8. Carpel solitary, included in calyx tube, style filiform, brush-like at apex. Fruiting calyx tube 4-angled or winged, enclosing 1 achene.

1. S. diandra (Hook. f.) Nordborg; *Poterium diandrum* Hook. f. Dz: *Jadum*. Fig. 37n—q

Stems 30—90cm, glabrous. Leaves 7—30cm, leaflets 4—8 pairs, elliptic or suborbicular 1—3×0.7—2.5cm, apex rounded, base cordate, margin serrate, glabrous above, sparsely pubescent beneath, petiolules up to 1.5cm. Flowers bisexual. Calyx tube 2—3mm, lobes ovate-lanceolate ± as long, reddish green or purple. Stamens (1−)2(−4) exserted 2−2.5mm. Style c1mm purplish. Fruiting heads 1−2cm diameter, calyx tube 5−8mm glabrous or finely pubescent, wings 1.5−2mm broad, ± crenate.

Bhutan: C—Thimphu district, N—Upper Mo Chu, Upper Pho Chu, Upper Bumthang Chu and Upper Kulong Chu districts; **Sikkim.** Steep open slopes, 3400−4120m. June−July.

Two varieties have been recognised in this area; var. **diandra** with dark purple flowers and completely glabrous fruiting calyces and var. **villosa** Purohit & Panigrahi with green or reddish green flowers and finely pubescent fruiting calyces.

2. S. filiformis (Hook. f.) Handel-Mazzetti; *Poterium filiforme* Hook. f.

Similar to *S. diandra* but smaller, 5−25(−35)cm; leaves 2.5−10cm, leaflets 7−9, ± suborbicular 2−10mm long and broad, ± crenately dentate; flower heads 6−8mm diameter, lower flowers male, upper ones female; male flower calyx tubes 1.5−2mm, lobes 4, ovate 2.5−3×1.5mm, white; stamens (3−)5−7(−8), ± as long as calyx lobes; female flower calyx tubes c1mm, lobes absent, style exserted c 0.5mm; fruiting calyx c 2mm, 4-angular.

Bhutan: C—Ha, Tongsa, Bumthang and Sakden districts, N—Upper Mo Chu and Upper Kulong Chu districts; **Sikkim.** In damp meadows and marshes, 2750−4050m. May September.

17. ROSA L.

by A. J. C. Grierson

Erect or scrambling shrubs, stems bearing prickles. Leaves pinnate, stipules adnate to petioles. Flowers solitary or several, ± corymbose, showy. Calyx tube turbinate, lobes 4−5. Petals 4−5 or more. Stamens numerous. Carpels numerous, free, sessile within calyx tube; styles free or connate above. Fruiting calyx tube (hip) fleshy containing numerous hairy ellipsoid achenes.

1. Calyx lobes and petals 4(−5) .... **1. R. sericea**
+ Calyx lobes and petals 5 or more 2

2. Styles connate ± club shaped 8−10mm .... **2. R. brunonii**
+ Styles free (short and ± concealed by white hair in *R. macrophylla*) ... 3

3. Calyx tube densely covered with spinous bristles .... **3. R. roxburghii**
+ Calyx otherwise, sometimes densely covered with glandular bristles .. 4

| 4. | Stipules entire, ciliate with subsessile glands .... **4. R. macrophylla** |
| + | Stipules toothed, pectinate or lacerate, sometimes glandular .... 5 |

5.	Prickles usually in pairs below leaves; plants native .... **5. R. lyellii**
+	Prickles without apparent relation to leaves; plants cultivated
	6. R. hybrida

1. R. sericea Lindley. Dz: *Sew Shing;* Med: *Sewai Metog;* Nep: *Sisi Chungchung* (131)

Erect shrub 1−4m, stems naked or bearing ± straight paired or scattered, slender or broad prickles. Leaves 3−9cm, leaflets 7−11(−17), oblong or narrowly obovate 0.5−2.5×0.3−1cm, acute or rounded, base cuneate, margins usually serrate near apex otherwise entire, glabrous or pubescent above, usually finely sericeous beneath; stipules 0.8−1.5×1−2mm. Flowers solitary on short lateral shoots, 4(−5)-merous. Calyx tube turbinate 4−5mm, lobes lanceolate 0.7−1.5cm, ± sericeous. Petals yellow or creamy white, broadly obovate 1.5−2cm long and broad. Styles free c 1.5mm. Hips obovoid or subglobose 0.7−1cm, orange-red, pedicels sometimes becoming thick and fleshy.

Bhutan: S—Chukka district, **C**—Ha, Thimphu, Bumthang and Tashigang districts, **N**—Upper Bumthang Chu district; **Sikkim.** On open dry hill slopes, 1220−3800m. April−June.

A very variable species in which several varieties, the characters of which overlap to some extent, are recognised: plants with fruiting pedicels slender and green, var. **sericea;** fruiting pedicels thickened and fleshy, reddish, var. **omeiensis** (Rolfe) Rowley; main stems with broad prickles (c 2cm at base). var. **pteracantha** (Franchet) Bean; stems with small prickles, bristles and glands, var. **hookeri** Regel.

Flowers used medicinally; fruit edible (34).

2. R. brunonii Lindley; *R. moschata* sensu F.B.I. non Herrmann. Dz: *Taktsher*

Scrambling shrub 3−10m, stems bearing scattered recurved prickles. Leaves 8−15cm, leaflets 5−9, ovate-elliptic, 2−6×1−3cm, acute or acuminate, base rounded, margins finely serrate, pubescent especially beneath and sparsely glandular; stipules 1−1.5cm adnate to petiole along most of their length, free and subulate at apex. Flowers fragrant in large terminal corymbs. Calyx tube turbinate 5−8mm finely pubescent and densely covered with shortly stalked glands, lobes lanceolate, 1.5−2cm, acuminate, sometimes with 2−3 pairs of pinnate lateral appendages, becoming reflexed. Petals creamy-white, rarely pink, obovate 1.5−3×1−2.5cm. Styles united club-shaped 8−10mm. Hips reddish obovoid 1−1.3×0.8−1cm, teeth deciduous.

Bhutan: S—Chukka, Gaylegphug and Deothang districts, **C**—Thimphu,

Punakha, Tongsa and Mongar districts, N—Upper Mo Chu district. Scrambling over shrubs and small trees in moist forest, 1370–2550m. May–July.

3. R. roxburghii Trattinnick; *R. microphylla* Lindley non Desfontaines

Erect shrub 1–3m, stems usually bearing paired straight upwardly pointing prickles c5mm below each leaf. Leaves 5–8cm, leaflets 7–11(–15), elliptic or obovate 1–2.5×0.7–1.5cm, obtuse or acute, base cuneate, margin finely serrate, glabrous. Flowers usually solitary on short lateral shoots, fragrant, usually double, buds globose 2–2.5cm diameter. Calyx lobes obovate 1–1.3×0.7–1cm densely covered with 2–4mm long spinous bristles. Petals usually numerous, sometimes 5, obovate 2–2.5×1.5–2cm, often emarginate, outer ones white or pale pink, inner ones deep rose red. Hips subglobose c 2cm diameter, prickly.

Sikkim: Darjeeling. 2100m. June–September.

Cultivated. Native of China.

4. R. macrophylla Lindley Fig. 37 1 & m

Erect shrub 1–5m, stems sometimes bearing paired straight prickles below leaves. Leaves 7–20cm, leaflets 7–11, ovate-elliptic 1.5–7×1–3cm, acute, base rounded, margins finely serrate, glabrous above, sparsely pubescent beneath; stipules oblong-elliptic 10–20×2–5mm, purple, ciliate with subsessile glands. Flowers 1–2, terminal on short lateral shoots. Calyx tube ellipsoid c 1cm, purplish, densely covered with short stalked glandular bristles, lobes lanceolate, 2–3cm, acuminate but broadening slightly near apex. Petals obovate 2.5–3×2–2.5cm, deep pink. Styles 5–7 free, almost concealed by white hairs. Hips obovoid 3–3.5×2cm reddish, calyx teeth persistent.

Bhutan: C—Ha, Thimphu, Punakha, Bumthang and Mongar districts, N—Upper Mo Chu and Upper Pho Chu districts; **Sikkim; Chumbi.** Hillsides and forest margins, 2100–3800m. June–July.

5. R. lyellii Lindley; *R. involucrata* Roxb.

Scrambling shrub, stems usually bearing a pair of straight downwardly pointing prickles below each leaf, densely pubescent. Leaves up to 8cm, leaflets 5–7, elliptic 15–30×5–15mm, acute, base cuneate, margin finely serrate, midrib pubescent beneath; stipules c 1cm, lacerate and glandular at apex. Flowers in terminal corymbs, pedicels pale tomentose and stipitate-glandular. Calyx tube obconical c 5mm, lobes ovate c 1.3cm acuminate, tomentose and glandular. Petals white, obovate, c 2×1.5cm. Styles free, up to 3mm.

Sikkim: locality unknown.

6. R. hybrida Hortorum

There are a number of roses cultivated in E Himalaya apart from *R. roxburghii* which are of hybrid origin. They are included under this collective name as it is not within the scope of this flora to deal with such plants.

18. COTONEASTER L.

by A. J. C. Grierson

Shrubs or small trees. Leaves simple, entire; stipules small, deciduous or subpersistent. Flowers solitary or several to many in terminal corymbose cymes. Calyx tube turbinate, lobes 5 persistent. Petals 5, erect or spreading. Stamens 10−20. Carpels 2−5, united, inferior, borne within and adnate to calyx tube. Fruit (pome) ± fleshy containing 2−5 1-seeded stones.

It has been shown that the majority of *Cotoneaster* species are triploid but some diploids and tetraploids are known. Most species appear to be facultative apomicts regularly producing offspring that closely resemble the maternal parent but variants do occur among the progeny, thus cross pollination and therefore hybridisation sometimes do appear to take place. This breeding pattern has led to the recognition of numerous 'species' differing from each other by some minor characters. What is offered here is an account of the apparent major species and the associated taxa that may be included in them to form aggregate species.

1. Petals suborbicular, spreading, usually white 2
+ Petals oblong or obovate, erect, pink or reddish (white or pink tinged in *C. acuminatus* and *C. simonsii*) 7

2. Flowers solitary; leaves 1cm or less; low-growing shrubs up to 15cm .. 3
+ Flowers 2 or more in cymes; leaves 1−8cm; shrubs 1−2m 4

3. Leaves elliptic or obovate .... **1. microphyllus**
+ Leaves broadly ovate or suborbicular .... **2. C. rotundifolius**

4. Cymes 2−10-flowered; leaves 1−2.5cm or less 5
+ Cymes many-flowered; leaves 4−13cm 6

5. Leaves elliptic or obovate, 0.7−2.0cm .... **3. C. sherriffii**
+ Leaves broadly elliptic or ovate, 1.5−2.5cm .... **4. C. racemiflorus**

6. Leaves 4−6(−9)cm; fruit blackish purple .... **5. C. bacillaris**
+ Leaves 7.5−13cm; fruit red .... **6. C. frigidus**

7. Flowers usually solitary; leaves small 0.5–3cm, obtuse, emarginate or acute 8
+ Flowers (1–)2–5 in cymes; leaves usually larger, 1–8cm, acute or acuminate 11

8. Branches spirally arranged; leaves obtuse or emarginate
   .... **7. C. sandakphuensis**
+ Branches distichously arranged; leaves ± acute never emarginate 9

9. Branches stiffly hirsute; stipules subpersistent .... **8. C. nitidus**
+ Branches softly pubescent; stipules early deciduous 10

10. Plants erect; calyx glabrous .... **9. C. sanguineus**
+ Plants prostrate; calyx pubescent .... **10. C. rubens**

11. Leaves (2–)3–8cm, acuminate .... **11. C. acuminatus**
+ Leaves 1–2cm, acute .... **12. C. simonsii**

1. C. microphyllus Lindley; including *C. congestus* Baker, *C. thymifolius* Baker and *C. integrifolius* (Roxb.) Klotz. Med: *Japho Tsi Tsi;* Nep: *Brush jhar* (34)

Low growing much branched shrub c 15cm. Leaves elliptic or obovate, 5–10×2.5–5mm, coriaceous, obtuse or subacute, base cuneate, margins ± inrolled, glabrous above, appressed pubescent beneath; petioles up to 3mm. Flowers solitary. Calyx including lobes c 4mm, pubescent. Petals white or tinged pink, suborbicular c 3mm diameter. Fruit subglobose 7–10mm, scarlet.

Bhutan: C—Ha, Thimphu, Mongar and Tashigang districts, N—Upper Mo Chu, Upper Pho Chu and Upper Kulong Chu districts; **Sikkim.** Rock faces and gravel banks, 2100–4570m. April–July.

Fruit used medicinally. Valued as an ornamental shrub in Europe.

2. C. rotundifolius Lindley

Similar to *C. microphyllus* but leaves broadly ovate or suborbicular, subacute or obtuse, base broadly cuneate or rounded.

Bhutan: C—Punakha and Tongsa districts, N—Upper Mo Chu and Upper Pho Chu districts. On cliffs and boulders, 2300–3600m.

3. C. sherriffii Klotz; including *C. ludlowii* Klotz

Shrub 1–2m. Leaves elliptic or obovate 7–20×4–10mm, acute or obtuse, base cuneate, glabrous above, sparsely pubescent beneath. Flowers 2–5(–10) in small cymes. Calyx including lobes 4–5mm, sparsely pubescent. Petals white, suborbicular 3–4mm diameter. Fruit ellipsoid or subglobose 5–7mm diameter, red.

Bhutan: C—Thimphu district (Tsalimaphe). On open dry hillsides, 2133m.

Possibly not distinct from *C. racemiflorus*.

4. C. racemiflorus (Desfontaines) Koch; *C. nummularia* sensu F.B.I. non Fischer & Meyer. Fig. 37d−g

Similar to *C. sherriffii* but sometimes taller up to 4m; leaves broadly elliptic or ovate 1.5−2.5×1−1.5cm, obtuse or subacute, base rounded; flowers 5−10 in cymes; calyx including lobes 3−4mm, subglabrous.

Bhutan: C—Thimphu district (Paro Chu Valley). On open dry hillsides, 2140m. April.

5. C. bacillaris Lindley; including *C. obtusus* Lindley, *C. gamblei* Klotz, *C. griffithii* Klotz and *C. cooperi* Marquand

Small tree 3−7m. Leaves elliptic or obovate 4−6(−9)×1.5−3(−5)cm, acute or obtuse, base cuneate, glabrous above, sparsely pubescent beneath; petioles up to 5mm. Flowers 10−20 in cymes, pedicels pubescent. Calyx including lobes 3−4mm, pubescent. Petals white, suborbicular, 3−4mm diameter. Fruit subglobose 7−8mm diameter, blackish purple.

Bhutan: C—Thimphu, Punakha, Tongsa, Bumthang and Tashigang districts, **N**—Upper Kuru Chu district; **Sikkim.** Riverbanks, 1620−3000m. May−June.

Wood used to make walking-sticks and tent-pegs (16).

6. C. frigidus Lindley. Nep: *Charu* (131)

Similar to *C. bacillaris* but leaves larger, 7.5−13cm, ± tomentose beneath when young; cymes many-flowered; fruit globose scarlet.

Bhutan: C—Punakha district (between Ritang and Ratsoo) (71), **N**—Upper Mo Chu district (between Tamji and Gasa) (71); **Sikkim.** 2000−2800m. April−May.

Young shoots dried and used to make tea (131)

7. C. sandakphuensis Klotz

Suberect evergreen shrub up to 0.5m, irregularly branched, young shoots minutely warted. Leaves obovate, 5−10×4−8mm, obtuse or emarginate, base cuneate, margins somewhat inrolled, subglabrous. Flowers solitary. Calyx including lobes c 5mm, ciliate. Petals erect c 3mm, red. Fruit ellipsoid 6−8mm, red.

Sikkim: between Sandakphu and Phallut. 3350−3650m. May−June.

8. C. nitidus Jacques; *C. distichus* Lange, *C. cavei* Klotz

Prostrate shrub or up to 1m, branches and leaves distichously arranged, young shoots stiffly hirsute. Leaves elliptic or obovate 7−15×5−12mm, obtuse or subacute, mucronate, base rounded, subglabrous; stipules linear-lanceolate 4−5mm, ± persistent. Flowers solitary, rarely 2. Calyx lobes obtuse. Petals erect, red. Fruit obovoid or subglobose, 8−10mm, red.

Bhutan: C—Tashigang district (Diri Chu); **Sikkim:** Kalapokri. On rocks in exposed situations, 2250−3200m. May−June.

9. C. sanguineus Yu; including *C. bakeri* Klotz

Deciduous erect shrub 2−3m, branches distichously arranged, young shoots pale pubescent. Leaves ovate-elliptic 10−30×7−15mm, obtuse or acute, base cuneate or rounded, appressed pale pubescent beneath; petioles 3−5mm; stipules c 2mm, early deciduous. Flowers solitary. Calyx including obtuse lobes 4−5mm, glabrous. Petals oblong or obovate, erect, 2−3mm, red. Fruit obovoid 7−8mm.

Bhutan: N—Upper Mo Chu district (near Laya) and Upper Bumthang Chu district (Pangotang). Streamsides, 3300−3800m. June.

10. C. rubens W. W. Smith

Similar to *C. sanguineus* but prostrate; leaves densely pale pubescent beneath; calyx pubescent.

Bhutan: C—Thimphu district (between Barshong and Dotena), N—Upper Kulong Chu district (Me La). Glacier moraines and rocks, 3300−3750m. June.

Records of *C. horizontalis* Decaisne and *C. adpressus* Bois from E Himalaya probably refer to this species.

11. C. acuminatus Lindley; including *C. mucronatus* Franchet and *C. wallichianus* Klotz

Shrub 1−3m. Leaves ovate (2−)3−8×1.25−3.5cm, acuminate, base rounded, pubescent especially beneath. Flowers solitary or 3−4. Calyx including lobes 6−7mm, pubescent. Petals erect, oblong or obovate 3−4mm, pink or white tinged pink. Fruit obovoid or subglobose, 8−10mm, red.

Bhutan: C—Thimphu, Punakha, Bumthang and Tashigang districts, N—Upper Mo Chu and Upper Bumthang Chu districts; **Sikkim.** In Conifer and Rhododendron forests, 2750−3350m. April−August.

Wood used to make walking sticks (48).

12. C. simonsii Baker

Similar to *C. acuminatus* but more regularly branched; leaves 1−2×0.7−1.5cm, acute; fruit obovoid 7−8mm.

Bhutan: C—Thimphu district (Dochu La and Semi La); **Sikkim.** Margins of Spruce forest, 2300−3800m. July.

19. PYRACANTHA Roemer
by A. J. C. Grierson

Evergreen spinous shrubs. Leaves simple; stipules minute, caducous. Flowers in corymbose cymes terminal on short lateral shoots. Calyx tube turbinate, lobes 5. Petals 5. Stamens c 20. Carpels 4–5, partly adnate to calyx tube, each with 2 fertile ovules; styles 4–5 free. Fruit (pome) ± fleshy, stones 4–5, free.

1. P. crenulata (D. Don) Roemer; *Crataegus crenulata* (D. Don) Roxb.

Shrub 2–4m, branches bearing stout spines 0.75–1.5cm. Leaves oblong or obovate 1–4×0.5–1.7cm, obtuse, base cuneate or attenuate, margin shallowly crenate-serrate, glabrous or pubescent only at base, subsessile or petioles c 5mm. Cymes 5–10-flowered. Calyx tube c 2mm; lobes triangular c 1mm. Petals white, obovate 3–5×2–3mm. Pomes globose 6–7mm diameter, red.

Bhutan: C—Punakha and Tashigang districts, N—Upper Kuru Chu district. Streamsides, 1200–1800m. April–May.

Wood used to make walking sticks (48).

20. CRATAEGUS L.
by A. J. C. Grierson

Small deciduous trees or shrubs, branches bearing spines (modified lateral shoots). Leaves simple, deeply pinnatifid into 3–5 lobes, coarsely serrate; stipules small, deciduous. Flowers in terminal corymbose racemes, bracts usually deciduous. Calyx tube campanulate, lobes 5. Petals 5, obovate, usually white or pink. Stamens numerous. Carpels (1–)3(–5), inferior, united at least at base, each with 1 sterile and 1 fertile ovule. Fruit obovoid or subglobose, ± fleshy, often red, containing a stone usually 3-celled, each cell 1-seeded.

1. C. sp. Eng: *Hawthorn*

A species of unknown identity and origin, possibly imported from Europe, has been recorded (34) as cultivated in Darjeeling gardens.

21. SORBUS L.
by D. G. Long

Deciduous trees. Leaves alternate (often ± clustered on short lateral shoots), simple or 1-pinnate, margins mostly serrate; stipules deciduous or persistent. Flowers in terminal cymes or corymbs. Calyx tube funnel-shaped or obconic, lobes 5, reflexed, persistent or deciduous. Petals 5. Stamens 20 or more. Ovary inferior or semi-inferior, 2−5-celled, adnate to calyx tube; styles 2−5, free or united below. Fruit globose or pyriform, flesh smooth or granular, 2−5-celled, each cell 1−2-seeded.

Many species of *Sorbus* are valued as ornamental trees.

1. Leaves simple (sometimes lobed) 2
+ Leaves 1-pinnate 9

2. Leaves deeply 3-lobed .... **8. S. bhutanica**
+ Leaves unlobed or with many shallow lobes 3

3. Leaves white or greyish tomentose beneath (sometimes brown on veins); calyx lobes persistent in fruit (except in *S. griffithii*) 4
+ Leaves glabrous or brownish pubescent or tomentose beneath, never white; calyx lobes deciduous in fruit 7

4. Midrib and lateral veins brown tomentose beneath ; leaf margins lobed and serrate .... **3. S. hedlundii**
+ Midrib and lateral veins white tomentose beneath; leaf margins simply or doubly serrate, not lobed 5

5. Leaves thickly woolly tomentose beneath; petals 4 −5mm; calyx lobes c 2mm rounded, deciduous .... **4. S. griffithii**
+ Leaves closely appressed tomentose beneath; petals 6−7mm; calyx lobes 3−5mm, persistent in fruit 6

6. Leaf margins sharply irregularly and often doubly serrate; lateral veins 11−14 pairs; styles 2−3(−4) .... **1. S. thibetica**
+ Leaf margins regularly crenate-serrate; lateral veins 6−11 pairs; styles 3−5 .... **2. S. vestita**

7. Leaves 10−15cm; lateral veins 12−14 pairs; petioles 12−18mm; fruit not spotted .... **7. S. rhamnoides**
+ Leaves 6−12cm; lateral veins 7−10 pairs; petioles 4−10mm; fruit spotted 8

8. Leaves ovate-elliptic, 6−8×2−3cm, acute or acuminate, brownish pubescent beneath becoming glabrous; calyx lobes obtuse; fruit c10mm
 .... **5. S. thomsonii**
+ Leaves broadly elliptic, 6−12×3−5cm, abruptly acuminate, brownish tomentose beneath; calyx lobes acute; fruit 12−15mm. **6. S. ferruginea**

74. ROSACEAE

9. Leaflets 7–17 pairs, usually serrate in upper ¼–½ or to base 10
+ Leaflets 4–8 pairs, entire, obscurely crenulate throughout or serrate only in upper ¼ (in *S. kurzii* and *S. oligodonta* sometimes serrate in upper ⅓–½) 15

10. Leaflets 9–17 pairs, 5–22×2–8mm, serrate almost to base 11
+ Leaflets 7–12 pairs, 25–55×7–14mm, serrate in upper ¼–½ (to below middle in *S. himalaica*) 13

11. Leaflets (12–)14–17 pairs, 5–10×2–4mm .... **11. S. rufopilosa**
+ Leaflets 9–12 pairs, 14–22×5–8mm 12

12. Stipules of uppermost leaves subulate or lanceolate, not leafy, 3–4mm
9. S. microphylla
+ Stipules of uppermost leaves green, herbaceous, 5–10mm
10. S. prattii

13. Leaflets 25–35×8–12mm, obtuse or subacute, often serrate to below middle; stipules linear-subulate, pubescent, early-caducous; petals c 4mm, red or pink .... **13. S. himalaica**
+ Leaflets 25–55×7–14mm, mucronate, acute or acuminate; serrate in upper ¼–½; stipules lanceolate, ovate or suborbicular, herbaceous at least in part, bifid or serrate, glabrous, mostly persistent; petals 2–3mm, white or pale pink 14

14. Stipules large, leafy, broadly ovate or suborbicular, 5–15mm broad; fruits 9–10mm .... **12. S. arachnoidea**
+ Stipules small, lanceolate or ovate, 2–3mm broad; fruits 7–8mm
14. S. foliolosa

15. Leaflets 60–90×15–25mm, obscurely crenulate .... **16. S. insignis**
+ Leaflets 18–50×7–17mm, entire or serrate 16

16. Shoots and leaf undersides densely white sericeous when young; leaflets entire or with a few small teeth at apex .... **15. S. wallichii**
+ Shoots and leaf undersides glabrous or minutely puberulous when young; leaflets distinctly serrate in upper ¼–½ 17

17. Leaflets petiolulate, 20–30×7–10mm; fruits 5–6mm ...... **17. S. kurzii**
+ Leaflets sessile, 35–50×13–17mm; fruits 8–10mm ... **18. S. oligodonta**

1. S. thibetica (Cardot) Handel-Mazzetti. Dz: *Chasokey*

Tree 4–10m. Leaves elliptic, occasionally obovate, 10–15×5–10cm, shortly acuminate or acute, base cuneate, margins sharply and often doubly serrate, closely white appressed tomentose beneath, becoming glabrous above, lateral veins 11−14 pairs, prominent beneath except when young; petioles 5−12mm; stipules linear, c15mm, deciduous. Cymes corymbose, fragrant, 3−6cm long and broad, pedicels white tomentose. Calyx tube funnel-shaped, tomentose, lobes lanceolate c4mm. Petals creamy, elliptic, 5−7mm, with a few white hairs within. Anthers bright crimson. Styles 2−3(−4), united only at base. Fruit orange or yellow, globose, 15mm diameter, 3−5-seeded; calyx lobes persistent.

Bhutan: C—Thimphu, Punakha, Bumthang and Mongar districts, N—Upper Mo Chu and Upper Bumthang Chu districts; **Sikkim:** Jongri and Lachung. Spruce, Hemlock and Fir forests, 2740−3340m. May−June.

2. S. vestita (G. Don) Loddiges; *S. cuspidata* (Spach) Hedlund, *Pyrus vestita* G. Don. Nep: *Tenga* (34)

Similar to *S. thibetica* but leaves regularly, never doubly, crenate-serrate, lateral veins 6−11 pairs; petals woolly within; styles 3−5.

Sikkim: Darjeeling district, Takdah (34, 69, 80). Rhododendron forests, 2440−3050m. April−May.

Fruit edible (48). The records of this species from Bhutan (71) probably refer to *S. thibetica;* some of those from Sikkim may also belong there.

3. S. hedlundii Schneider. Nep: *Sanu Tenga, Lekh Mehel* (34)

Similar to *S. thibetica* but leaf margins coarsely doubly serrate, often shallowly lobed; white tomentose beneath but brown tomentose on midrib and veins; lateral veins 12−17 pairs; styles 3−5.

Bhutan: N—Upper Kuru Chu district (Nanpe near Shawa) and Upper Kulong Chu district (Tobrang); **Sikkim:** Tonglu, Laghep, Rechi La etc. Cool broad-leaved forests, 2560−2860m. April−May.

4. S. griffithii (Decaisne) Rehder; *Pyrus griffithii* (Decaisne) Hook. f., *Micromeles griffithii* Decaisne, *Sorbus sikkimensis* Wenzig var. *microcarpa* Wenzig

Tree to 15m. Leaves obovate-elliptic, 12−20×5.5−8cm, acuminate, base cuneate, margins regularly serrulate, thickly white-woolly tomentose beneath, more thickly on veins, glabrous above except when young, lateral veins 10−15 pairs; petioles up to 1cm, densely white tomentose; stipules deciduous. Corymbs 4−5cm, densely woolly. Calyx tube white tomentose, lobes c2mm, rounded-triangular, woolly outside, deciduous. Petals white, obovate, 4−5mm. Styles 2. Fruit globose, 6−7mm, 1−2-seeded.

Bhutan: C—Punakha district (Thinleygang), N—Upper Kuru Chu district (Nashima) and Upper Kulong Chu district (Tobrang); **Sikkim:** Lachung, Rungbi, Cho La etc. (34). Cool broad-leaved forests, 2100−2750m. March−May.

74. ROSACEAE

5. S. thomsonii (Hook. f.) Rehder; *S. verticillata* Merrill, *Pyrus thomsonii* Hook. f., *Micromeles thomsonii* (Hook. f.) Schneider. Nep: *Pasi* (34)

Tree to 15m. Leaves membranous, ovate-elliptic, 6−8×2−3cm, shortly acuminate, base cuneate, serrulate towards apex, brownish pubescent beneath when young, becoming glabrous, lateral veins 7−10 pairs, brownish tomentose beneath when young; petioles 4−8mm, brownish tomentose. Corymbs 2−4cm, pubescent or glabrous, fragrant. Calyx lobes rounded-triangular, c1.5mm, deciduous. Petals creamy, obovate-elliptic, c4mm, thinly hairy at base within. Styles 2−3. Fruit globose, c1.2cm diameter, with a few pale spots.

Bhutan: C—Thimphu district (Paro and Gunisawa) and Mongar district (Pangkhar near Lhuntse), N—Upper Kulong Chu district (Tobrang); **Sikkim:** Tonglu, Yak La, Senchal etc. Moist coniferous forests, 2400−3050m. April−May.

6. S. ferruginea (Wenzig) Rehder; *S. sikkimensis* Wenzig var. *ferruginea* Wenzig, *Pyrus ferruginea* (Wenzig) Hook. f., *Micromeles ferruginea* (Wenzig) Schneider

Very close to *S. thomsonii* but leaves broadly elliptic, 6−12×3−5cm, abruptly acuminate, brownish tomentose beneath; calyx lobes acute; fruit 12−15mm diameter.

Bhutan: N—Upper Mo Chu district (Gasa). 2285m.

Endemic to Bhutan; Griffith's type locality is unknown. A poorly known plant which may not be specifically distinct from *S. thomsonii*.

7. S. rhamnoides (Decaisne) Rehder; *S. sikkimensis* Wenzig var. *oblongifolia* Wenzig, *Micromeles rhamnoides* Decaisne, *Pyrus rhamnoides* (Decaisne) Hook. f. Nep: *Lahara Pasi* (34)

Similar to *S. thomsonii* but a small tree, often epiphytic; leaves coriaceous, ovate-elliptic, 10−15×3−5cm, pubescent beneath when young, becoming glabrous; lateral veins 12−14 pairs, very prominent beneath; petioles 12−18mm; corymbs pubescent at first; petals glabrous; fruit 6mm, smooth, not spotted, 4-seeded.

Sikkim: Darjeeling, Tonglu, Sandakphu, Karponang, Singalela and Sureil. 2100−2750m. May−July.

8. S. bhutanica (W. W. Smith) Balakrishnan; *Pyrus bhutanica* W. W. Smith

Tree. Leaves 3−5-lobed to middle, 3.5−6×2.5−4cm, base cuneate; lobes ovate or elliptic, subacute, crenate-serrate, with a few hairs above, rugulose beneath, veins ± pubescent; petioles 1.5−2cm, pubescent; stipules linear-lanceolate 5−10mm, entire, subpersistent. Flowers 4−8 in corymbs on lateral branchlets, pedicels tomentose. Calyx tomentose, tube urn-shaped, lobes triangular, 3mm, obtuse. Petals orbicular, 7−8mm, white hairy within. Ovary 5-celled; styles 5, hairy at base. Fruit unknown.

Tibet: Lhalung.
Known only from the original collection by J. C. White; the small 3-lobed leaves suggest that this species may be wrongly placed in *Sorbus;* it is somewhat similar to *Crataegus* except for the small entire stipules and lateral inflorescences; fruit is needed to resolve the problem.

9. S. microphylla Wenzig; *Pyrus microphylla* (Wenzig) Hook. f. Dz: *Tsema Shing;* Nep: *Sanu Pasi* (34)

Shrub 2–3m or small tree to 10m. Leaves borne on short lateral shoots, pinnate, 8–13cm, rachis winged and glandular; leaflets 9–12 pairs, oblong-elliptic, 14–22×5–8mm, acute or mucronate, base obliquely rounded, sessile, margins sharply serrate almost to base; glabrous or white to pale brown pubescent on both surfaces; stipules subulate, lanceolate or bifid, 3–4mm, persistent or deciduous. Corymbs 4–7cm across, 15–50-flowered; pedicels brownish pubescent. Calyx lobes broadly triangular, 1–1.5mm. Petals rose pink or white tinged pink, 3–4.5mm. Stamens pinkish. Styles 5. Fruit globose, 8–12mm, white or pink.

Bhutan: C—Thimphu, Tongsa and Tashigang districts, N—Upper Mo Chu and Upper Bumthang Chu districts; **Sikkim.** Common in Fir, Hemlock and Spruce forests, 3050–4100m. June–July.

A variable species in development of stipules, number of flowers, petal size, and fruit colour and size. This may in part be due to hybridization. The following two species are poorly-known and of doubtful status.

10. S. prattii Koehne

Similar to *S. microphylla* but stipules of uppermost leaves herbaceous, green, ovate-lanceolate, 5–10mm, bifid or toothed.

Bhutan: C—Ha district (between Damthang and Yatung, Chumbi), N—Upper Mo Chu district (Lingshi) and Upper Kulong Chu district (Shingbe); **Sikkim:** Gamothang. In Juniper/Rhododendron scrub, 3600–4000m. June–August.

11. S. rufopilosa Schneider

Similar to *S. microphylla* but leaflets more numerous, (12–)14–17 pairs, ovate-lanceolate, 5–10×2–4mm, obtuse or acute with fewer teeth; corymbs 5–10-flowered; fruit white (34).

Bhutan: C—Thimphu district (Pajoding and Barshong) and Tongsa district (Maruthang, Rinchen Chu), N—Upper Mo Chu district (Soe); **Sikkim:** Sandakphu, Phalut, Megu and Tonglo. Fir/Rhododendron forests, 3000–3650m. June–July.

12. S. arachnoidea Koehne; *S. foliolosa* auct. p.p. non (Wall.) Spach. Nep: *Pasi*

Tree 5–10m. Leaves borne on moderately stout branchlets or short lateral shoots, pinnate, 12−20cm, rachis winged and glandular; leaflets 7−9(−11) pairs, pale green beneath, oblong, 25−55×7−15mm, apex acute, mucronate, base obliquely cuneate, margins finely serrate in upper ¼−½, sessile, brownish pubescent at least on veins beneath; stipules large, leafy, persistent, broadly ovate or suborbicular, 5−15mm across, entire or serrate, glabrous. Corymbs large, 4−8cm diameter, many-flowered; pedicels brown tomentose. Calyx lobes bluntly triangular, 1−1.2mm. Petals white or pink, ovate, 2−3mm. Stamens pinkish. Styles 5, free. Fruit globose, 9−10mm, pink.

Bhutan: C—Thimphu district (Jato La) and Bumthang district (Bumthang and Yuto La), **N**—Upper Bumthang Chu district (Shimitang) and Upper Kulong Chu district (Shingbe); **Sikkim:** Tonglo, Sandakphu, Changu, etc.; **Chumbi.** Fir and Hemlock forests, 2900−4400m. May−July.

13. S. himalaica Gabrielian

Similar to *S. arachnoidea* but branchlets more slender; leaflets proportionately shorter and broader, 25−35×8−12mm, obtuse or subacute, often serrate to below middle; stipules linear-subulate, 6−9mm, pubescent, early caducous, never green; petioles and rachises narrowly winged; petals larger, c 4mm, pink or red; fruits white or pink 10−11mm.

Bhutan: N—Upper Pho Chu district (Lunana). Montane conifer forests, 3650m. June.

14. S. foliolosa (Wall.) Spach; *S. ursina* (G. Don) Shauer, *Pyrus foliolosa* Wall., *P. ursina* G. Don, *S. ursina* var. *wenzigiana* Schneider

Similar to *S. arachnoidea* and *S. himalaica* but branchlets often stout, leaflets proportionately longer and narrower, 25−40×7−11mm, mucronate or shortly acuminate, serrate in upper ¼−½; stipules of upper leaves ovate or lanceolate, 5−8mm, bifid or sharply serrate into cuspidate teeth; petioles and rachises narrowly winged; petals white or creamy, rarely pinkish, c 3mm; fruits 7−8mm, white or flushed with pink.

Bhutan: C—Thimphu district (Pumo La); **Sikkim:** Tonglo and Sandakphu. Montane conifer forests, 3050−3350m. May−June.

15. S. wallichii (Hook. f.) Yu; *S. foliolosa* auct. p.p. non (Wall.) Spach, *Pyrus wallichii* Hook. f.

Tree up to 10m; young shoots densely white sericeous. Leaves pinnate, 10−15cm; rachises winged, white pubescent; leaflets 5−8 pairs, oblong, 25−33×8−12mm, subacute or rounded and mucronate, base oblique, sessile, entire or with a few teeth near apex, densely white sericeous beneath at first becoming glabrous; stipules lanceolate, 3.5−4mm. Corymbs dense, 2−5cm diameter, fragrant. Calyx woolly, lobes triangular, 1mm. Petals white, 2.5−3mm. Stamens white. Styles 3. Fruit globose, 8−10mm, pink.

Bhutan: C—Mongar district (Dengchung, Khoma Chu), **N**—Upper Mo

Chu district (Gasa and Lingshi) and Upper Kulong Chu district (Tobrang); **Sikkim:** Lachen, Tonglo etc. Evergreen oak forests, 2100−2800m. April−May.

Two specimens, possibly juvenile or hybrid, from Senchal, have leaflets finely serrate to the middle.

16. S. insignis (Hook. f.) Hedlund; *Pyrus insignis* Hook. f.

Similar to *S. wallichii* but shoots brownish tomentose when young; leaves 15−25cm; leaflets 4−6 pairs, 60−90×15−25mm, apex acute, margins obscurely crenulate throughout, revolute when dry, brownish tomentose beneath when young, becoming glabrous; stipules ovate, leafy, up to 2cm; corymbs large, 9−13mm diameter, brown tomentose at first; fruits red, c 8mm.

Sikkim: Cho La, Phallut, Tonglo and Senchal. Rhododendron forests, 2750−3350m. April−May.

17. S. kurzii (Prain) Schneider; *S. wattii* Koehne, *Pyrus kurzii* Prain

Similar to *S. wallichii* but subglabrous; leaves 7−11cm; leaflets 4−5 pairs, elliptic or somewhat obovate, 20−30×7−12mm, apiculate, base obliquely rounded, distinctly petiolulate (petiolules 1−1.5mm), margins finely toothed in upper ½; corymbs narrow, 2−3cm diameter; petals white; styles 3−4; fruits red, 5−6mm.

Sikkim: Kalapokri, Phallut and Sandakphu. Rhododendron forests, 3350−3650m. May.

18. S. oligodonta (Cardot) Handel-Mazzetti; *S. hupehensis* Schneider var. *oligodonta* (Cardot) Yu

Similar to *S. wallichii* and *S. insignis* but subglabrous; leaves 10−20cm; leaflets 4−6 pairs, elliptic-obovate, 35−50×12−17mm, obtuse or subacute, base rounded, slightly oblique, serrate only in uppermost ¼, glabrous except for hairs at base of midrib beneath; corymbs broad, 5−15cm diameter; petals white; styles 3−4; fruits cherry red, 8−10mm.

S.E. Tibet: Trimo, Nyam Jang Chu. Moist coniferous forests, 3050m. May−June.

22. PHOTINIA Lindley

by A. J. C. Grierson

Evergreen or deciduous shrubs or trees. Leaves simple, stipules minute, deciduous. Flowers in terminal corymbose panicles. Calyx cup-shaped with 5 lobes. Petals 5. Stamens numerous (about 20). Ovary 2(−3)-celled, inferior; styles 2(−3) ± united at base. Fruit (pome) somewhat fleshy, 2−3-celled bearing persistent calyx lobes at apex, 1−2-seeded.

1. Leaves entire; inflorescence branches smooth without lenticels
   .... **1. P. integrifolia**
+ Leaves serrate; inflorescence branches smooth or bearing conspicuous lenticels 2

2. Deciduous; petioles 2−4cm; inflorescence branches always smooth
   .... **2. P. griffithii**
+ Evergreen; petioles up to 1cm; inflorescence branches bearing conspicuous lenticels especially when fruiting 3

3. Inflorescence 40−50-flowered, branches glabrous or sparsely pubescent, lenticels always visible .... **3. P. beauverdiana**
+ Inflorescence up to 12-flowered, branches pubescent or tomentose at first, lenticels visible later .... **4. P. arguta**

1. P. integrifolia Lindley. Nep: *Kursimla* (34), *Phalame* (34), *Hare, Saro Kat*
   Evergreen shrub or tree up to 12m. Leaves elliptic or oblanceolate, 7−15×2−5cm, acuminate, base cuneate, margin entire, glabrous; petioles 0.5−1.5cm. Inflorescence branches glabrous. Calyx turbinate 2−3mm including lobes. Petals white, ± as long as calyx. Styles 2. Pomes subglobose c 5mm diameter, reddish brown at first becoming bluish glaucous.
   Bhutan: S—Chukka, Gaylegphug and Deothang districts, **C**—Thimphu, Punakha, Tongsa, Mongar and Tashigang districts, **N**—Upper Kuru Chu district; **Sikkim.** In Warm broad-leaved forest, 1500−2350m. April−May.

2. P. griffithii Decaisne
   Similar to *P. integrifolia* but deciduous; greyish pubescent or tomentose at first on young shoots and on midribs of leaves beneath, leaf margins finely serrate especially near apex; petioles 2−4cm, greyish brown tomentose at first; inflorescence branches smooth and ± tomentose.
   Bhutan: C—Punakha district (Thinleygang and near Chuzomsa) and Mongar district (Lingmachey). Warm broad leaved-forests, 1200−1400m. April.

3. P. beauverdiana Schneider
   Evergreen shrub up to 8m. Leaves elliptic or ovate 5−10×2−4cm, acute or acuminate, base rounded, margin finely serrate, pubescent at first beneath; petioles 5−10mm. Corymbs 4−8cm, 40−50-flowered, branches glabrous or sparsely pubescent, warted with pale rounded or oblong lenticels up to 1mm. Calyx including broadly triangular lobes c 3mm, glabrous. Petals suborbicular 4−5mm diameter, sparsely white pubescent at base. Styles 2−3 connate near base. Pomes globose 5−8mm diameter, red.
   Bhutan: C—Tashigang district (Sherpang). Margin of cultivated land, 1825m. April.

4. P. arguta Lindley; *P. mollis* Hook. f., *Pourthiaea arguta* (Lindley) Decaisne

Similar to *P. beauverdiana* but leaves elliptic or oblanceolate 5−12×1−4cm, acuminate, base cuneate, ± glabrous; corymbs 5−8cm, up to 12-flowered, branches pubescent or tomentose at first, strongly warted with lenticels; pomes globose, 5−10mm diameter.

Sikkim: Jalpaiguri Duars. Streamsides in Terai forests.

The above description refers to var. **hookeri** (Decaisne) Vidal which has larger corymbs than in the typical variety.

23. ERIOBOTRYA Lindley

by A. J. C. Grierson

Evergreen trees. Leaves simple, coriaceous, lateral veins ± prominent and extending to leaf margin; stipulate. Flowers in conical terminal panicles. Calyx tube cup-shaped, lobes 5. Petals 5. Stamens numerous (15−40). Ovary inferior, 2−5-celled; styles 2−5, connate at base. Fruit (pome) fleshy, 2−5-celled, 1(−2) seeded; calyx lobes persistent.

1. Leaves subsessile or with petioles up to 1cm .... **Species 1 & 2**
+ Leaves with petioles 2−5cm .... **Species 3 & 4**

1. E. hookeriana Decaisne

Tree up to 10m. Leaves elliptic or oblanceolate, 15−25×3−8cm, acuminate, base attenuate to thick petiole up to 1cm, margins coarsely serrate, glabrous above, sparsely pubescent beneath along the 15−30 pairs of veins; stipules semi-lunar c1cm, deciduous. Panicles spreading 10−15cm, tomentose. Calyx tube 3−4mm, lobes 1.5−2mm. Petals broadly elliptic c3mm, white or pink. Stamens c 20. Styles 2, pubescent at base. Pomes ellipsoid 1.5−1.8×0.7cm, yellowish, 1−2-seeded.

Bhutan: S—Chukka district (near Chukka), **C**—Tongsa district (near Tashiling); **Sikkim.** Dry hill slopes, 1525−2300m. October−November.

2. E. dubia (Lindley) Decaisne. Nep: *Maya* (34)

Similar to *E. hookeriana* but leaves smaller, oblanceolate 7−15×1.5−4cm, with 10−16 pairs of veins, glabrous; stipules spathulate 3−4mm deciduous; panicles smaller 5−10cm, ± tomentose at first; calyx including lobes c 3mm; stamens 15−20; styles 2(−3); pomes c 1cm.

Bhutan: S—Chukka district (Sinchu La and near Chukka); **Sikkim.** In moist forest on exposed slopes, 1200−2100m. September−November.

3. E. petiolata Hook. f. Nep: *Maya* (34)

Tree up to 15m. Leaves oblong-elliptic, 15−30×7−10cm, shortly acuminate, base cuneate, margin entire or shallowly serrate near apex, brownish tomentose at first becoming glabrous, lateral veins 15−18 pairs; petioles 4−5cm; stipules lanceolate 1.5cm deciduous. Panicles c15cm tomentose. Calyx tube 3−4mm, lobes rounded c 2mm. Petals obovate, 7−8×4mm. Stamens c 20. Styles 4−5. Pomes subglobose 1−1.5cm diameter.

Bhutan: S—Chukka and Gaylegphug districts, **C**—Tongsa and Mongar districts; **Sikkim**. Steep hillsides in Warm broad-leaved forests, 1700−2200m. March−May.

Timber useful (48).

4. E. bengalensis (Roxb.) Hook. f. Nep: *Maya* (34)

Similar to *E. petiolata* but leaves elliptic or lanceolate 10−20×4−8cm, acuminate, base attenuate, margin serrate, glabrous, lateral veins 8−14 pairs; petioles 2−4cm; panicles 8−12cm, fragrant; calyx including lobes 3−4mm; petals obovate 4−5×3−4mm, white; styles 2−3; pomes ovoid c15×10mm.

Sikkim: Sivoke, Lopchu and Dumsong. Warm broad-leaved forest slopes, 1000−1500m. November−February.

Merits cultivation for its showy flowers (16).

24. CHOENOMELES Lindley

by A. J. C. Grierson

Deciduous shrub or small tree, branches spiny. Leaves simple; stipules foliaceous. Flowers 2−3 on short lateral branches, bisexual. Calyx tube obconical, lobes 5. Petals 5, ± clawed at base. Stamens 30−40. Ovary 5-celled, each cell with many ovules; styles 5, connate at base. Fruit a pome, ovoid or oblong.

1. C. lagenaria (Loiseleur) Koidzumi; *Cydonia cathayensis* Hemsley. Sha: *Khomang Shing;* Eng: *Quince*

Shrub or tree 1−4m, branches usually bearing stout spines c1cm. Leaves elliptic 4−8×1.5−2.5cm, obtuse or acute, base cuneate, margin entire or finely serrate, glabrous or pubescent beneath; stipules suborbicular 1−1.5cm, oblique at base, deciduous. Flowers subsessile or peduncles c 5mm. Calyx tube c1cm, lobes ovate or spathulate 4−5mm, rounded, pubescent within. Petals obovate 1.5−2×1−1.5cm, pink. Stamens c1cm; styles slightly longer, broadened at apex. Pomes up to 15×10cm, yellowish, seeds numerous.

Bhutan: C—Cultivated as a decorative plant e.g. at Paro and near Punakha, 2400m. April.

In the typical var. **lagenaria** the leaves are pubescent beneath at first but later glabrous; in var. **wilsonii** Rehder the leaves are persistently densely brown pubescent beneath. Both varieties of this Chinese species are cultivated in Bhutan.

25. DOCYNIA Decaisne

by A. J. C. Grierson

Small deciduous trees; sometimes spiny when young. Leaves simple, sometimes lobed in young trees; stipules subulate. Flowers 1−3 at branch ends, unfolding with young leaves. Calyx tube turbinate, lobes 5. Petals 5, clawed at base. Stamens numerous. Ovary 5-celled, each cell with 3−10 ovules, styles 5−6, connate at base, ± as long as stamens. Fruit (pome) ovoid or subglobose, 5-celled usually with 3 seeds in each; calyx lobes persistent.

1. D. indica (Wall.) Decaisne; *D. griffithiana* Decaisne, *Eriolobus indicus* (Wall.) Schneider. Dz: *Tong Shing;* Sha: *Thung Kakpa;* Nep: *Mehel* (34).Fig. 38 i−1

Tree 4−10m, branches sometimes spiny. Leaves ovate 9−15×3.5−6cm, acuminate, base rounded, margin entire or shallowly serrate, glabrous above, densely white tomentose beneath at first becoming greyish pubescent later; petioles 1.5−2.5cm; stipules 2−4mm deciduous. Peduncles short, thick up to 1cm. Flowers fragrant. Calyx tube c 8mm, lobes lanceolate, 8−15mm, white tomentose. Petals white flushed pink 1.5−3×2cm, obovate. Pomes 2.5−4×2.5cm, greenish yellow with red spots, c 15-seeded.

Bhutan: C—Punakha, Tongsa, Mongar and Tashigang districts; **Sikkim:** Kalimpong. In open forest, 1300−2440m. March−May.

Fruit edible (34).

26. MALUS L.

by A. J. C. Grierson

Deciduous trees or shrubs. Leaves simple, unlobed; stipules deciduous. Flowers in terminal corymbs on short lateral shoots. Calyx tube ellipsoid, lobes 5. Petals 5. Stamens numerous (20 or more). Ovary 4−6-celled, inferior; styles 4−6, very shortly united at base. Fruit (pome) fleshy, pithy not granular, 3−6-celled, cells cartilaginous within, 1−2-seeded; calyx lobes persistent or deciduous.

1. Pedicels slender 3−4cm; fruit thinly fleshy 7−15×4−10mm; forest shrubs or trees .... **Species 1 & 2**
+ Pedicels ± stout 1−2cm; fruit thickly fleshy 40−80×25−70mm; cultivated or near habitation .... **3. M. pumila**

1. M. baccata (L.) Borkhausen; *Pyrus baccata* L. Dz: *Khomang Shing*
Shrub or tree 3−5m. Leaves elliptic-lanceolate, rarely ovate 3.5−8×1.5−5cm acute or acuminate, base cuneate or ± rounded, margin finely serrate, sparsely pubescent along the veins and sometimes along lateral veins mostly beneath; petioles 1−4cm; stipules subulate 3−5mm. Corymbs 3−7-flowered, pedicels 2−3.5cm, slender. Calyx tube 3−4mm, constricted above ovary, lobes lanceolate ± as long. Petals obovate 10−15×7−10mm white, often pink tinged, glabrous. Styles 4−5, pubescent at base. Pomes ellipsoid 7−8×4−5mm, thinly fleshy, red, floral parts deciduous.

Bhutan: C—Ha, Thimphu, Punakha, Tongsa and Bumthang districts, N—Upper Mo Chu district (Gasa). In scrub amongst Blue Pine forest, 1800−3050m. April−May.

2. M. sikkimensis (Wenzig) Koehne; *Pyrus sikkimensis* (Wenzig) Hook. f. Dz: *Mindu Shing*
Similar to *M. baccata* but undersides of leaves, petioles, pedicels and calyces ± densely white pubescent at first; corymbs 4−20-flowered; petals pubescent at base; styles 5−6, glabrous; fruit ellipsoid c 1.5×1cm, calyx lobes ± persistent.

Bhutan: C—Ha, Thimphu, Punakha, Tongsa and Mongar districts; **Sikkim.** In open forest, 2450−3050m. April−June.

3. M. pumila Miller; *Pyrus malus* L. Eng: *Apple*
Shrub or tree to 10m. Leaves elliptic 3.5−8×2−5cm, acuminate, base rounded, margin serrate, pubescent especially beneath at first, later ± glabrous; petioles 2−3cm. Corymbs 4−7-flowered; pedicels ± stout 5−12mm. Calyx tube 4−5mm, lobes triangular ± as long, pubescent. Petals

FIG. 38. **Rosaceae** and **Connaraceae**. Rosaceae. a & b, *Rubus biflorus:* a, flowering shoot; b, fruit. c−e, *Prunus cerasoides:* c, flowering shoot; d, fruit; e, fruit in longitudinal section. f−h, *Spiraea arcuata:* f, flowering shoot; g, flower with two petals removed; h, mature follicles. i−l, *Docynia indica:* i, flowering shoot; j, fruit; k, fruit in longitudinal section; l, fruit in cross section. **Connaraceae.** m−r, *Connarus paniculatus:* m, flowering shoot; n, flower; o, flower dissected, ovary removed; p, ovary; q, fruit; r, seed with basal aril. Scale: a, i, j, k, l, m, q, r × ½; b, c, d, e, f × 1; n, o × 3; g, p × 4; h × 6.

white or pink tinged, elliptic 1−2×0.7−1cm, glabrous. Styles pubescent at base. Fruit ovoid-ellipsoid or subglobose 4−8×2.5−7cm, thickly fleshy, green, yellow or red when mature, floral parts ± persistent.

Bhutan: C—Cultivated e.g. at Paro. 2400m. April.

Probably SW Asiatic in origin and possibly of hybrid derivation; widely cultivated for its edible fruit.

27. PYRUS L.

by A. J. C. Grierson

Deciduous trees. Leaves simple, unlobed; stipules deciduous. Flowers in corymbs unfolding before or with young leaves. Calyx tube oblong, broadening above, lobes 5. Petals 5. Stamens numerous (20−30). Ovary inferior; styles 3−5, free. Fruit (pome) ellipsoid or obovoid, flesh granular, 3−5-celled, each cell cartilaginous, 1(−2)-seeded, calyx lobes deciduous or ± persistent.

1. P. pashia D. Don. Dz: *Lee;* Sha: *Litong;* Nep: *Naspati*

Tree 5−12m. Leaves ovate 4.5−9×2.5−4cm, acuminate, base rounded or shallowly cordate, rarely cuneate, margin finely serrate, sparsely white pubescent at first; petioles 1.3−2.5cm; stipules 4−10mm, subulate, adnate to petiole in lower half. Corymbs 3−8-flowered; pedicels glabrous, 9−20mm. Calyx tube 4−6mm, lobes ± as long, brownish pubescent within. Petals obovate 8−15×6−9mm, white. Pomes ellipsoid 2−3cm long and broad, brownish with pale spots, calyx lobes deciduous.

Bhutan: C—Ha, Thimphu, Punakha, Bumthang, Mongar and Tashigang districts; **Sikkim**: Darjeeling district (48). Cultivated or growing near habitation, 2130−3000m. April.

Fruit edible. The Bhutanese plants differ from the typical *P. pashia* which has crenately serrate leaf margins; because of their sharper serrations they have sometimes been incorrectly identified as *P. serrulata* Rehder, a Central Chinese species.

2. P. communis L. Sha: *Lise;* Nep: *Lishi, Naspati;* Eng: *Pear*

Similar to *P. pashia* but leaves broadly ovate up to 6cm broad, apex usually abruptly acute, base rounded, margin crenately serrate or subentire; petioles up to 6cm; corymbs 5−11-flowered borne on short much-scarred shoots, pedicels glabrous, 12−30mm; pomes obovoid or subglobose 4−6cm diameter, calyx lobes ± persistent.

Bhutan: S—Sarbhang district (Dara Chu), **C**—Thimphu district (Thimphu); **Sikkim**. Cultivated, 2150−2700m. April.

Native of Europe and W Asia, widely cultivated for its edible fruit (17).

Family 75. CONNARACEAE

by D. G. Long

Climbing shrubs or small trees. Leaves alternate, pinnately trifoliate or unevenly 1-pinnate; exstipulate; leaflets entire, pinnately veined. Flowers in terminal and axillary panicles, actinomorphic, bisexual. Sepals 5, very shortly connate at base. Petals 5, free. Stamens 10, connate at base into a short tube, filaments alternating long and short. Ovary superior, 1-celled; style slender with capitate stigma; ovules 2, ± basal. Capsules pod-like, oblique, slightly compressed, stalked, 1-seeded, seed with basal aril.

1. CONNARUS L.

Description as for Connaraceae.

1. C. paniculatus Roxb. Fig. 38 m−r

Large woody climber. Leaflets 3−7, coriaceous, subopposite, oblong-elliptic, 10−21×4−7.5cm, shortly and bluntly acuminate or apiculate, base rounded, glabrous, lateral veins 4−6 pairs; petioles 4−6mm, thickened. Panicles 10−25cm, red-brown tomentose. Sepals ovate, c 2.5mm. Petals white or creamy, narrowly spathulate, 5.5−6.5mm, lower half erect, upper half spreading, pubescent. Longer stamens distinctly exserted. Capsules obliquely compressed-obovoid, 3−3.5×1.8cm, finely striate, with obscure sub-apical beak. Seeds black with yellow aril.

Bhutan: S—Gaylegphug district (Gaylegphug) (117). Subtropical forests, 270m. February−March.

This literature record requires confirmation as no specimens have been located from the E Himalaya.

Family 76. LEGUMINOSAE

by A. J. C. Grierson & D. G. Long

Trees, shrubs or herbs, sometimes twining or climbing. Leaves alternate, sometimes simple but usually with few to many leaflets arranged trifoliately, pinnately, bipinnately or digitately; stipules usually present. Flowers zygomorphic and bisexual, or actinomorphic and sometimes polygamous; inflorescence often terminal, axillary or supra-axillary. Calyx campanulate or tubular, 5-toothed, the posterior pair of teeth often ± connate. Petals mostly 5 or sometimes fewer, free or some or all connate, equal or unequal.

Stamens commonly 10, rarely only 3 or 7 fertile and fully developed, sometimes numerous; filaments variously united, rarely free. Ovary monocarpellate, superior, unilocular or sometimes falsely plurilocular, ovules 1 or more borne on posterior suture; style simple, filiform or capitate. Fruit a pod (legume), valves usually dry rarely fleshy, usually dehiscent along both sutures or breaking transversely into 1-seeded segments, or indehiscent. Seeds often pea- or bean-like, rarely conspicuously arillate.

The family is naturally divisible into 3 subfamilies:

1. Caesalpinoideae (Genera 1–8)
Leaves evenly pinnate or bipinnate, sometimes simple or bilobed. Flowers ± zygomorphic, medium-sized to large and showy. Sepals often united at base forming a receptacular cup. Petals 5 or fewer, somewhat unequal, patent. Stamens usually 10, sometimes fewer, filaments usually free.

2. Mimosoideae (Genera 9–18)
Leaves evenly bipinnate, rarely reduced to a broadened simple leaf-like petiole (phyllode). Flowers regular, small, in spikes or globose heads, often polygamous. Sepals and petals usually united at base. Stamens 10 to numerous rarely 4–8, conspicuously exserted, filaments usually united at base.

3. Papilionoideae (Genera 19–85)
Leaves simple or with 2 to many leaflets arranged digitately or more commonly, pinnately. Flowers zygomorphic, small or large. Calyx often somewhat unequal, 4–5-toothed, the upper (posterior) 2 ± connate. Petals usually 5, the uppermost (standard) generally longer than the others, the 2 lateral petals (wings) usually parallel to each other and the 2 lowest often partially connate to form the keel to which the wings often adhere. Stamens usually 10, filaments free or more usually variously united: monadelphous with all 10 connate into a tube, or diadelphous where 9 are connate into a split tube and one (posterior) free or, more rarely, where the filaments are united in two bundles of 5. Style usually upwardly curved, hooked or sometimes coiled.

The sequence of subfamilies and genera follows that published by Polhill and Raven (166a) and although tribes and subtribes are not indicated, allied genera may be found adjacent to one another.

The key is divided into four groups as follows:
Leaves 1-foliate (sometimes deeply 2-or 3-lobed) .... **Group 1.**
Leaves 2-or 3-foliate .... **Group 2.**
Leaves simply pinnate .... **Group 3.**
Leaves bipinnate .... **Group 4.**

Group 1.
(Leaves 1-foliate, sometimes deeply 2- or 3-lobed)

1. Leaves unlobed, lanceolate, oblong, ovate or elliptic, lamina longer than broad 2
+ Leaves 2−3-lobed or lamina as broad as or broader than long 11

2. Trees; leaves (in fact, phyllodes) elliptic, ± falcate, leathery, veins several, parallel; flowers mimosoid, in spherical heads 7−8mm diameter .... **13. Acacia** *(A. melanoxylon)*
+ Herbs or shrubs, leaves not as above; flowers caesalpinioid or papilionoid 3

3. Stipels absent; pods not jointed 4
+ Stipels present; pods jointed, segments 1-seeded 9

4. Tendrillar climbing shrub; flowers caesalpinioid
    7. Bauhinia *(B. scandens)*
+ Erect herbs or shrubs without tendrils; flowers papilionoid 5

5. Keel tapering into a distinct beak distally; pods 6−many-seeded 6
+ Keel acute or rounded distally, not beaked; pods 1−2-seeded 7

6. Calyx campanulate, distinctly 5-toothed; keel sharply hooked below upwardly pointing beak .... **81. Crotalaria**
+ Calyx spathe-like, teeth minute; keel gently incurved below outwardly pointing beak .... **85. Spartium**

7. Prostrate herb, sometimes woody at base; flowers minute, c 3mm, scarlet .... **29. Indigofera** *(I. linifolia)*
+ Shrubs ± erect; flowers small to medium-sized, 7−12mm, white, pink or purple 8

8. Simple-stemmed or little-branched shrubs; leaves less than 1cm broad
    60. Eriosema
+ Branching shrubs; leaves more than 1.5cm broad .... **58. Flemingia**

9. Calyx papery, striate or with prominent parallel veins; segments of pods ± symmetrical on both sides .... **33. Alysicarpus**
+ Calyx herbaceous, not as above; segments of pods generally more convex on lower side 10

10. Segments of pods folded on top of one another or coiled in a circle; pedicels becoming hooked distally .... **31. Uraria***
+ Segments of pods remaining in a ± straight line; pedicels not becoming hooked distally .... **30. Desmodium***

11. Trees or woody climbers; leaves 2-lobed; flowers caesalpinioid
7. Bauhinia
+ Subshrubs or herbs sometimes with woody rootstocks; leaves 3-lobed or lamina as broad as or broader than long; flowers papilionoid 12

12. Leaves 3-lobed; petals dark purple, c2.5cm .... **84. Thermopsis**
+ Leaves not or scarcely lobed; petals whitish 3−7mm 13

13. Weak prostrate subshrub; leaves reniform; petals 3−4mm
30. Desmodium *(D. renifolium)*
+ Erect herb; leaves bat-wing-like, often white streaked; petals 6−7mm
32. Christia

Group 2.
(Leaves 2−3-foliate)

1. Leaves 2-foliate 2
+ Leaves 3-foliate 3

2. Climbing herbs; leaf rachis ending in a branched tendril; flowers large, 2−3.5cm, solitary or few on long peduncles forming short racemes
74. Lathyrus
+ Erect or spreading herbs; leaf rachis ending blindly; flowers small, 3−5mm, each concealed by a pair of medifixed bracts forming elongated racemes .... **63. Zornia**

3. Leaves digitately 3-foliate (i.e. petiolules equal) 4
+ Leaves pinnately 3-foliate (i.e. petiolules of terminal leaflet longer than others) 10

4. Slender prostrate or trailing herbs 5
+ Erect herbs or shrubs 7

5. Flowers 15−22mm, blue, solitary or in pairs .... **77. Parochetus**
+ Flowers 4−15mm, white or pink, few or numerous in short racemes or globose heads 6

*It is difficult to distinguish specimens of *Uraria* which lack pods from those of *Desmodium*. In general, *Uraria* has ovate, ± obtuse leaves and dense racemes. The leaves of *Desmodium* are variable and the inflorescence is less densely racemose or ± loosely paniculate.

6. Flowers in globose pedunculate heads; leaflets obovate, denticulate
   80. Trifolium
+ Flowers in ± sessile racemes; leaflets oblanceolate or spathulate, entire
   29. Indigofera *(I. trifoliata)*

7. Stipules deciduous leaving prominent annular scars on stems; flowers 2–3cm .... **83. Piptanthus**
+ Stipules persistent or if deciduous than not leaving conspicuous scars; flowers less than 1.5cm 8

8. Stems usually less than 1m tall, simple, ± unbranched; petioles 1–2mm
   35. Lespedeza
+ Stems more than 1m tall (except some *Crotalaria spp.*), branched; petioles 7–40mm 9

9. Flowers yellow; keel strongly hooked and tapering upwards into a beak; pods more than 2-seeded (except *C. trifoliastrum*) .... **81. Crotalaria**
+ Flowers pink or purplish (but pale yellow in *F. bhutanica*); keel upwardly curved but not beaked; pods 1–2-seeded .... **58. Flemingia**

10. Prostrate or erect herbs, shrubs or trees, never twining; lateral leaflets ± symmetrical at base 11
+ Woody or herbaceous twiners with elongated internodes; lateral leaflets usually very asymmetric at base (except *Shuteria* and *Dumasia*), rarely straggling or suberect (in *Vigna, Phaseolus* and *Rhynchosia*) 22

11. Flowers 2.5–5cm, red or orange (sometimes creamy); trees or robust perennial herbs 12
+ Flowers less than 2cm, colours various often blue or purple; herbs or shrubs (rarely trees e.g. *Desmodium oojeinense*) 13

12. Stipels knob-like; flowers red, standard longer than and sheathing other petals; pods 3–8-seeded .... **36. Erythrina**
+ Stipels subulate; flowers yellow or orange, keel longer than other petals, stongly curved; pods oblong containing one distal seed
   38. Butea

13. Prostrate herbs with yellow flowers in subcapitate heads c 0.7cm diameter on peduncles 1–2.5cm .... **80. Trifolium**
+ Erect or scrambling herbs or shrubs; flowers solitary or in elongated racemes, panicles or fascicles, colours various 14

14. Leaflets denticulate or finely serrate 15
+ Leaflets entire 16

15. Flowers minute, c3mm in elongated racemes; pods 3–4mm, usually 2-seeded .... **78. Melilotus**
+ Flowers small, 5–15mm, solitary, axillary or in short racemes or umbels; pods 10–100mm, few (usually more than 2)–many-seeded
79. Trigonella

16. Terminal leaflet broader than long, bat-wing-like .... **32. Christia**
+ Terminal leaflet longer than broad, not bat-wing-like 17

17. Calyx stiff, papery, parallel-veined, ± concealing flowers and pods
33. Alysicarpus
+ Calyx ± herbaceous, veins inconspicuous or net-like, not concealing flowers and pods 18

18. Leaves exstipellate; petioles less than 1cm; pods 1-seeded
34. Campylotropis
+ Leaves stipellate; petioles more than 1cm (except *Desmodium microphyllum* and *D. triflorum*); pods 2 or more seeded 19

19. Petals yellow or reddish 1.7–2cm; pods oblong, 1.5–7cm, obliquely grooved between the seeds .... **56. Cajanus**
+ Petals whitish, pink or purple (rarely yellowish, e.g. *Desmodium duclouxii*), 4–15mm; pods jointed or, if unjointed, not grooved between seeds 20

20. Annuals (cultivated); racemes 5–8-flowered; pods not jointed
44. Glycine
+ Perennial herbs or shrubs; racemes 10 or more flowered (except *Desmodium microphyllum* and *D. triflorum*); pods jointed 21

21. Segments of pod folded on top of one another; flowers in dense spike-like racemes (except *U. sinensis*) .... **31. Uraria**
+ Segments of pod remaining in a line, not folded; racemes or panicles ± diffuse (densely clustered in *D. oojeinense*) .... **30. Desmodium**

22. Pedicels borne on short, often glandular swellings on inflorescence axis
23
+ Pedicels not borne on swellings on inflorescence axis 31

23. Flowers less than 1cm, in branched terminal panicles; pods oblong, containing one distal seed .... **38. Butea** *(B. parviflora)*
+ Flowers usually more than 1cm (except some *Vigna* spp.), borne in axillary or terminal racemes; pods 2 or more seeded 24

24. Flowers 3−5.5cm; pods 1.5−5.5cm broad, ± terete and thickly coriaceous or woody, sometimes thinly coriaceous but then covered with irritant hairs 25
+ Flowers less than 3cm; pods 0.5−1.5cm broad (except *Lablab* and *Mastersia* up to 3cm broad), valves thin or membranous 27

25. Petals, or some of them, dark purple (turning black when dry), standard shorter than wings; pods either constricted between seeds or covered with irritant hairs .... **37. Mucuna**
+ Petals pink or purplish but not dark purple, standard longer than wings; pods neither constricted between seeds nor covered with irritant hairs 26

26. Calyx strongly bilabiate, upper lip larger, rounded and emarginate; keel ± straight; pods compressed, glabrous, strongly ribbed along lower suture .... **41. Canavalia**
+ Calyx scarcely bilabiate, lowest tooth longer than others; keel strongly incurved; pods terete, densely brown tomentose .... **50. Dysolobium**

27. Keel curled almost in a circle; pods 0.5cm or less broad, ± terete **54. Vigna**
+ Keel not curled; pods 0.6cm or more broad, ±compressed 28

28. Keel hook-shaped; pods sublunate bearing persistent style base, minutely warted along both sutures .... **51. Lablab**
+ Keel ± straight or incurved; pods narrowly oblong, 6−30mm broad . 29

29. Terminal leaflet rhombic, broader than long with 3−7 broad sharp teeth; pods 10−15mm broad .... **42. Pachyrhizus**
+ Terminal leaflet elliptic or ovate, longer than broad, entire or shallowly to deeply 3-lobed; pods 6−10mm or 25−30mm broad 30

30. Pods 6−10mm broad, seeds elliptic arranged parallel to pod axis; terminal leaflet entire or 3-lobed .... **43. Pueraria**
+ Pods 25−30mm broad, seeds oblong arranged transversely to pod axis; terminal leaflet always entire .... **46. Mastersia**

31. Leaflets bearing yellow or reddish glands on lower surface (sometimes sparse in *Cajanus* and *Rhynchosia*); flowers always yellowish 32
+ Leaflets without glands beneath; flower colours various 34

32. Pods 4−11-seeded, valves not grooved between seeds ..... **57. Dunbaria**
+ Pods either 3−9-seeded with valves grooved between seeds or 2-seeded 33

33. Pods 2-seeded; leaflets either 1−2cm or 18−23cm .... **59. Rhynchosia**
+ Pods 3−9-seeded; leaflets 2−9cm .... **56. Cajanus**

34. Calyx obliquely truncate; pods torulose, much constricted between seeds .... **48. Dumasia**
+ Calyx with distinct teeth; pods ± compressed, sometimes depressed but not constricted between seeds 35

35. Petals 5−6mm; pods bearing thickened persistent hooked style
    45. Teramnus
+ Petals 8−25mm; pods without thickened hooked style 36

36. Inflorescences up to 2.5cm (rarely 5cm); standard with 2 appendages near midline 37
+ Inflorescences 3−45cm; standard without appendages near midline 38

37. Flowers solitary or few (2−3, rarely 5) subsessile, fasciculate, rarely subracemose with peduncles up to 1.5cm; standard bearing lamelliform appendages on either side of midline .... **53. Macrotyloma**
+ Flowers in a distinct raceme; standard bearing hooked appendages on either side of midline .... **52. Dolichos**

38. Keel and style coiled distally through 360° 39
+ Keel and style straight or incurved by no more than 90° 40

39. Stipules and stipels absent or caducous .... **40. Cochlianthus**
+ Stipules and stipels present, persistent .... **55. Phaseolus**

40. Standard c4cm, placed lowermost in flower, much exceeding other petals .... **49. Clitoria** *(C. mariana)*
+ Standard 1−2.5cm, placed uppermost in flower, not much longer than other petals 41

41. Flowers borne singly on inflorescence axis; claw of wing and keel petals longer than blade .... **47. Shuteria**
+ Flowers borne in clusters of 3−6 on inflorescence axis; claw of wing and keel petals shorter than blade .... **43. Pueraria**

Group 3
(Leaves simply pinnate)

1. Leaves even-pinnate, rachis sometimes produced as a bristle or tendril but not ending in a leaflet 2
+ Leaves odd-pinnate, rachis ending in a leaflet 11

2. Leaf rachis and sometimes stipules persistent and stiffly spinescent; shrubs .... **65. Caragana**
\+ Leaf rachis and stipules never spinescent; habit various 3

3. Leaf rachis ending in a simple or branched tendril; climbing herbs 4
\+ Leaf rachis ending blindly or in a short simple point; habit various but not climbing herbs 6

4. Stipules 3−5cm, larger than leaflets; cultivated vegetable .... **75. Pisum**
\+ Stipules 0.4−2.5cm, usually smaller than leaflets; native herbs or cultivated ornamental 5

5. Small native climbers up to 50cm; stipules 0.4−1.2cm .... **73. Vicia**
\+ Large cultivated ornamental climber 1.5−2m; stipules 1.5−2.5cm
   74. Lathyrus

6. Slender twining shrubs; flowers and fruits borne on glandular swellings on upper part of peduncle .... **23. Abrus**
\+ Trees, shrubs or herbs, never twining; flowers not borne on glandular swellings 7

7. Stipules medifixed, produced below point of insertion; herbs with sensitive leaves 8
\+ Stipules basifixed, persistent, minute or caducous; trees, shrubs or herbs, some with sensitive leaves... 9

8. Leaflets 3−8 pairs; pod segments folded on each other within calyx; plants of dry stony ground with thin woody rootstocks .... **62. Smithia**
\+ Leaflets 20−40 pairs; pods linear not enclosed by calyx; plants of marshy ground with thick pithy stems and rootstocks
   61. Aeschynomene

9. Shrubs or trees; flowers purple with yellow wings; pods linear, elongate, terete, 9−23×0.3−0.4cm, shallowly constricted between seeds
   28. Sesbania
\+ Herbs, shrubs or trees; flowers yellow or red but not purple; pods compressed or terete but not as above 10

10. Prostrate herbs; leaves with 2 pairs of leaflets; flowers papilionoid, solitary or few, axillary; pods oblong, cylindrical, developing underground at end of gynophore .... **64. Arachis**
\+ Trees, shrubs or ± erect herbs; leaves with more than 2 pairs of leaflets (except sometimes in *Cassia tora*); flowers caesalpinioid, mostly in racemes or panicles, seldom solitary or few axillary; pods terete or compressed, developing above ground 11

11. Trees; petals and fertile stamens 3; pods terete or cylindrical covered with a dry brittle brown skin without obvious sutures; seeds surrounded by juicy pulp .... **8. Tamarindus**
+ Trees, shrubs or herbs; petals 5; fertile stamens 7–10; pods compressed or terete, valves membranous, coriaceous or woody, sutures conspicuous; seeds not embedded in pulp .... **6. Cassia**

12. Herbs, ± erect, sometimes woody at base but never climbing or twining 13
+ Trees or shrubs, erect, climbing or twining, sometimes ± herbaceous (*Apios*) 24

13. Leaflets sharply serrate; flowers solitary, axillary .... **76. Cicer**
+ Leaflets entire; inflorescence racemose (rarely 1-flowered in *Chesneya*) sometimes subcapitate or umbellate 14

14. Racemes elongate, densely spreading-hirsute; slender erect herbs ... 15
+ Racemes or umbels short or subcapitate, racemes rarely elongate but then not spreading-hirsute; cushion-forming or tufted bushy herbs (except *Astragalus floridus*) 16

15. Perennial; leaflets narrowly oblong-lanceolate, usually variegated with a pale median band .... **31. Uraria** *(U. picta)*
+ Annual; leaflets elliptic-obovate, not variegated
    29. Indigofera *(I. astragalina)*

16. Acaulescent herbs often with thick woody rootstocks 17
+ Herbs with distinct stems (10cm or less in *Gueldenstaedtia*) 20

17. Flowers (5–)10–20 in pedunculate racemes; leaf rachises sometimes persistent, pale coloured .... **68. Oxytropis**
+ Flowers solitary or few, ± sessile, axillary 18

18. Leaf rachises always persistent, dark brown .... **66. Spongiocarpella**
+ Leaf racimes not persistent 19

19. Flowers yellow; pods dehiscent, sutures and valves smooth, 10–15-seeded .... **67. Astragalus** *(A. acaulis)*
+ Flowers purple; pods dividing transversely into 2–4 1-seeded segments, sutures and mid-line of valves toothed .... **71. Stracheya**

20. Leaflets, at least the upper ones, obovate or oblanceolate; flowers solitary or up to 8 in an umbel or very condensed raceme; pods transversely septate 21

+ Leaflets oblong, elliptic, ovate or lanceolate, rarely obovate or oblanceolate; racemes 3–30-flowered (flowers rarely solitary in *Astragalus balfourianus* and *A. donianus*); pods either compressed and divided into 1–3 1-seeded segments or linear-oblong, ± inflated or terete, wholly or partially 2-celled by a longitudinal septum, never transversely septate 22

21. Leaflets 9–17, emarginate; flowers solitary or 2–3, dark purple rarely white .... **69. Gueldenstaedtia**
+ Leaflets 4–5, upper 3 crowded at apex of leaf rachis and 1 or 2 stipule-like at its base, at least upper ones obtuse or subacute; flowers 4–8, petals yellow or red-tipped .... **72. Lotus**

22. Wing petals with basal appendage ± as long as claw; pods compressed and divided into 1–3 elliptic, 1-seeded segments .... **70. Hedysarum**
+ Wing petals with basal appendage much shorter than claw; pods ± turgid, oblong, ovoid or ellipsoid, dehiscent 23

23. Keel with a small but distinct beak distally .... **68. Oxytropis**
+ Keel unbeaked, rounded distally .... **67. Astragalus**

24. Herbaceous twiners 25
+ Distinctly woody trees or shrubs, erect or climbing 26

25. Lateral leaflets asymmetric at base; flowers in racemes; standard shorter than keel and uppermost in flower; style coiled through 360°
39. Apios
+ Lateral leaflets symmetric at base; flowers solitary, axillary; standard longer than other petals and lowermost in flower; style ± straight
49. Clitoria

26. Leaflets in part at least alternately arranged on rachis 27
+ Leaflets opposite (sometimes subopposite in *Cladrastis* and *Sophora*) 28

27. Inflorescence paniculate; flowers less than 1cm; pods oblong, compressed .... **22. Dalbergia**
+ Inflorescence racemose; flowers 1.3–1.7cm; pods ± terete, strongly constricted between the seeds .... **21. Sophora**

28. Leaflets 15–18cm, coriaceous, broadly elliptic, borne on a narrowly winged rachis; petals white, 1.7–1.9cm; pods ± fleshy, 1-seeded
82. Euchresta
+ Leaflets usually less than 15cm (sometimes longer in *Millettia* but then membranous), mostly narrowly elliptic or oblong, rachis never winged; petals often less than 1.5cm; pods 1–12-seeded, never fleshy 29

29. Leaflets 0.5−4cm (rarely up to 6cm in *I. atropurpurea*); flowers 4−12mm; pods linear, inflated, ± terete, septate within . **30. Indigofera**
+ Leaflets 4−18cm (sometimes 3cm in *Tephrosia*); flowers 0.9−2.5cm; pods compressed or, if terete, then torulose, rarely broadly ellipsoid in *Ormosia* 30

30. Calyx 10−12mm; pods ± terete, c1cm broad, strongly constricted between the seeds .... **21. Sophora**
+ Calyx 2−7mm; pods compressed (rarely terete or torulose in *Millettia* but then 2.5−5cm broad) 31

31. Leaflets narrowly elliptic or oblong, 0.8−1.8cm broad; petals 2−2.3cm; pods narrowly oblong, 7−8mm broad .... **26. Tephrosia**
+ Leaflets elliptic, ovate or obovate, 2−7cm broad; petals 0.9−1.5cm (up to 2.5cm in *Millettia pachycarpa*); pods broader 32

32. Filaments free or coherent only at base; pods membranous or thickly woody 33
+ Filaments monadelphous or diadelphous; pods leathery or woody, never membranous 34

33. Leaflets 5−7 pairs; flowers in terminal panicles; pods narrowly oblong, 7−8×1cm, membranous .... **20. Cladrastis**
+ Leaflets 3−4 pairs; flowers in axillary racemes; pods 5−7.5×3cm, thickly woody .... **19. Ormosia**

34. Inflorescence a true raceme, flowers borne singly on rachis; pods unwinged, glabrous .... **27. Gliricidia**
+ Inflorescence paniculate, often racemiform but flowers borne in clusters or several on short lateral branches; pods winged or pubescent 35

35. Stipules usually linear-lanceolate, deciduous (but broadly ovate-triangular, persistent in *M. pachycarpa*); stipels present or absent; calyx distinctly short-toothed; pods unwinged, pubescent **25. Millettia**
+ Stipules shortly triangular lozenge-shaped, appressed, ± thickened, persistent (often inconspicuous); always exstipellate; calyx ± truncate; pods winged along one or both sutures, glabrous .... **24. Derris**

Group 4
(Leaves bipinnate)

1. Rachises of leaves and pinnae with rounded or oval, sessile or raised glands; stamens mostly numerous but 10 in *Leucaena* 2

+   Rachises of leaves and pinnae without rounded or oval glands; stamens 5–10 8

2. Shoots bearing stipular spines or scattered prickles 3
+   Shoots unarmed 4

3. Leaves with one pair of pinnae, each with one pair of leaflets
    17. Pithecellobium
+   Leaves with 4–26 pairs of pinnae, each with 20–70 pairs of leaflets
    13. Acacia

4. Stamens 10; petals free .... **12. Leucaena**
+   Stamens numerous, 15–50; petals connate 5

5. Petals 12–15mm; pods fleshy .... **15. Samanea**
+   Petals 1–10mm, but sometimes up to 12mm in *Albizia sherriffii*; pods not fleshy 6

6. Central flower of head unlike others: slightly larger, staminal tube exserted from corolla, filaments more intensely coloured ..... **14. Albizia**
+   All flowers of head apparently alike, though some sometimes unisexual
    7

7. Leaflets generally increasing in size towards apex of pinnae; filaments connate in lower half; pods coiled .... **18. Archidendron**
+   Leaflets generally larger near middle of pinnae; filaments free or connate only at base; pods straight or curved but not coiled .. **13. Acacia**

8. Leaflets always less than 10mm broad 9
+   Leaflets in part at least more than 10cm broad 13

9. Flowers small, ± sessile in rounded heads 1–1.5cm diameter; pods divided transversely into 1-seeded segments, sutures persistent after fragmentation ....**11. Mimosa**
+   Flowers small, medium-sized or large, pedicellate in racemes or panicles; pods not dividing transversely into 1-seeded segments 10

10. Rachis of leaves bearing recurved prickles; pods 1-seeded, samaroid with a distal wing .... **5. Pterolobium**
+   Rachis of leaves unarmed; pods several to many-seeded, not samaroid but sometimes with narrow wing along dorsal suture 11

11. Petals 5–6cm; pods thickly woody, seeds cylindrical or squarish in section, borne in chambers between valves .... **3. Delonix**
+   Petals 0.7–2.5cm; pods coriaceous, dehiscent and continuous within or indehiscent and winged; seeds compressed or subglobose 12

12. Lowest sepal larger than others, hooded and enclosing bud, upper petal differing in size and shape from others; pods oblong, narrowly winged along dorsal suture only .... **4. Caesalpinia**
+ Sepals and petals all similar; pods elliptic, broadly (5−7mm) winged along both sutures ....**2. Peltophorum**

13. Flowers in globose heads; pods 10−12cm, thickened along both sutures
16. Calliandra
+ Flowers in spikes or racemes; sutures of pods unthickened, except in *Entada* but then pods 20−60cm more 14

14. Flowers less than 5mm in dense spikes or racemes; pods transversely jointed or constricted between the seeds 15
+ Flowers 1−7cm in lax racemes; pods compressed, not jointed or constricted between seeds 16

15. Leaves with 2 pairs of pinnae, rachis ending in a branched tendril, leaflets 3−6cm; pods 20−60cm or more, transversely jointed, breaking away from sutures into 1-seeded segments ....**10. Entada**
+ Leaves with 3−5 pairs of pinnae, rachis ending blindly; pods linear, 15−20cm, constricted between the rounded seeds, not jointed
9. Adenanthera

16. Petals narrowly elliptic, up to 3mm broad; stamens 5; pods membranous, seeds 15−20 .... **1. Acrocarpus**
+ Petals elliptic or obovate, more than 3mm broad; stamens 10; pods thickly coriaceous or if membranous then 1-seeded .... **4. Caesalpinia**

1. ACROCARPUS Arnott

Tall unarmed deciduous trees. Leaves evenly bipinnate; leaflets entire; stipels absent; stipules small, caducous. Flowers precocious in terminal racemes. Sepals 5, inserted on margin of campanulate receptacle. Petals 5, narrow. Stamens 5, free, exserted. Pods linear, compressed, narrowly winged along dorsal suture, long-stalked.

1. A. fraxinifolius Arnott. Nep: *Mandane*. Fig. 40 d−f
Tree 20−40m. Leaves 50−60cm, pinnae 2−4 pairs with an odd terminal one, leaflets 6−8 pairs, ovate-elliptic, 5−12×2−6cm, acuminate, base

rounded or cuneate, sparsely pubescent at first. Racemes 30−40cm, densely flowered, pedicels 2−3mm. Sepals triangular, 4−5×2−3mm, pubescent, inserted on margin of receptacular cup 6−7mm. Petals yellowish, oblanceolate, 10−12×2−3mm, pubescent. Stamens exserted c 20mm, filaments red. Pods 10−15×1.5−2cm including narrow wing 3−5mm, thinly woody; seeds 15−20, ellipsoid, 6×4mm, brown.

Bhutan: S—Throughout southern foothills, **C**—Tongsa to Tashigang districts: **Sikkim**. Subtropical and Warm broad-leaved forests, 220−1500m. February−March.

The wood is used for making tea-boxes and furniture (48).

2. PELTOPHORUM (Vogel) Bentham

Unarmed deciduous trees. Leaves evenly bipinnate; leaflets numerous, entire; stipels absent; stipules minute, caducous. Flowers showy in terminal panicles, bracts deciduous. Sepals 5, subequal, connate at base, reflexed in flower. Petals 5, ± crumpled or wavy. Stamens 10, free. Style filiform, stigma peltate. Pods elliptic, compressed, winged, indehiscent, 1−4-seeded.

1. P. pterocarpum (DC.) Heyne; *P. ferrugineum* Bentham

Tree 10−15m. Leaves 20−35cm, pinnae 8−10(−15) pairs, 8−12cm; leaflets 8−18 pairs, oblong, 10−20×4−8mm, emarginate, base oblique, truncate or rounded, sessile, glabrous. Panicle branches brown pubescent. Sepals oblong-ovate, c10×7mm, brown outside, yellowish within. Petals yellow, broadly obovate, 15−20×15−20mm, narrowed to a broad claw at base, brown lanate along midline in lower half. Stamens c15mm, brown lanate at base. Pods leathery, 6−11×2.5−3cm, including wing 5−7mm broad around both sutures.

Bhutan: S—Phuntsholing and Gaylegphug districts. 200−300m. April−May.

Native of Australia and SE Asia, cultivated as a decorative wayside tree.

3. DELONIX Rafinesque

Unarmed trees, partly or completely deciduous. Leaves evenly bipinnate; leaflets small, numerous, entire; stipels absent; stipules bipinnate, leaf-like, deciduous. Flowers showy in terminal or axillary corymbose racemes, pedicels jointed near apex, bracts persistent. Sepals 5, subequal, valvate, shortly connate at base. Petals 5, tapering gradually below into a claw, upper one somewhat larger than other petals, and variegated. Stamens 10, free, filaments long. Pods large, linear-oblong, compressed; seeds transverse, borne in chambers in woody endocarp.

76. LEGUMINOSAE

1. D. regia (Hooker) Rafinesque; *Poinciana regia* Hooker. Eng: *Flamboyant, Flame Tree, Gold Mohur* (34)

Tree 10–15m. Leaves 20–60cm, pinnae 8–12(–25) pairs, leaflets 12–25 pairs, oblong, 8–12×3–4mm, obtuse or subacute, base obliquely rounded, ± sessile; stipules 1–3cm. Racemes 8–12-flowered, pedicels 3–5cm, bracts ovate c8×5mm. Sepals oblong-lanceolate, 20–25×7–8mm, reddish within. Petals obovate, 4–6×2.5–5cm, red, upper one streaked yellow. Stamens 3.5–4cm. Pods 30–70×4–6cm; seeds 20–50, oblong, 2×0.7cm, brown.

Bhutan: S—Phuntsholing, Sarbhang, Gaylegphug and Deothang districts. May–June.

Native of Madagascar. Cultivated in gardens and on roadsides for its decorative habit and colourful flowers.

4. CAESALPINIA L.

Climbing shrubs or small trees, usually armed with recurved spines. Leaves evenly bipinnate, rachis often bearing recurved spines beneath; pinnae opposite, leaflets few or numerous; stipels rarely present; stipules present or absent. Racemes or panicles axillary, supra-axillary or terminal, pedicels jointed or not. Flowers bisexual or sometimes unisexual; flower parts borne on margin of a concave receptacle which is persistent in fruit. Sepals 5, the lowest one longer than the other 4, ± hooded and enclosing bud, other sepals subequal. Petals 5, upper one usually differing in shape and size from other petals. Stamens 10, free, usually decurved in open flower. Pods dehiscent or indehiscent, ± compressed, winged along dorsal suture or wingless, membranous, coriaceous or woody, valves sometimes armed with stiff bristles; seeds 1–10.

1. Leaflets narrowed to a point at apex, acute, acuminate or mucronate .. 2
+ Leaflets rounded or emarginate at apex 4

2. Leaflets ± sessile or petiolules less than 0.5mm; pods with densely bristled valves .... **3. C. bonduc**
+ Leaflets with petiolules 2–4mm; pods with smooth valves 3

3. Leaflets 3.5–10×2–5cm; upper petal fan-shaped, bilobed; pods membranous, oblong-lanceolate, 7–10×2.5–3cm, with a wing c 0.5cm along upper suture .... **1. C. cucullata**
+ Leaflets 2–6×1–3.5cm; upper petal suborbicular; pods woody, elliptic, 4–7×2.5–3.5cm, unwinged ....**2. C. crista**

4. Pinnae with 12–30(–40) or more pairs of narrow leaflets 2–5mm broad
   4. C. tortuosa
+ Pinnae with 5–12 pairs of leaflets usually more than 5mm broad 5

5. Stipels absent; filaments 1−1.5cm; pods 2.5−3cm broad
   5. C. decapetala
+ Stipels present, subulate, c 0.5mm; filaments 5−7cm; pods 2−2.5cm broad .... **6. C. pulcherrima**

1. C. cucullata Roxb.; *Mezoneuron cucullatum* (Roxb.) Wight & Arnott. Dz: *Tse Hein;* Nep: *Bokshi Khanra.* Fig. 39e

Climbing shrub. Leaves 15−35cm, rachis bearing paired hooked spines, pinnae 2−6 pairs, 10−15cm, leaflets 4−5 pairs, ovate 3.5−10×2−5cm, acuminate, base rounded, glabrous; petiolules 2−4mm; stipules absent. Panicles terminal, 30−50cm; pedicels 5−12mm, jointed near middle. Lowest sepal twice as long as others, 8−12×4mm, other sepals 5×3mm. Petals yellow, upper one fan-shaped, bilobed, 6−8×10−12mm, other petals with claws c 0.5mm, oblong or broadly elliptic, 5−8×4−6mm. Stamens exserted, filaments 20−25mm. Pods membranous, indehiscent, oblong-lanceolate, 7−10×2.5−3cm including wing c 0.5cm broad, brown; seeds 1(−2) in middle of pod.

Bhutan: S—Samchi, Phuntsholing, Sarbhang, Gaylegphug and Deothang districts; **Sikkim.** Subtropical terai forests, 200−390m. November−December.

2. C. crista L.; *C. nuga* (L.) Aiton

Similar to *C. cucullata* but leaf pinnae 2−4(−5) pairs, 2.5−8(−10)cm, leaflets 2−3(−5) pairs, ovate-elliptic, 2−6×1−3.5cm, acute, base rounded, upper surface ± glossy, dull and glabrous beneath; petiolules 2−4mm; flowers on pedicels 7−15mm, jointed near apex; lowest sepal 8×4mm, others 6−8×2−3mm, reflexed in flower; upper petal suborbicular, c 5mm, borne on claw c 5×2mm, other petals 7−9×4mm, on claws c 1mm; stamens c10mm, filaments lanate to above middle; pods indehiscent, woody, elliptic, 4−7×2.5−3.5cm, acute, base narrowed to a stalk 3−5mm; seeds 1−2.

Sikkim: Terai. Banks of streams. January−April.

3. C. bonduc (L.) Roxb.; *C. bonducella* (L.) Fleming. Eng: *Fever Nut* (126), *Nicker Tree* (126). Fig. 39f

Climber. Leaf rachis 15−80cm, armed with recurved spines at least at base of pinnae; pinnae 6−11 pairs, 8−20cm, leaflets elliptic 6−12 pairs, 1−4×0.5−2.5cm, acute, mucronate, base rounded, pubescent especially beneath; petiolules c 0.5mm; stipules pinnate, 3−5-foliate, similar to leaflets. Racemes 30−60cm, supra-axillary, inserted up to 2cm above leaf axils; bracts longer than buds, lanceolate, c 10mm reflexed, caducous; pedicels 7−8mm, jointed near apex. Flowers unisexual, anthers sterile in female flowers, ovary rudimentary in male flowers, otherwise similar. Sepals almost equal, oblong, 7−10×2−3mm, reflexed, brown pubescent.

Petals yellow, claws c 3mm, upper one ovate, 4−7×4mm, reflexed, other petals oblanceolate c 7×3mm. Stamens 6−10mm. Ovary in female flowers ovoid, 7−8×3mm, densely bristly. Pods ellipsoid, 6.5−9×3.5−4.5cm, covered with spiny bristles 5−10mm; seeds 1−2, subglobose, c 2×1.5×1.2cm, grey with fine dark parallel lines concentric around hilum.
Sikkim: Badamtam. September.
The seeds are used medicinally (126).

4. C. tortuosa Roxb.; *C. cinclidocarpa* Miquel

Small tree or shrub up to 10m, sometimes climbing. Leaves 30−45(−60)cm, rachis with recurved spines, pinnae 7−20 pairs, 5−16cm, leaflets 12−30(−40) pairs, oblong, 9−20×2−5mm, obtuse or rounded, base oblique truncate; stipules absent. Racemes supra-axillary, 20−60cm, pedicels 8−15mm, not jointed. Lowest sepal deeply hooded, 12−14×12mm, other sepals 10×8mm. Petals with broad claws, 3−4mm, upper one broadly elliptic, c 10×7mm, reflexed, other petals suborbicular, 1.3−1.8cm. Stamens 10−14mm, somewhat exserted, lanate. Pods oblong-elliptic, 3.5−9×2−3.5cm, indehiscent, woody, often twisted and constricted between the seeds; seeds subglobose, c10mm.
Sikkim: Sivoke Terai. August.

5. C. decapetala (Roth) Alston; *C. sepiaria* Roxb. Dz: *Tatse Tsang, Tsangi Metog*. Fig. 39d

Climber or shrub, branchlets and leaf rachises with recurved spines or sometimes unarmed. Leaves 10−35cm, pinnae 3−10 pairs, 2.5−7cm, leaflets 5−12 pairs, oblong or obovate, 1.2−2×0.5−1cm, obtuse or emarginate, base rounded, finely pubescent and minutely glandular beneath; stipules ovate, 8−10×4−5mm. Racemes 15−30cm, pedicels 15−35mm, jointed near apex. Lowest sepal ± hooded, 8−10×3−4mm, others oblong, 6−8×3−4mm. Petals yellow often with red veins, upper one suborbicular, 7−8mm, borne on claw 5−6mm, other petals obovate, 14−15×10−12mm on claws 1−2mm. Stamens c15mm, lanate to middle. Pods oblong, 6−8(−10)×2.5−3cm, woody, narrowly (2−3mm) winged along dorsal suture, sharply beaked with style remnant c 0.75cm; seeds 4−9.
Bhutan: C—Punakha and Mongar districts. Streamsides and banks in forest, 1200−1850 m. May−September.

FIG. 39. **Leguminosae.** a, *Cassia fistula:* leaf, inflorescence and pod partly dissected to show seeds. b, *Cassia lechenaultiana:* portion of flowering shoot. c, *Bauhinia variegata:* portion of flowering shoot. d, *Caesalpinia decapetala:* flower, e, *Caesalpinia cucullata:* pod. f, *Caesalpinia bonduc:* pod. g, *Pterolobium macropterum:* portion of leaf and pod. Scale: a × ⅓; c × ½; b, e × ⅔; f × ¾; d, g × 1.

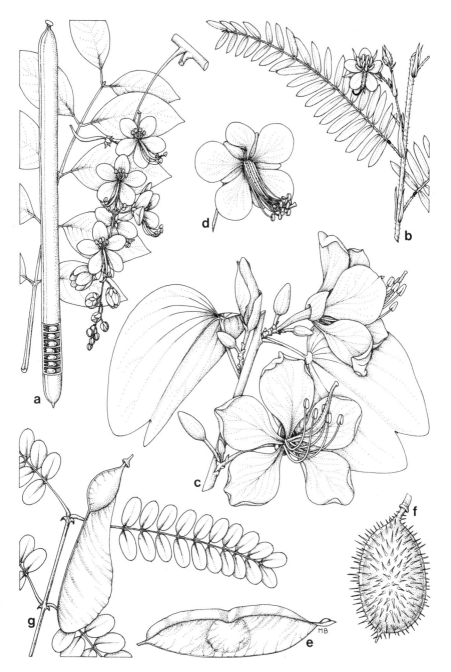

6. C. pulcherrima (L.) Swartz

Shrub or small tree c5m, branchlets unarmed or with a few prickles. Leaves 20−40cm, rachis unarmed or with a few weak prickles, pinnae 5−9 pairs, 5−12cm, leaflets 6−12 pairs, oblong or obovate, 10−30×5−15mm, obtuse or emarginate, base obliquely rounded, glabrous; petiolules 1−2mm; stipels subulate c 0.5mm; stipules subulate c 2mm, caducous. Racemes axillary or terminal, 20−30cm, pedicels 3−8(−10)cm, jointed near apex. Lowest sepal deeply hooded, c 15×5mm, others oblong, 10−13×4−5mm. Petals reddish fading to orange or yellow, upper one suborbicular, c 5mm borne on claw c 1.5cm, other petals obovate or suborbicular, 2−2.5×1.2−1.8cm, margins wavy. Stamens 5−7cm, lanate at base. Pods oblong, 6−10×1−2cm, beaked, somewhat woody, dehiscent; seeds 8−10, ± rectangular 8−10×5−8×2−3mm, blackish, separated by pithy septa.

Bhutan: locality unknown.

Native of S America, cultivated as a handsome ornamental and sometimes becoming naturalised.

5. PTEROLOBIUM Wight & Arnott

Climbing shrubs, stems and leaf rachises armed with recurved spines. Leaves evenly bipinnate, leaflets numerous, entire; stipels absent; stipules minute. Flowers in axillary or terminal racemes or panicles, bracts caducous. Sepals 5, shortly connate at base, lowest one larger than others. Petals 5, uppermost one slightly broader than others and lanate near base. Stamens 10, free. Style filiform, stigma concave, ciliate. Pods samaroid, compressed, basally 1-seeded, membranously winged distally.

1. P. hexapetalum (Roth) Santapau & Wagh; *P. indicum* sensu F.B.I. non A. Richard

Leaves 12−18cm, with 4−8 pairs of pinnae 3−5cm, leaflets 6−8 pairs, oblong, 8−12×3−5mm, obtuse or emarginate, mucronate, base obliquely rounded, sparsely pubescent. Inflorescence branches pubescent. Sepals oblong or oblanceolate, lowest one c 5×2mm, others c 3×1.5mm. Petals oblanceolate or elliptic, uppermost one yellow, c 3.5×2mm, others white 4×1.25mm. Stamens c 5mm. Pods ± sessile, crimson, fertile part obovate-elliptic, 15×10mm, sparsely pubescent, wing obliquely elliptic 20−25mm, 10−15mm broad near middle.

Bhutan: C—Tashigang district (Gamri Chu Valley). Dry slopes, 1080m.

2. P. macropterum Kurz; *P. indicum* A. Richard var. *macropterum* (Kurz) Baker. Fig. 39g

Similar to *P. hexapetalum* but leaves with 7−8 pairs of pinnae 4−5cm, leaflets 7–10 pairs, oblong, 8–12×5–7mm; pods with basal stalk 2–3mm, pubescent, fertile part 20×12mm, wing 30–40mm, 13–18mm broad near middle.

Bhutan: C—Tongsa district (Mangde Chu Valley). Rock faces in mixed forest, 1325m.

6. CASSIA L.

Trees, shrubs or herbs. Leaves even-pinnate; petioles and leaf rachises sometimes glanduliferous; leaflets few or numerous, entire, sometimes sensitive to contact; stipels absent; stipules deciduous or ± persistent. Flowers solitary, axillary, or in axillary or terminal racemes or panicles. Sepals 5, subequal. Petals 5, subequal. Stamens 4–10, perfect and subequal or very unequal in size and some sterile or absent, anthers opening by pores or by longitudinal slits. Pods linear, flat or ± cylindric, membranous, leathery or woody, dehiscent or indehiscent, usually septate within; seeds few or many.

1. Petioles and leaf rachises glandless; trees or (*C. alata*) shrubs 2
+ Petioles and leaf rachises bearing one or more ± prominent glands (scarcely prominent in *C. mimosoides*); mostly herbs or shrubs but *C. surattensis* usually a small tree 6

2. Shrub; leaflets 7–15×4–7cm, obtuse; pods membranous, winged
   .... **5. C. alata**
+ Trees; leaflets 2–9×1–3cm, ± acute (obtuse in *C. grandis* but then leaves 2–5×1–1.5cm); pods unwinged, ± woody 3

3. Stamens all equally long (3 staminodes also present); filaments shorter than anthers .... **4. C. spectabilis**
+ Three stamens at least twice as long as the others; filaments many times longer than anthers (staminodes present or not) 4

4. Inflorescence short, 6–15cm, erect, bracts remaining attached as long as flowers .... **3. C. javanica**
+ Inflorescence 20–40cm, pendent, bractless when in full bloom 5

5. Leaflets 3–4(–8) pairs, ± acute, glabrescent; petals 2cm or more, yellow .... **1. C. fistula**
+ Leaflets 10–20 pairs, obtuse, pubescent beneath; petals c1.5cm, pink
   2. C. grandis

6. Leaflets narrow (0.5–)1–2mm broad; sepals acute or acuminate, glands on leaf rachis cushion-like or cup-shaped 7
+ Leaflets 1–2.5cm broad; sepals obtuse, glands on petiole or leaf rachis erect, linear, ovoid or club shaped 9

7. Small herbs, stems 10–45cm, sometimes prostrate; stamens 4
    8. C. hochstetteri
+   Tall herbs or subshrubs, 0.5–1.5m; stamens 9–10 8

8. Leaf rachis smooth in side view, middle leaflets 6–12mm
    6. C. lechenaultiana
+   Leaf rachis serrate in side view, middle leaflets 4–8mm
    7. C. mimosoides

9. Leaves with a gland near base of petiole only 10
+   Leaves with a gland between lowermost pairs of leaflets and sometimes with other glands between upper leaflets 11

10. Leaflets 4–9 pairs; bracts ovate, 4–5mm, obtuse; pods ± terete
    9. C. sophera
+   Leaflets 3–5 pairs; bracts ovate, acuminate, 1–1.5(–2)cm; pods ± compressed .... **10. C. occidentalis**

11. Leaflets acuminate .... **11. C. floribunda**
+   Leaflets obtuse, subacute or emarginate 12

12. Leaflets ovate-elliptic; shrub or small tree .... **12. C. surattensis**
+   Leaflets obovate; annual herb .... **13. C. tora**

1. C. fistula L. Sha: *Donka Sey, Donko Shing;* Med: *Donga;* Nep: *Raj Birse, Sunalo, Rajbriksh* (34), *Bandarlata* (34). Fig. 39a

Deciduous tree, 10–20m. Leaves 15–40cm, leaflets 3–4(–8) pairs, ovate, 7–15×4–7cm, acute, base rounded, glabrous but greyish pubescent beneath at first, petiolules 4–5mm; stipules deltoid 1–2mm, deciduous. Racemes axillary, pendent, 10–40cm, pedicels 4–6cm. Sepals elliptic, 8–10×4–5mm, becoming reflexed. Petals yellow, obovate, 2–3× 1–1.5cm. Stamens 10, longest 3 with filaments 3cm, 4 medium-sized with filaments 8–10mm and 3 smaller stamens with filaments 5–6mm and poorly developed anthers. Style ± as long as longest stamens. Pods terete, 20–60×1.5×cm, woody, indehiscent, black, transversely septate into many chambers. Seeds ovate, 12–13×7mm, glossy, brown with a darker line from hilum to top of seed on one side.

Bhutan: S—Phuntsholing and Sarbhang districts, **C**—Tashigang district; **Sikkim**. On dry hillsides, 300–1200m. May–June.

A native possibly of Malaysia, it is widely cultivated as a decorative wayside tree, occasionally becoming naturalised. The pulp found in the pods is a very strong purgative (16).

2. C. grandis L. f.

Similar to *C. fistula* but flowers precocious; leaves 15−25cm, leaflets 10−15(−20) pairs, oblong-elliptic, 2.5−5×0.8−1.5cm, rounded at apex and base, mucronate, sparsely pubescent above, densely brown pubescent beneath; racemes 15−25cm; petals red, upper one spotted yellow c1.5cm; stamens 10, 3 longest with filaments c 2cm, 5 medium-sized with filaments 7−9mm and 2 small with filaments c 2mm; pods 30−65(−90)×3−4.5cm, not or slightly compressed, transversely rugose, blackish, 70−80-seeded.

Bhutan: S—Phuntsholing district (Phuntsholing) (117). Cultivated, 220m. April.

Native of tropical S America, widely planted as a decorative wayside tree.

3. C. javanica L.

Similar to *C. fistula* but tree up to 40m; leaflets 5−12(−20) pairs, ovate-elliptic, 6−9×2−3.5cm, acuminate, base rounded, minutely pubescent beneath, petiolules 2−4mm; racemes 5−15cm, corymbose, ± erect, bracts lanceolate, 8−10(−18)×2mm, pedicels 2−3.5cm; sepals ovate, 4−5mm, green; petals elliptic, 1.5−2×0.75cm, pink at first later yellowish; stamens 10, 3 longest with filaments 1.5−2cm, thickened at middle, 4 medium-sized with filaments 10−12mm, 3 small with much reduced anthers; pods 30−70×1−2cm, terete, annulate, dark brown.

Bhutan: S—Samchi district (Samchi). Cultivated, 500m. April−June.

Native of Malaysia, planted as an ornamental wayside tree.

The above description refers to subsp. **nodosa** (Roxb.) K. & S. Larsen which differs from the typical variety in its acute (not obtuse) leaflets, its green (not dark red) sepals and its shorter petals which turn yellowish (not dark red).

4. C. spectabilis DC.

Similar to *C. fistula* but a small tree 8−10m; leaves 30−40cm, leaflets 10−15 pairs, ovate, 4−10×1.5−4.5cm, acute, mucronate, base rounded, brownish pubescent especially beneath, petiolules 2−4mm; racemes 10−15cm, pedicels 2−2.5cm, bracts small, caducous; sepals petaloid, broadly ovate, 5−7×3.5−5mm, obtuse; petals yellow, obovate, 1.5−2×1−1.5cm, distinctly veined; stamens 7, subequal, anthers 5×2.5mm on filaments 2−3mm, staminodes 3, bilobed 1.5×2.5mm; pods subterete, 18−25×1−1.5cm, annulate, 50−70-seeded.

Bhutan: C—Punakha district (Punakha Dzong). Cultivated, 1270m. September.

Native of Central America, cultivated for its handsome flowers.

5. C. alata L. Fig. 40k

Shrub 1−5m. Leaves 30−60cm, leaflets 8−12(−15) pairs, oblong or obovate, 4−12(−19)×2.5−7(−12)cm, obtuse or emarginate, mucronate, base rounded, glabrous; petiolules 2−4mm; stipules triangular 5−7mm long and broad, persistent. Racemes 40−60cm, bracts ovate or obovate, 2−2.5×1−1.5cm, orange, deciduous. Sepals obovate, 12−15×4−7mm. Petals yellow, oblong or obovate, 15−18×10mm, borne on a claw 2−3mm. Stamens 10, 2 lateral ones larger than others with anthers 12×3−4mm, on filaments c 4mm, other stamens with anthers 4×1.5mm on filaments c 2mm, except one (the lowest) on filament c 4mm, 2−3 upper stamens much reduced (sterile?). Pods linear, tetragonous, 12−15(−19)×1.5cm, both valves bearing a crenate wing c 7mm broad, 50−60-seeded; seeds flat, rhombic, c 10×8mm.

Bhutan: S—Phuntsholing district (Phuntsholing) and Gaylegphug district (Tori Bari). Subtropical forest margins and waste ground, 200−400m. May.

Native of S. America, now a pantropical weed. A paste made from the leaves is an effective remedy for ringworm (126).

6. C. lechenaultiana DC.; *C. mimosoides* L. var. *wallichiana* (DC.) Baker and var. *auricoma* Bentham. Fig. 39b

Herb, sometimes woody at base, 20−100(−150)cm. Leaves 7−8×2cm, leaflets 10−25(−40) pairs, sensitive to contact and then becoming folded together lengthwise, obliquely oblong-lanceolate, 6−12×1.5−2mm, acute, mucronate, base rounded, midrib excentric, closer to upper margin, ciliate, rachis smooth, unsculptured; petioles 3−7mm bearing an ovoid sessile gland (c 0.5mm) just below lowest pair of leaflets; stipules lanceolate, 6−8×1.5mm. Racemes few-flowered, 3−10mm, supra-axillary, inserted 3−8mm above axils, bracts stipule-like. Sepals ovate-lanceolate, 7−8×2−4mm, pubescent on midrib, becoming reflexed. Petals yellow, broadly obovate, 8−10×5−10mm. Stamens 10, 2 or 4 of them with anthers 5−6mm, others 2.5−4mm, filaments of all c 1mm. Pods linear-oblong, compressed, 3−4.5×0.4cm, transversely grooved between the 10−12 black, glossy seeds.

Bhutan: C—Punakha, Mongar and Tashigang districts. Open hillsides and cultivated ground, 1350−2400m. May−August.

Specimens with yellowish hirsute stems have been segregated as var. **auricoma** (Bentham) de Wit.

7. C. mimosoides L.

Similar to *C. lechenaultiana* but leaves 4−10×0.7−1cm, leaflets 15−40(−70) pairs, 4−8×0.8−1.3mm, rachis broadened between leaflets, ± serrate in side view; petioles c 3mm bearing at lowermost pair of leaflets a discoid scarcely prominent gland.

Sikkim: Terai. July−August.

Records from Bhutan (117) require confirmation.

8. C. hochstetteri Ghesquiere; *C. mimosoides* L. var. *dimidiata* (Roxb.) Baker

Similar to *C. lechenaultiana* but prostrate or suberect, 10−40cm; leaflets 4−8×1−2mm, leaf rachis not broadened between leaflets; petioles 2−3mm bearing an ovoid cushion-like gland between lowermost leaflets; sepals lanceolate, 5−6×1.5mm, reddish-green; petals 4−7×2−5mm; stamens 4, anthers 3.5−4mm, subsessile.

Bhutan: C—Thimphu, Punakha and Tongsa districts; **Sikkim.** Roadsides, 1500−2370m. August−September.

9. C. sophera L. Dz: *Heyduk*

Shrub 1−3m. Leaves 10−20cm, leaflets 4−9(−12) pairs, ovate-lanceolate, 2.5−7×1−2cm, acuminate, base rounded, finely ciliate, otherwise glabrous; petioles 3−4cm, bearing an erect club-shaped gland (c 1.5mm) 5−10mm from its base; stipules ovate c 6mm, caducous. Racemes axillary, 4−10-flowered, bracts ovate 4−5mm, acute, deciduous. Sepals obovate, 5−10×4−6mm. Petals elliptic or obovate 10−15×5−7mm, yellow or orange. Stamens 10, 3 larger with anthers 5−6mm, on filaments 3−4mm, 3−4 smaller stamens with anthers 4−5mm on filaments c 2mm and 3−4 minute c 3mm, sterile. Pods subterete, 8−10×0.7−1cm, straight or curved, brown with a broad pale band along each suture, septate; seeds 30−50, ovate, 4−5×3−4mm, greyish.

Sikkim: Sivoke Terai. June.

10. C. occidentalis L.

Similar to *C. sophera* but leaflets 3−5 pairs, ovate, 3−9(−12)×1.5−3(−4)cm, acuminate, base rounded, ciliate and minutely glandular pubescent beneath; petioles 3−6cm with a globose or ovoid gland (c 2mm) very close (1−5mm) to base; stamens 9−10, unequal, 2 much larger than rest, 4 medium-sized and 3−4 very small; pods linear, ± compressed, 10−12×0.7−1cm, brown with a pale band along both sutures.

Bhutan: S—Phuntsholing district (near Phuntsholing), **C**—Punakha district (near Punakha and Wangdu Phodrang); **Sikkim.** River banks and waste ground, 200−1450m. May−September.

Possibly S. American in origin, now a pantropical weed. The dried seeds, ground to powder, make a good substitute for coffee (126).

11. C. floribunda Cavanilles; *C. laevigata* Willdenow

Similar to *C. sophera* but leaflets 3−5 pairs, ovate, 4−9(−11)×2−3.5cm, acuminate, base rounded, glabrous, rachis bearing a linear gland (c 2mm) between all except uppermost pair of leaflets; sepals obovate, 5−10×3−5mm, obtuse; petals obovate, 1−1.5×1cm, yellow; stamens 10, 3 large with anthers 6−8mm, 2 of them on filaments 8−10mm, the third on

filament c4mm, 4 smaller stamens with anthers 4—5mm on filaments c 2mm, 3 other stamens much reduced or sterile; pods 8—10×1.5cm, terete, transversely septate, c 50-seeded.
Sikkim. 900m. May-August.

12. C. surattensis Burman f.
Similar to *C. sophera* but usually a small tree 2—7m; leaflets 4—7 pairs, 3.5—8(—10)×1.5—4(—5)cm, subacute, emarginate, base rounded, sparsely pubescent beneath, rachis with a linear gland (c 2mm) between the lower 2—3(—4) pairs of leaflets; stipules linear, falcate 7—10mm, ± persistent; racemes axillary ± corymbose; sepals broadly ovate, 5—12×5—10mm, obtuse; petals obovate, 2—3×1—1.5cm; stamens 10, all with anthers 4—5mm on filaments 1—2mm except one filament 3—4mm; pods linear, compressed, 8—15×1—1.5cm, 10—30-seeded.
Bhutan: S—Phuntsholing and Sarbhang districts, **C**—Tongsa district. Cultivated, 200—400m. September—May.
Native of S.E. Asia planted as a decorative wayside tree.

13. C. tora L.
Annual 30—100(—120)cm. Leaves (4—) 8—10cm, leaflets 2—3 pairs, obovate, 2—5×1.5—2.5cm, obtuse, base rounded or cuneate, pubescent beneath, rachis bearing a linear gland (2—3mm) between all except uppermost pair of leaflets; stipules linear 4—5mm, ± persistent. Flowers axillary, solitary or 2-flowered. Sepals obovate, 5—7×3—4mm. Petals yellow, obovate, 8—12×6—8mm. Stamens usually 7, 3 longer with anthers c4mm on filaments 4mm, 4 smaller stamens with anthers 1.5mm on filaments c1mm, 3 staminodes present or absent. Pods terete or subtetragonous, 15—20×0.4—0.5cm, 20—30-seeded.
Bhutan: S—Phuntsholing and Gaylegphug districts; **Sikkim.** Roadsides and river banks, 200—300m. May—August.
A paste made from the leaves of this plant is an effective remedy for ringworm (126).

7. BAUHINIA L.

Trees or shrubs, erect or large tendrillar climbers, bisexual or dioecious. Leaves simple, usually bilobed at apex, palmately veined; stipules usually small, caducous. Flowers in axillary or terminal racemes or panicles. Sepals borne on margin of receptacular tube, free or ± united, calyx sometimes spathe-like. Petals 5, subequal, usually clawed at base. Fertile stamens 3, 5 or 10, staminodes sometimes present. Pods stalked, oblong, linear or elliptic, compressed, leathery or ± woody, dehiscent.

7. BAUHINIA

1. Trees or erect shrubs, branches without tendrils; flowers mostly large, petals 3–5cm but small (c 8mm) in *B. malabarica* 2
+ Climbing shrubs, branches bearing tendrils; flowers small to medium-sized, petals 3–25mm 4

2. Flower buds 7–10mm, ± rounded at apex; sepals spreading; petals 8–10mm; stamens 10 .... **1. B. malabarica**
+ Flower buds 1.5–3cm, acute at apex; sepals connate, calyx becoming spathe-like; petals 3–5cm; stamens 3 or 5 3

3. Flower buds pentagonous with 5 dark longitudinal ridges; petals 3–3.5cm; fertile stamens 3 .... **2. B. purpurea**
+ Flower buds smooth, without ridges; petals 4.5–5cm; fertile stamens 5
3. B. variegata

4. Flowers small, petals c 3mm .... **4. B. scandens**
+ Flowers medium-sized, petals 12–25mm 5

5. Leaves densely brown pubescent beneath, 11–13-veined. ...... **5. B. vahlii**
+ Leaves glabrous or sparsely pubescent beneath, 7–9-veined 6

6. Flowers in corymbose panicles .... **6. B. glabrifolia**
+ Flowers in elongate racemose branches .... **7. B. wallichii**

1. B. malabarica Roxb.; *Piliostigma malabaricum* (Roxb.) Bentham. Nep: *Amil Tanki* (34)

Dioecious tree 4–15m. Leaves suborbicular in outline, 4.5–9×6–12cm, apex ⅕ or ⅙ bifid, lobes rounded, base truncate, rounded or shallowly cordate, veins 9–11, sparsely pubescent on veins beneath; petioles 2.5–4cm. Racemes slender, 4–7cm, buds ellipsoid 7–10mm, rounded or subacute at apex, pedicels 10–25mm. Sepals 4–5mm, spreading, acute, inserted on receptacular cup c4mm. Petals oblanceolate 8–9×2–3mm. Male flowers with 10 stamens, filaments 3–4mm, ovary vestigial. Female flowers with pubescent ovary, stigma capitate; staminodes minute. Pods narrowly oblong, 15–30×1.5–2cm, leathery; seeds 20–30 oblong, c 7×4mm.

Bhutan: S—Sankosh and Deothang districts; **Sikkim.** Subtropical forests, 200–500m. September.

2. B. purpurea L. Sha: *Pegpeyposhing*; Nep: *Tanki* (34)

Erect shrub or tree 2–12m. Leaves broadly elliptic, 16–18×5–15cm, apex deeply (⅓–½) bilobed, lobes subacute or obtuse with a subulate point 2–3mm in the sinus between them, base truncate or cordate, glabrous or minutely puberulous beneath, veins 9–11. Racemes 10–12-flowered,

633

axillary or terminal, buds 2—2.5cm, 5-ridged in upper half, ridges dark coloured. Sepals elliptic, c 20×3—4mm, connate, calyx spathe-like at anthesis, receptacular tube 9—12mm. Petals pink or mauve, elliptic, 3—4×1.5—2cm, borne on a claw c1cm. Fertile stamens 3, filaments 3—4cm, shorter hair-like staminodes also present. Pods linear-oblong, 15—25×2—2.5cm; seeds ellipsoid, c 1.5×1cm.

Bhutan: S—Samchi, Phuntsholing, Chukka and Gaylegphug districts, **C**—Tashigang district; **Sikkim.** Subtropical forests, 250—1500m. October—November.

The bark is used in dyeing and tanning. The leaves are used for cattle fodder (126).

3. B. variegata L. Nep: *Koerlo, Koiralo* (34), *Taki* (13). Fig. 39 c

Similar to *B. purpurea* but leaves broadly ovate, 5—18×5—18cm, apex ⅕—⅓ bifid, lobes obtuse, base cordate, veins 9—11; racemes 1—3cm, flowers ± sessile, receptacular tube 2.5—3cm, buds ovoid, smooth; sepals 2—3cm; petals elliptic, 4—5×2.5—3cm, on claws c1.5cm, white or pink, posterior one purple; fertile stamens 5, 4.5—5cm, pods 15—20×2—3cm.

Bhutan: S—Phuntsholing and Gaylegphug districts, **C**—Tongsa and Tashigang districts; **Sikkim.** Warm and subtropical forests, 250—1200m. February—May.

Originally possibly a native of China, widely cultivated and naturalised throughout warm and subtropical regions. Bark, leaves and flowers are used medicinally (13).

4. B. scandens L.; *B. anguina* Roxb., *Lasiobema scandens* (L.) de Wit. Nep: *Nagbaele* (34)

Climber with paired opposite tendrils on young shoots, old stems flattened (up to 15cm broad) but with alternate concavities from opposite sides. Leaves ovate, 6—11×5—8cm, apex usually ⅕—⅓ bifid, sometimes entire, lobes acute, base shallowly cordate, veins 7—9, ± glabrous. Flowers in slender racemes 15—18cm, sometimes forming terminal panicles, pedicels 3—5mm. Calyx c 3mm including short (c 1mm) teeth. Petals suborbicular, c 3×3mm, yellowish. Fertile stamens 3, 4—5mm; staminodes 2, minute. Pods elliptic, 3—4×2—2.5cm, 1—2-seeded.

Bhutan: S—Chukka district (Marichong); **Sikkim.** Subtropical forests, 1200m. October.

This species is represented here by var. **horsfieldii** (Miquel) Larsen & Larsen which differs from the typical variety by its smaller pods.

5. B. vahlii Wight & Arnott; *Phanera vahlii* (Wight & Arnott) Bentham. Nep: *Bhorla* (34)

Large densely brown pubescent climber, young shoots bearing simple tendrils in opposite pairs. Leaves suborbicular, 18—40×20—50cm, apex ⅓ bifid, lobes obtuse, base cordate, brown pubescent especially beneath; petioles 5−20cm; stipules oblong, c10×4mm, deciduous. Racemes densely corymbose; pedicels 3.5−4cm bearing a pair of small (c 2mm) bracteoles near apex; bracts lanceolate 6×1.5mm; buds ovoid c 12mm. Calyx smooth, sepals connate, free at tips becoming split into 2 parts at flowering time, becoming reflexed, receptacular tube 1−1.5cm. Petals white, 2−2.5×0.8−1.5cm, densely pubescent on the back. Fertile stamens 3, filaments 2.5−3cm, red. Stigma capitate. Pods oblong, 25−30×6−8cm, woody, brown felted, seeds 8−12, oblong, 2−3.5×1.5−2.5cm.

Sikkim: Duars. 300m. April−June.

Leaves and seeds are used medicinally (13); fibres from the bark are valued for rope-making (126).

6. B. glabrifolia (Bentham) Baker

Similar to *B. vahlii* but less densely brown pubescent; leaves 10−18×7−15cm, apex usually bifid to ⅓, lobes acuminate, base cordate, veins 9, upper surface glabrous, pubescent on veins beneath at first; petioles 5−9cm; flowers in branched corymbs; pedicels c 25mm bearing a pair of bracteoles c 1.5mm near middle; buds subglobose c 5mm, finely brown pubescent; receptacular tube 3−4mm; calyx splitting into 2−3 reflexed segments; petals white, obovate or suborbicular, 12−15×12−15mm, margins undulate; fertile stamens 3, filaments 2.5−3cm; pods oblong 14−35×4−5cm, seeds 5−15, oblong, 21×12mm.

Bhutan: S—Deothang district (N of Deothang). Warm broad-leaved forest, 1425m. June.

7. B. wallichii MacBride; *B. macrostachya* Baker non Bentham

Similar to *B. vahlii* but subglabrous; leaves broadly ovate, 8−12×8−12cm, apex usually ⅕ bifid, lobes acute, base cordate, sparsely pubescent beneath; racemes elongate; receptacular tube 6−8mm bearing ovate obtuse sepals c 2.5mm, appressed greyish brown pubescent; petals obovate, 12−15×7mm, greyish pubescent on the back; fertile stamens 3, filaments c 2cm; pods oblong 10−15×5cm, tomentose, 2−4-seeded.

Darjeeling: Sukna. 300m. April.

8. TAMARINDUS L.

Semi-evergreen trees. Leaves even-pinnate, leaflets entire; stipules minute, caducous. Flowers small in drooping axillary or terminal racemes, buds enclosed at first by deciduous bracts and bracteoles. Sepals 4, reflexed from margin of receptacular cup. Petals 3, with vestiges of other 2. Fertile stamens 3, filaments united in lower half with vestiges of other 6 stamens appearing as points at margin of staminal sheath. Pods oblong, subcylindric, curved, indehiscent, covered by a scurfy brittle brown skin; seeds embedded in sticky pulp.

1. T. indicus L. Dz: *Tengsey*; Nep: *Titiri, Titri* (34); Hindi: *Himli*; Eng: *Tamarind*. Fig. 40 a–c

Tree 5–20m. Leaves 6–15cm, leaflets 10–20 pairs, oblong, 0.8– 2.5 (–3)×0.4–1cm, obtuse, subacute or emarginate, base obliquely rounded, glabrous. Racemes 3–15cm, pedicels 1–1.5cm. Sepals oblanceolate, 10×4mm, reddish outside, yellow within, receptacular cup obconical c 3mm. Petals yellow streaked red, the two lateral ones oblong, c 12×4mm, the upper one oblanceolate, c 12×3mm, concave. Fertile stamens 12–15mm. Pods 6–15×3cm; seeds 5–10, oblong c 2×1cm, compressed.

Bhutan: S—Phuntsholing district (Phuntsholing); **Sikkim**. Cultivated, 200m. June–July.

Planted around towns and villages for its edible fruit which are used in jam and curries; dried or preserved pulp is often exported (16).

9. ADENANTHERA L.

Unarmed deciduous trees. Leaves bipinnate, pinnae evenly paired, odd-pinnate; leaflets alternate; stipules small, deciduous. Flowers numerous in spike-like axillary or terminal racemes or panicles. Calyx campanulate, 5-toothed. Petals 5, connate at base, subequal. Stamens 10, anthers bearing a deciduous apical gland. Pods linear, curved or coiled, leathery, compressed between the suborbicular biconvex seeds.

1. A. pavonina L.

Tree 4–20m. Leaf rachis 15–25(–40)cm, pinnae 3–5 pairs 7–20cm, leaflets 11–19 per pinna, ovate-oblong, 2.5–4×1.5–2cm, obtuse or emarginate, base rounded, sparsely appressed pubescent on both surfaces, pale beneath. Panicle branches 8–15cm. Calyx c1mm, pubescent. Petals yellowish or orange, c 3mm, elliptic. Stamens 4–5mm. Pods 15–20×1.5cm, seeds 10–15, red, glossy 7–8mm diameter.

Sikkim: foothills. 1200m. May–June

The timber is employed in construction work and the wood may be pulverised to yield a red dye (126).

FIG. 40. **Leguminosae.** a–c, *Tamarindus indicus:* a, flowering shoot; b, flower; c, pods, one bisected to show seeds immersed in pulp. d–f *Acrocarpus fraxinifolius:* d, part of leaf; e, part of inflorescence; f, bisected flower. g–i, *Albizia gamblei:* g, flowering shoot; h, flower; i, pod with one valve partly dissected away to show seeds and elongated funicles. j, *Mimosa himalayana:* mature pod, one segment separated from persistent sutures. k, *Cassia alata:* winged pod, Scale: c × ¼; i × ⅓; d, k × ⅖; g × ½; a, e × ⅔; j × ¾; f × 1½; b × 2; h × 2½.

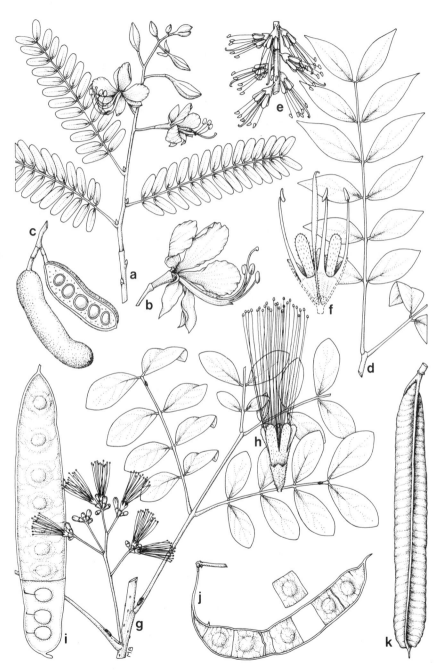

637

10. ENTADA Adanson

Large woody climbers. Leaves evenly bipinnate, rachis usually ending in a branched tendril; stipules small. Flowers small, numerous in crowded axillary spikes, polygamous. Calyx campanulate, obscurely 5-toothed. Petals 5, free, subequal. Stamens 10 free, anthers with a deciduous apical gland. Pods oblong, compressed, woody, sutures thickened, valves transversely jointed, breaking away from sutures into 1-seeded, indehiscent segments.

1. E. rheedii Sprengel; *E. pusaetha* DC., *E. scandens* sensu F.B.I. p.p. non (L.) Bentham. Sha: *Kolokpo*; Med: *Ngingshosha*; Nep: *Pangro, Pangra* (34), *Taktokhajim* (13)

Leaves 12−25cm; pinnae 2 pairs, 7−12cm, each bearing 3−4 pairs of elliptic or narrowly obovate leaflets 3−6×1.5−3cm, obtuse or subacute, base rounded, glabrous, petiolules 2−3mm; stipules subulate, 4−7mm. Spikes 10−25cm, flowers mostly male, bisexual flowers usually in upper parts of spike, rachis brownish tomentose. Calyx c1mm, pubescent. Petals yellowish, elliptic, 2−3×0.75mm. Stamens 4−5mm. Pods oblong, somewhat curved, 20−60(−75)×8−10cm, segments 4−15, ± rectangular, 4−6cm long; seeds suborbicular or ovoid, 4−5cm diameter, c 1.75cm thick, brown.

Bhutan: S—Phuntsholing and Deothang districts, **C**—Mongar and Tashigang districts; **Sikkim.** Warm and Subtropical forests, 600−1500m. March−June.

The E. Himalayan plants belong to subsp. **sinohimalensis** (Grierson & Long) Panigrahi (*E. "entity C"* Brenan, *E. laotica* Gagnepain, *E. pusaetha* DC. subsp. *sinohimalensis* Grierson & Long).

The seeds and bark are used medicinally (13).

11. MIMOSA L.

Herbs or shrubs, scrambling or prostrate, armed with recurved spines. Leaves evenly bipinnate or pinnae digitately arranged, often sensitive to contact, leaflets numerous; stipules persistent. Flowers polygamous, sessile, forming globose heads (lower flowers of heads functionally male, upper one or more female), axillary, few or forming terminal racemiform inflorescences. Calyx tubular, campanulate, minutely toothed. Petals 4, connate in lower half. Stamens 4 or 8, long exserted. Pods oblong, compressed, papery, dividing transversely into 1-seeded segments which separate at maturity from the persistent sutures.

1. M. himalayana Gamble; *M. rubicaulis* sensu F.B.I. p.p., *M. rubicaulis* Lamarck subsp. *himalayana* (Gamble) Ohashi. Nep: *Arere Khanra, Arare* (34). Fig. 40j

Scrambling shrub 4−5m. Leaves evenly bipinnate, 12−25cm, pinnae 6−9 pairs, 2.5−6cm, leaflets 16−20 pairs, oblong, asymmetric, 5−12×1.5−3.5mm, acute, base rounded, sparsely appressed pubescent beneath, midrib closer to upper margin and without lateral veins on that side; stipules subulate, 5−8mm. Flower heads pink, c1.5cm diameter, in axillary clusters of 3−4 on peduncles 2−4cm forming a terminal leafy racemiform inflorescence. Calyx c0.5mm. Petals oblong c 3mm. Stamens 8, 6−8mm. Pods 1−4 from each head, 7−11×1−1.5cm, fragmenting into squarish segments; seeds ovoid, compressed, 5−6mm, brown.

Bhutan: S—Samchi, Phuntsholing, Sarbhang, Gaylegphug and Deothang districts; **Sikkim.** Subtropical forest, 200−750m. May−October

M. rubicaulis Lamarck, a S. Indian species, has been erroneously recorded from Bhutan. It has 4−6 pairs of pinnae and leaflets with lateral veins on both sides of midrib.

2. M. pudica L. Nep: *Booarey, Bohari Jhar* (34), *Lajunia* (34). Fig. 41 a−c

Similar to *M. himalayana* but an annual or perennial subshrubby herb; stems densely hispid and usually with fewer spines, sometimes prostrate; pinnae of leaves 4, arranged digitately on petioles 3−6cm, bearing long bristles; leaflets 20−26 pairs, oblong, 10−13×2−3mm, acute, base rounded, margin ciliate with bristles, lateral veins along both sides of midrib; stipules lanceolate, 9−14mm, bristly ciliate; flower heads bearing narrowly triangular bracteoles 1.8−2.2mm, longer than buds, margins bearing a few bristles; flowers pink or bluish-purple; stamens 4; pods 4−12 developing from each head, 20×3−4mm, fragmenting into 2−5 rounded segments, sutures bearing long (c 3mm) bristles.

Bhutan: S—Samchi, Phuntsholing and Deothang districts; **Sikkim.** Roadsides, 200−350m. May.

The above description refers to var. **hispida** Brenan to which all the Himalayan material seen hitherto belongs.

12. LEUCAENA Bentham

Unarmed shrubs or small trees. Leaves even-pinnate, rachis bearing a gland near lowest and near uppermost pinnae; stipules minute, caducous. Flowers small, numerous in globose heads on axillary peduncles, or terminal and racemosely arranged. Calyx campanulate, 5-toothed. Petals 5, free. Stamens 10, free, exserted. Pods linear-oblong, compressed, membranous or thinly coriaceous, dehiscent.

1. L. leucocephala (Lamarck) De Wit; *L. glauca* Bentham non L.

Shrub or tree 0.5–12m. Leaf rachises 15–20cm, pinnae 3–8 pairs, 3–9cm, leaflets 7–20 pairs, obliquely oblong-ovate, 6–15×2–5mm, acute, base cuneate, sparsely ciliate otherwise glabrous. Heads c1.5cm diameter, creamy white, on peduncles 2–5cm. Calyx c 2mm. Petals oblong, 4×0.75mm, pubescent. Stamens 5–6mm. Pods 10–20×1.5–2.5cm, borne on stalks up to 3cm, glabrous; seeds 15–25, ovate, c 8×4–5mm, brown, glossy, ± transversely arranged.

Bhutan: S—Samchi district (Samchi). Cultivated, 500m. March.

Native of Tropical America, cultivated as a forage plant for cattle and goats but not for horses, donkeys and pigs which are poisoned by it. Timber is valuable for firewood (130).

13. ACACIA Miller

Trees or shrubs, sometimes climbers, shoots unarmed or bearing prickles or stipular spines. Leaves usually evenly bipinnate, rarely reduced to a leaf-like petiole (phyllode), leaflets usually numerous, petioles and rachises often glanduliferous; stipules, if not spiny, small and deciduous. Flowers bisexual or polygamous, ± sessile in spikes or globose heads, solitary, paired or fascicled in leaf axils or in axillary or terminal racemes or panicles. Calyx campanulate, 4–5-lobed. Petals as many as calyx lobes, connate in lower half. Stamens numerous (more than 20), exserted, free or connate at base. Pods oblong, compressed or terete, membranous, leathery or woody, straight or coiled; seeds suborbicular or ellipsoid, funicles often twisted.

1. Flowers in globose heads 2
+ Flowers in spikes 7

2. Leaves reduced to simple elliptic petioles (phyllodes)
 .... **7. A. melanoxylon**
+ Leaves bipinnate 3

3. Leaves with glands at least on petioles 4
+ Leaves without glands on petioles or rachises 6

4. Branches unarmed .... **8. A. decurrens**
+ Branches bearing scattered prickles 5

5. Leaves with 12–26 pairs of pinnae .... **1. A. pennata**
+ Leaves with 4–5 pairs of pinnae .... **2. A. rugata**

6. Branches with paired stipular spines .... **3. A. farnesiana**
+ Branches bearing scattered prickles .... **4. A. gageana**

7. Leaves with 20−25 pairs of pinnae .... **5. A. catechu**
+ Leaves with 3−4 pairs of pinnae .... **6. A. lenticularis**

1. A. pennata (L.) Willdenow. Nep: *Arare* (34)
Climber or tree 3−8m, shoots bearing scattered recurved prickles. Leaves bipinnate, 15−20cm, pinnae 12−26 pairs, 3−7(−10)cm; leaflets 60−70 pairs, linear-oblong 3−7×0.7−1.5mm, obtuse or subacute, base truncate, asymmetric, attached very obliquely, finely ciliate; petioles bearing an oval gland 4−6mm and rachis usually bearing 1−3 smaller glands at base of upper pinnae. Flower heads globose, c1cm diameter, forming terminal panicles. Calyx tube 2−2.5mm, 5-toothed. Petals oblong 2.5−3mm. Stamens 4−5mm, yellow. Pods leathery, 10−16×2.5−3.5cm, produced from upper flower in head; seeds 6−12, oblong c 1.2×6mm, funicle twisted.
Bhutan: S—Phuntsholing, Chukka and Sarbhang districts; **Sikkim.** Subtropical forests, 900−1000m. June−July.
Stems and fruit used to poison fish (146).

2. A. rugata (Lamarck) Voigt; *A. concinna* (Willdenow) DC. Nep: *Arare Khanra, Arare* (34)
Similar to *A. pennata* but pinnae of leaves 4−5 pairs, 3−8cm, leaflets 12−16 pairs, 8−15×2.5−4.5mm, obtuse, mucronate, ± glabrous; flower heads in leafy racemes; sepals and petals subequal, 2.5−3mm, sepals purplish; stamens whitish 4−5mm; pods 8−12×1.7−2.5cm, sinuate along both sutures, fleshy but becoming very wrinkled when dry, 5−10-seeded.
Bhutan: S—Sarbhang district (Phipsoo Khola); **Sikkim.** Subtropical forests, 300m. March.

3. A. farnesiana (L.) Willdenow
Shrub or tree up to 7m, branches bearing whitish stipular spines 0.5−2cm. Leaves bipinnate, 5−8cm, rachis with a minute circular gland on petiole and sometimes with another near the apex of the rachis; pinnae 4−6 pairs, 1.3−2.5cm; leaflets oblong, 12−20 pairs, 3.5−5.5×1mm, subacute, base rounded or truncate, slightly oblique, ± glabrous. Flower heads 1.3−1.5cm diameter, in axillary clusters, peduncles 2−3cm. Calyx c2mm, shallowly lobed. Petals 2.5−3mm. Stamens 4−5mm, orange yellow. Pods 5−8×1−1.5cm, terete or subcompressed, woody, seeds 10−16, ellipsoid, 7×5mm, in 2 ranks embedded in pith.
Bhutan: S—Samchi district (Dorokha); **Sikkim.** 700m. December−March.
It can be used as a host for the lac insect and the bark possesses tannin (16).

76. LEGUMINOSAE

4. A. gageana Craib. Nep: *Arari Khanra, Arare* (34)

Similar to *A. farnesiana* but stems bearing small scattered prickles; leaves 10–20cm, pinnae 9–11 pairs, 4–8cm, leaflets 15–30 pairs, 6–10×2–2.75mm, obliquely acute, base symmetrically truncate, finely ciliate; flower heads c1cm diameter in terminal racemes; calyx c1mm; petals c 2mm; stamens c 4mm, creamy white; pods 10–15×2cm, compressed, thinly woody, 6–8-seeded.

Bhutan: S—Phuntsholing district, **C**—Mongar district; **Sikkim.** Warm and subtropical forests, 200–1100m. July.

Records of *A. intsia* (L.) Willdenow (*A. caesia* (L.) Wight & Arnott) probably refer to this species.

5. A. catechu (L.f.) Willdenow; *A. catechuoides* (Roxb.) Bentham. Nep: *Khair*

Tree 2–7m, shoots unarmed or with paired stipular spines. Leaves bipinnate, 15–20cm, rachis with a large (2–3mm) rounded gland at base of lowest pair of pinnae and 2–4 smaller glands at base of upper ones, pinnae 20–25 pairs, 3–5cm, leaflets linear-oblong, crowded, (20–)30–45 pairs, 3–6×0.5–0.7(–1)mm, acute or subacute, base rounded or truncate, ± glabrous. Spikes solitary, axillary, narrowly cylindrical, 8–10cm, densely flowered. Calyx c1.5mm, shortly toothed. Petals oblong 2.5–3mm. Stamens c 4mm, yellow. Pods oblong, 5–7×1–1.5(–2)cm, compressed, leathery; seeds 6–8, suborbicular, c 5mm diameter.

Bhutan: S—Samchi, Phuntsholing, Sarbhang and Deothang districts; **Sikkim.** Subtropical forests by river banks, 230–550m. May–June.

The timber is durable and a crystalline extract of it, *cutch* or *catechu,* is used for chewing with betel nut and *Piper betle* leaves (126).

6. A. lenticularis Bentham. Nep: *Kakur*

Similar to *A. catechu* but leaves with 3–4 pairs of pinnae, rachis bearing a rounded gland at base of each pair; pinnae 5–12cm, leaflets 5–10 pairs, obovate 1.5–3×1–1.5cm, obtuse or emarginate, base rounded, glabrous, glaucous; spikes 6–7cm; calyx c 3mm; petals 3.5–4mm; stamens 5–6mm; pods woody, thickened along both sutures, 8–18×2.5–3cm, 3–8-seeded.

Sikkim: Darjeeling terai. April.

7. A. melanoxylon R. Brown

Tree 5–15m. Leaves reduced to elliptic flattened petioles (phyllodes) 8–12×0.5–2.5cm, leathery, glabrous, usually curved. Flower heads globose, c 0.75cm diameter, 2–5 in short axillary or terminal racemes. Calyx c1mm. Petals c 2mm. Stamens c3mm. Pods 5–10×0.5–0.7cm, leathery, ± compressed, curved or coiled; seeds 5–10, ellipsoid, 5×3mm, black, glossy.

Sikkim: Darjeeling. Cultivated.

Native of S Australia, planted for its valuable dark brown timber (16, 48).

8. A. decurrens Willdenow

Tree c 10m, branches prominently angled, ± pubescent. Leaves bipinnate 10−15cm, pinnae 8−15 pairs, leaflets linear, numerous, 30−40 pairs, 2−12×0.75−1.25mm, subacute, glabrous or finely pubescent, rachis bearing a gland at base of each pair of pinnae. Flower heads globose, c 5mm diameter, in axillary racemes or terminal panicles. Sepals c0.5mm. Petals c 1mm. Stamens c 2.5mm. Pods linear-oblong, 8−12×0.7cm, much compressed and narrowed between the 7−12 ellipsoid seeds, each 4×3mm, black.

Sikkim: Darjeeling. Cultivated, 2100m.

Native of S Australia, grown for its timber and for its bark which is used for tanning (48). This record is based on a single sterile specimen. The above description is drawn up from Australian material of *A. decurrens* but the Sikkim plant may prove to be a related species.

14. ALBIZIA Durazzini

Unarmed shrubs or trees, rarely climbers. Leaves evenly bipinnate, rachises usually with oval or circular glands; stipules usually small (but large in *A. chinensis*), caducous or deciduous. Flowers small in globose heads, in fascicles or forming axillary or terminal panicles. Calyx tubular-campanulate, obscurely toothed. Petals 5, united at least to middle into a tubular corolla. Stamens numerous (more than 15), exserted, filaments united below into a tube as long as or shorter than corolla. Central flower of head (terminating peduncle) usually different from others in being slightly larger, staminal tube exserted from corolla, filaments more intensely coloured; apparently such flowers never form pods. Pods coriaceous, sometimes thinly so, oblong, compressed, continuous within; seeds ovoid, oblong or suborbicular, funicles usually elongated and contorted.

1. Leaflets medium-sized to large, 2−15×0.5−7cm 2
+ Leaflets small, 4−20×1−7mm 5

2. Flower heads solitary or in axillary fascicles; peduncles simple, rarely forming racemes .... **1. A. lebbeck**
+ Flower heads in axillary or terminal panicles; peduncles much branched 3

3. Leaflets oblong, 5−10mm broad .... **2. A. odoratissima**
+ Leaflets elliptic or obovate, 1.3−7cm broad 4

4. Pinnae 1(−3) pairs; leaflets (1−)2(−7) pairs, 7−15×3−7cm
   3. A. lucidior
+ Pinnae 2−6 pairs; leaflets 2−12 pairs, 2−9×1.3−3.5cm ..... **Species 4 & 5**

5. Climber; shoots with prominent leaf scars; leaflets very numerous and small 4−8×1−1.5mm .... **6. A. myriophylla**
+ Trees or shrubs; shoots without prominent leaf scars; leaflets less numerous and larger, 5−20×1.5−6mm .... **Species 7−9**

1. A. lebbeck (L.) Bentham
Tree 10−30m. Leaf rachises with an oval or circular gland near base and apex, 8−15(−27)cm, pinnae 2−4 pairs, 7−17cm, leaflets 4−10 pairs, oblong or obovate, 2−6×0.9−3cm, rounded at apex and base, asymmetric, midrib closer to upper margin, glabrous or sparsely pubescent beneath, pale green, ± glaucous. Heads 30−40-flowered, solitary, axillary or several in supra-axillary lines rarely forming terminal racemes, peduncles 6−10cm, simple. Calyx c 3mm, pubescent. Corolla 7−8mm, toothed to middle. Stamens greenish-white, filaments 2−3cm. Pods coriaceous, 12−30× 3−4.5cm, seeds 5−15, suborbicular 9×7mm.
Bhutan: C—Tashigang district (Dangme Chu); **Sikkim**. River banks, 300−900m. April−June.
Non-fruiting specimens are closely similar to those of *Samanea saman* (q.v.). The wood is durable and polishes well (126). It is also valued for firewood (130).

2. A. odoratissima (L.f.) Bentham. Nep: *Karkur Siris* (34)
Similar to *A. lebbeck* but a tree 12−25m; leaves 10−20cm, pinnae 3−5 pairs, 7−15cm, leaflets oblong, 6−18 pairs, 1−3×0.5−1.3cm, acute or obtuse, base rounded, glabrous above, appressed pubescent beneath, leaf rachis with a gland near base and apex; panicles much branched, heads c 12-flowered; calyx c 1.5mm; corolla tubular c7mm; stamens white, filaments c 2cm; pods thinly coriaceous, 15−25×2.5−3cm, seeds 6−12, ovoid, compressed, 7−15×5−7mm.
Bhutan: C—Chukka district (Giengo), **C**−Tashigang district (Dangme Chu); **Sikkim**. Dry hillsides and banks of streams, 900−1200m. May−July.
The wood is used for making furniture (16).

3. A. lucidior (Steudel) Hara; *A. lucida* (Roxb.) Bentham *nom. illeg.* Nep: *Kalo Siris, Parke Siris, Rato Siris, Tata* (34), *Portka Siris* (34)
Tree 7−15m. Leaf rachises 3−7cm, with an elliptic gland near base and a circular one near apex, pinnae 1(−3) pairs, 1−5cm, leaflets (1−)2(−7) pairs, ovate-elliptic, 7−15×3−7cm, acute or acuminate, base cuneate or rounded, ± glabrous, upper surface ± glossy, reticulate. Panicles terminal, ± corymbose, heads 7−12-flowered. Calyx 2−3mm, whitish puberulous. Corolla 5−6mm, pubescent. Stamens yellow or cream-coloured, filaments c 2cm. Pods coriaceous 15−20×3−3.5cm, seeds 6−8, suborbicular, c 1cm diameter.

Bhutan: S—Samchi district (Dwarapani and Khana Bharti Khola); **W Bengal Duars:** Buxa; **Sikkim.** Hot dry slopes by rivers, 300−750m. April−September.

Non-fruiting specimens are closely similar to those of *Archidendron monadelphum* (q.v.).

4. A. gamblei Prain. Nep: *Patpate Siris, Harra Siris* (34). Fig. 40 g−i

Similar to *A. lucidior* but leaf rachises 6−15cm with a gland near base, pinnae 2−3 pairs, 4−12cm, leaflets 2−5 pairs, obovate-elliptic 3−12×1.5−4cm, acute or acuminate, base cuneate, glabrous or minutely pubescent on both surfaces; panicles 7−10cm, corymbose, ultimate peduncles 2−3cm, clustered, heads 10−12-flowered; calyx c 3mm, pubescent; corolla whitish, 6−7mm, pubescent; stamens creamy-white, filaments c 2cm; pods thinly coriaceous, 18×3cm, 6−8-seeded.

Bhutan: S—Samchi and Phuntsholing districts, **C**—Mongar and Tashigang districts; **Sikkim.** Warm and subtropical forest slopes, 200−1100m. June−August.

5. A. procera (Roxb.) Bentham. Nep: *Seto Siris* (34)

Similar to *A. lucidior* but leaf rachises 25−30cm with a large (5−10mm) oval gland near base and usually with one or more smaller ones at base of upper pinnae; pinnae 3−6 pairs, 10−20cm, leaflets ovate-elliptic, 4−12 pairs, 2−5.5×1.5−3cm, obtuse or emarginate, base rounded, appressed pubescent on both surfaces, dark green above, pale beneath; panicles elongated, up to 30cm, much branched, heads 12−20-flowered; calyx c 3mm, glabrous; corolla c6mm, teeth pubescent; stamens white, filaments c 10mm; pods thinly coriaceous, 12−18×2−2.5cm, 5−10-seeded.

Bhutan: S—Samchi district (Karpaga Hill), **C**—Mongar district (Lhuntse); **Sikkim.** River banks, 400−1200m. August.

6. A. myriophylla Bentham

Climbing shrub, shoots with leaf scars becoming enlarged and hooked. Leaf rachises 10−15cm with a prominent rounded gland 2−3mm near base, pinnae 8−16(−20) pairs, 2.5−7cm, leaflets linear, 25−50(−60) pairs, 4−8×1−1.5mm, obtuse or subacute, base truncate, obliquely attached to pinna. Panicles 6−10cm, heads 10−12-flowered. Calyx minute, c 0.5mm. Corolla 3−3.5mm. Stamens 8−10mm. Pods coriaceous, 10−15×2.5−3cm, seeds 12−15, oblong 8×14mm.

Sikkim. 900m. April−May.

7. A. julibrissin Durazzini

Tree 3−10m, branches pubescent. Leaf rachises 10−30cm, with an oval gland c 2mm near base and small ones at base of upper or all pinnae; pinnae 4−8 pairs, 5−15cm, leaflets 10−18(−25) pairs, oblong, 1−2×0.5−0.7cm, acute, base truncate, obliquely attached to pinna, asymmetric, midrib closer to upper margin, appressed pubescent on both surfaces. Heads 15–20-flowered, solitary or 2–4 axillary on simple peduncles 5–8cm, sometimes in terminal panicles. Calyx 3–3.5cm, pubescent. Corolla 9–10mm, pubescent. Filaments white at base, pink above, 3–3.5cm. Pods thinly coriaceous, 10–15×1.5–2.5cm; seeds 6–12, ovoid, 5×3mm.

Bhutan: C—Punakha and Tongsa districts, **N**—Upper Mo Chu district; **Sikkim.** Dry rocky slopes and river banks, 1500–2300m. April—June.

The above description refers to var. **mollis** (Wall.) Bentham which differs from the typical variety by being pubescent throughout and by its larger leaflets.

8. A. chinensis (Osbeck) Merrill; *A. stipulata* (Roxb.) Boivin. Nep: *Rato Siris* (34), *Kalo Siris* (34)

Similar to *A. julibrissin* but tree to 40m; leaf rachises 15–20cm with an oval ± prominent gland 2–3mm near base, pinnae (4–)7–12(–14) pairs, 4–10cm, leaflets 18–35 pairs, oblong-lanceolate, 6–8×1.5–2mm, acute, very asymmetric, midrib almost forming upper margin, base obliquely truncate, glabrous above, pubescent beneath, stipules ovate, 2–3× 1.5–2cm acuminate, base cordate, deciduous; panicles 15–20cm, heads 15–25-flowered; calyx 3.5–4mm, pubescent; corolla 7–8mm, pubescent; filaments 2.5–3cm, greenish or yellowish white; pods thinly coriaceous, 10–18×2–3.5cm, 9–12-seeded.

Bhutan: S—Samchi, Phuntsholing and Gaylegphug districts, **C**—Mongar and Tashigang districts; **Sikkim.** River banks, 450–1500m. April–July.

Used as a shade tree in tea plantations; its branches are also lopped for cattle fodder (48).

9. A. sherriffii E. G. Baker. Dz: *Lam Shing;* Nep: *Lahari Siris, Lebbek Siris*

Similar to *A. julibrissin* but tree 10–25m; leaf rachises 10–30cm with an oval gland c 3mm near base and another smaller one near apex; pinnae 9–18 pairs, 6–10cm; leaflets 16–27 pairs oblong 5–9(–12)×2–3(–4)mm, apex acute, upwardly curved, base truncate, sessile, midrib closer to upper margin, dark green glabrous above, pale appressed pubescent beneath; heads c 30-flowered, on simple peduncles 6–8cm in axillary clusters, sometimes in terminal panicles; calyx 5–6mm, brownish tomentose; corolla 10–12mm, whitish pubescent; filaments 3–3.5cm, creamy white; pods thinly coriaceous, 12–15×1.5–2cm, 15–22-seeded.

Bhutan: S—Chukka, Sarbhang and Gaylegphug districts, **C**—Tongsa and Mongar districts; **Arunachal Pradesh:** Nyam Jang Chu. Warm and Subtropical forests, 1000–2000m. April–June.

15. SAMANEA (DC.) Merrill

Large evergreen or semideciduous unarmed trees. Leaves evenly bipinnate, rachis and pinnae glanduliferous, leaflets opposite. Flowers small, numerous in globose heads ± clustered in terminal corymbs. Calyx campanulate, 5-toothed. Petals 5, connate to above middle forming a tubular corolla. Stamens many, exserted, filaments connate into a tube at base. Pods linear-oblong, straight or slightly curved, ± fleshy, septate between seeds, sutures thick, indehiscent.

1. **S. saman** (Jacquin) Merrill; *Pithecolobium saman* (Jacquin) Bentham, *Enterolobium saman* (Jacquin) Prain. Eng.: *Rain Tree* (34)

Tree 20(−50)m. Leaf rachises 15−25(−40)cm with a gland near insertion of each pair of pinnae; pinnae 2−6 pairs each with 3−9 pairs of leaflets with a small gland near insertion of each pair; leaflets rhombic-oblong, asymmetric, increasing in size from base to apex of pinna, 1.5−5×1−3.5cm, acute or obtuse, base cuneate, glabrous above, softly pubescent beneath. Heads c 20-flowered on peduncles 6−7cm. Calyx 6−8mm, tomentose. Corolla 12−15mm. Filaments 3−4cm, pink. Pods 10−20×1.5−2cm, 7−10mm thick, brown or blackish; seeds c 20, ovoid, c 8×5mm, brown, central part of each side paler and raised.

Bhutan: S—Phuntsholing district (near Phuntsholing); **Sikkim.** Cultivated, 250m. March−June.

Native of Tropical Central America planted as a wayside tree. The pods are edible and are relished by cattle and pigs. Non-fruiting specimens are very similar to those of *Albizia lebbeck* which differs in having less glandular leaf rachises and pinnae, almost glabrous or thinly pubescent leaflets, calyx and corolla shorter and filaments white or greenish yellow. It owes its English name to the fact that the leaflets fold together in a rain storm (16).

16. CALLIANDRA Bentham

Shrubs or trees. Leaves evenly bipinnate, pinnae 1 pair, leaflets several pairs increasing in size towards apex of pinnae; stipules persistent. Flowers polygamous, numerous in globose terminal or axillary heads. Calyx campanulate, 5-toothed. Petals 5, united into a tubular or campanulate corolla. Stamens numerous, filaments much exserted, united at base. Pods oblong-elliptic, narrowed at base, compressed, non-septate within, thickened along both sutures, 6−8-seeded.

1. **C. haematocephala** Hasskarl

Shrub c 2m. Petioles 10−15mm, pinnae 5−7cm, leaflets 4−5 pairs, ovate or obovate, often asymmetric, 2−5.5×1−2.5cm, obtuse or subacute, mucronate, base rounded or subcordate, sessile, ± glabrous above, pubescent beneath; stipules ovate-lanceolate, 10×4mm, striate. Flower heads c 7cm diameter, peduncles 3−4cm. Calyx 4−5mm, shallowly toothed, striate. Corolla tube 10mm, teeth 3mm, white. Stamens c 40, white, staminal tube c 7mm with a ring of short appendages at apex internally, filaments c 3cm. Pods (immature) 10−12×1−1.5cm.

Bhutan: S—Phuntsholing district (Phuntsholing). Cultivated, 250m. February.

Native of tropical S America; typically this has scarlet flowers, apparently the white form only is grown in Bhutan as an ornamental. We are grateful to Dr S.A. Renvoize of Royal Botanic Gardens, Kew for confirming the identity of this specimen.

17. PITHECELLOBIUM Martius

Trees or shrubs, shoots armed with stipular spines. Leaves evenly bipinnate; petioles bearing a minute gland between the one pair of pinnae; leaflets 1 pair with a gland between them. Flowers in globose heads, uniform, heads arranged in axillary or terminal racemes or panicles. Calyx campanulate, 5-toothed. Petals 5, united into a short tubular corolla. Stamens numerous (c50), filaments united below into a tube. Pods coiled, ± compressed, thinly coriaceous; seeds covered with a fleshy aril.

1. P. dulce (Roxb.) Bentham

Shrub or tree 3−15m; stipular spines 5−10mm. Petioles 2−3.5cm; rachises of pinnae 5−10mm; leaflets asymmetrically elliptic, 1.5−3.5 (−5)×0.7−1.5(−2)cm, obtuse, base rounded, glabrous. Flower heads c1cm diameter. Calyx 1−1.5mm, puberulous. Corolla 3−4mm, minutely pubescent. Stamens 8−10mm. Pods 10−12(−15)×1−1.5cm; seeds 4−10, ellipsoid, 10−12×7−8mm, 2−3mm thick, black, covered with a white or reddish aril.

Bhutan: S—Phuntsholing district (Phuntsholing). Cultivated, 200m. November−March.

Native of Tropical S America; planted as a wayside tree. The arils are edible.

18. ARCHIDENDRON F. v. Mueller

Unarmed trees. Leaves evenly bipinnate, rachis and pinnae glanduliferous, leaflets opposite. Flowers uniform in small subumbellate

FIG. 41. **Leguminosae.** a−c, *Mimosa pudica:* a, portion of flowering shoot; b, portion of leaf; c, fruiting head. d, *Archidendron monadelphum:* mature pod. e−j, *Parochetus communis:* e, portion of flowering shoot; f, standard; g, wing petals; h, keel; i, calyx opened and partly spread to show ovary and style; j, diadelphous staminal tube with posterior stamen free. k & l, *Cladrastis sinensis:* k, flower from above; l, calyx spread open and petals removed to show ovary surrounded by the almost completely free stamens. m, *Apios carnea:* inflorescence. n−r, *Phaseolus vulgaris:* n, flower; o, standard; p, wing; q, keel; r, ovary with style coiled terminally. s−u, *Erythrina arborescens:* s, portion of flowering shoot; t, flower; u, flower with standard removed to show keel and wing. Scale: s × ¼; d, t, u × ⅔; e × ¾; m × ⅘; a, f, g, h, i, j, k × 1½; c, n, o, p × 2; b, l × 2½; q, r × 3.

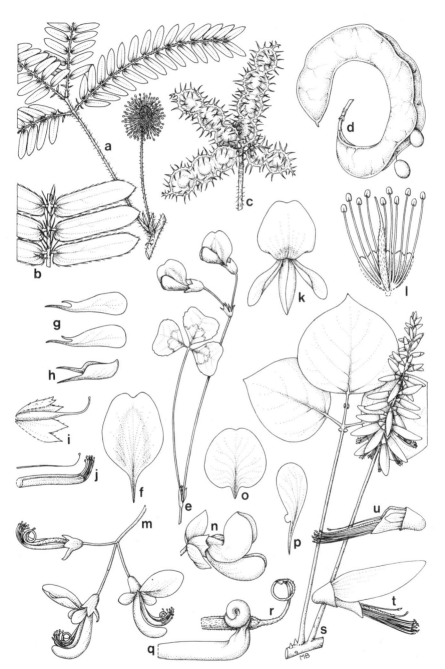

heads, arranged in axillary or terminal panicles. Calyx campanulate, 4−5-toothed. Petals 4−5 united into a tubular corolla. Stamens numerous, filaments united into a tube at base. Pods thinly coriaceous, ± compressed, coiled into a flat spiral; seeds ellipsoid, black, exarillate.

1. A. monadelphum (Roxb.) Nielsen; *Pithecellobium monadelphum* (Roxb.) Kostermans, *Pithecolobium bigeminum* sensu F.B.I. p.p., *Abarema monadelpha* (Roxb.) Kostermans. Nep: *Kalo Siris, Tikpi-Kung* (34). Fig. 41d

Tree c 10m, shoots rounded. Leaf rachises 1−3cm with a ± prominent cup-shaped gland near base, pinnae 1−2 pairs, 5−15cm bearing a gland at each node; leaflets 3 pairs, ± symmetrical, ovate-elliptic or obovate, 4−16×2.5−6cm, acute or acuminate, base cuneate, ± glabrous. Heads c 10-flowered. Calyx 2−3mm, brownish pubescent. Corolla white, 5−6mm. Filaments white, c 1.5cm. Pods 15−20×2cm, dehiscing along ventral suture, valves reddish within; seeds 6−8, black, 1.3×1cm, 0.5cm thick.

Bhutan: S—Chukka district (Marichong) and Gaylegphug district (Sher Camp); **Sikkim**. Subtropical forest, 850−1200m. March−May.

2. A. clypearia (Jack) Neilsen; *Pithecellobium clypearia* (Jack) Bentham, *P. heterophyllum* (Roxb.) MacBride, *Pithecolobium angulatum* Bentham, *Abarema clypearia* (Jack) Kostermans

Similar to *A. monadelphum* but shoots angular; leaf rachises 15−27cm bearing a stalked gland near base and at each node; pinnae 3−6(−10) pairs, proximal ones 2−5cm, distal ones 15−22cm; leaflets 3−10(−18) pairs, asymmetrically elliptic or trapeziform, 1.5−6×0.7−2cm, acute or acuminate, base cuneate, obliquely attached to pinna near lower margin, ± pubescent, pinna bearing a cup-shaped gland at each node; calyx c 3mm, pubescent; corolla 7−8mm; stamens c 1.5cm; pods up to 20−1cm, red or yellow outside, red within, seeds ellipsoid, 7−10×6−9mm, 6−7mm thick.

Sikkim: Rongbe and Endrango. Lower hill forests, 600−1200m. June.

19. ORMOSIA Jackson

Evergreen trees. Leaves odd-pinnate, leaflets coriaceous, entire; stipels absent; stipules minute. Racemes axillary or terminal. Calyx campanulate, bracteolate at base, teeth 5, upper 2 ± connate. Petals clawed, keel petals free. Stamens free. Pods ellipsoid, thick, dehiscent; seeds few.

1. O. glauca Wall.; *Federovia glauca* (Wall.) Yakovlev

Leaves 18−25cm, leaflets 7−9, elliptic, 6−11×1.5−5cm, bluntly acuminate, base cuneate, glaucous and minutely pubescent beneath. Racemes 9−12cm. Calyx tube 5−7mm, shortly brown pubescent, teeth ovate c 5mm. Petals white, c 1.8cm, standard suborbicular, wing and keel petals subequal, oblong. Pods 5−8×3cm, blackish, rugose, ribbed along each side of upper suture; seeds 2−4 red, 8−10×5−6mm, embedded in pithy lining of valves.

Sikkim: Sivoke Hills. 750m. May.

20. CLADRASTIS Rafinesque

Trees. Leaves deciduous, odd-pinnate, winter buds enclosed within inflated petiole bases, leaflets entire, stipels and stipules absent. Flowers in terminal panicles, bracts deciduous. Calyx campanulate, teeth short and broad. Petals clawed, standard recurved, wings and keel petals subequal and free. Stamens 10, ± free, shortly coherent at base, alternately long and short. Pods compressed, shortly stalked, membranous, tardily dehiscent, 4−6-seeded.

1. C. sinensis Hemsley. Fig. 41 k&l

Tree 3−12m. Leaves 15−30cm, leaflets 11−15, ovate-elliptic, 4.5−9×2−3.5cm, acute or acuminate, base cuneate or rounded, brown pubescent beneath especially on midrib, petiolules 3−5mm. Panicles 25−30cm. Calyx tube 5mm, brown pubescent, teeth ovate, 1.5−2mm. Petals white, claws 3−4mm, standard elliptic, 9−10×6−7mm, wings and keel petals roundly hastate, 8−10×4−5mm. Pods linear-oblong 7−8×1cm.

Bhutan: S—Chukka district (Bunakha), **C**—Punakha district (Ritang). In dense forest, 1800−2550m. July.

21. SOPHORA L.

Shrubs or trees. Leaves odd-pinnate, leaflets in part at least alternately arranged on rachis, entire; stipels absent; stipules subulate. Flowers in axillary or terminal racemes, bracts linear. Calyx campanulate, truncate or teeth 5, upper 2 connate. Standard narrowed in lower half, not clawed, wings and keel petals clawed. Stamens ± free, connate only at base. Pods cylindric, constricted between the few rounded seeds, indehiscent or tardily dehiscent.

1. S. wightii Baker subsp. **bhutanica** (Ohashi) Grierson & Long; *S. acuminata* sensu F.B.I. non Desveaux, *S. benthamii* van Steenis, *S. bhutanica* Ohashi. Fig. 43i

Shrub or tree 1−8m. Leaves 15−25cm, leaflets 11−15(−21), ovate-elliptic 4−9×1−2cm, acuminate, base cuneate, glabrous above, sparsely brownish pubescent beneath; stipules subulate, 6−10mm.

76. LEGUMINOSAE

Racemes 20–27cm, pedicels 8–10mm. Calyx oblique, pouched at base, 10–12mm, truncate, brown pubescent. Petals white, standard narrowly obovate, 16–17×5–6mm, recurved near apex, emarginate, wings with claw c 5mm, blade ovate 9×4mm, keel petals with claw c 7mm, blade oblong 7×4mm. Pods 7–12cm, 2–4-seeded, shortly pubescent. Seeds ellipsoid, 10–12×7–8mm, red.

Bhutan: C—Punakha district (between Rinchu and Mishichen). Warm broad-leaved forests, 1450–1600m. May–June.

2. S. velutina Lindley; *S. glauca* DC. non Salisbury

Similar to *S. wightii* subsp. *bhutanica* but shrub 1–2m; leaves 12–15cm, leaflets 15–17, elliptic, 2–5.5×0.7–1.5cm, acute, mucronate to a subulate point, base cuneate, densely brown pubescent beneath; stipules subulate 4–5mm; racemes 7–13cm; calyx tube 6–7mm, brown pubescent, teeth 2–3mm, acute; standard purplish, obovate, 12–13×5–6mm, wings and keel white, 12–13×3mm; pods 10–12cm, 3–6-seeded; seeds ellipsoid, 7×3.5–5mm, brown.

Bhutan: C—Mongar district (Lhuntse and near Mongar). Open dry hillsides, 1200–1750m. April–June.

22. DALBERGIA L.f.

Trees or shrubs sometimes twining. Leaves odd pinnate, leaflets entire, usually alternately arranged on rachis; stipels absent; stipules often deciduous. Flowers small, in axillary or terminal racemes or panicles. Calyx campanulate, teeth 5, upper 2 often connate. Petals clawed. Stamens 9 or 10, monadelphous, filaments connate into a sheath split along the upper side or diadelphous and united into 2 groups of 5. Pods compressed, ± coriaceous, conspicuously veined over the 1 or few seeds, indehiscent.

1. Leaves with few (3–9), large (3–13×2.5–6cm) leaflets ...... **Species 1–3**
+ Leaves with numerous (9–43), smaller (1–5×0.5–3cm) leaflets ......,.. 2

2. Leaflets 17–43, oblong, small, 1–4×0.5–0.8(–1.3)cm ...... **Species 4–6**
+ Leaflets (9–)11–23, ovate-elliptic to obovate, medium-sized, 2–5× 1.5–3cm .... **Species 7–9**

1. D. sissoo DC.; Nep: *Sissau* (34), *Sissoo*. Fig. 43h

Tree 10–25m. Leaves 10–18cm, rachis ± zigzag, leaflets (3–)4–6, suborbicular or broadly ovate, 3.5–8×2.5–5.5cm, abruptly acute or acuminate, base cuneate, appressed pubescent at first beneath, petiolules 5–7mm. Panicles axillary, 4–6.5cm, flowers ± sessile, bracts caducous. Calyx tube 3–3.5mm, teeth 1–2mm. Petals creamy white, claws c 3mm, standard obovate, blade 5−6×4mm, wings and keel oblong c 4×1mm. Stamens 9, filaments united into a sheath split along upper side. Pods elliptic 5−8×1−1.2cm, prominently veined over the 1−2(−3) seeds.

Bhutan: S—Phuntsholing, Sarbhang and Gaylegphug districts; **Sikkim.** River banks, 300m. March−April.

Sometimes cultivated as a decorative wayside or shade tree and also prized for its useful timber which is of value for all purposes where strength and elasticity are required (126).

2. D. latifolia Roxb. Nep: *Satisal* (34)

Similar to *D. sissoo* but leaves 15−22cm, rachis straight, leaflets (3−)5(−7), suborbicular, 3−6cm long and broad, obtuse or emarginate, base rounded; pedicels 3−4mm; calyx tube 3mm, teeth 2mm; petals with claws c 2mm, standard ovate, blade 4×3mm, wings and keel 3×1.5mm; pods oblong 3.5−8×1−1.5cm, 1−3-seeded.

Sikkim: Terai. September−October.

The timber is used largely for making furniture (126).

3. D. rimosa Roxb. Nep: *Tatebiri* (34)

Similar to *D. sissoo* but a shrub 2−7m, sometimes scandent; leaves 15−25cm, leaflets (3−)5−7(−9), ovate, elliptic or obovate, 3−13×1.5−5cm, acute or obtuse, mucronate, base rounded; panicles terminal or axillary 10−20cm, flowers ± sessile; calyx including teeth 2mm; petals yellow 3mm; pods elliptic, 5−7×2−2.5cm, strongly veined over the solitary seed.

Bhutan: S—Phuntsholing and Sarbhang districts; **Sikkim.** Dense subtropical mixed forest, 120−900m. April−May.

4. D. stipulacea Roxb. Nep: *Lahara Siris* (34)

Sprawling tree or scandent shrub. Leaves 15−20cm, leaflets 17−21, oblong, 2−4×0.6−1.3cm, obtuse, base rounded or cuneate, sparsely appressed pubescent beneath, petiolules 1−3mm; stipules oblong 10×2−3mm, deciduous. Panicles 8−12cm, branches bearing numerous oblong bracts 4−5×1mm in lower parts, bracteoles c 3×2mm. Petals pale blue, claws c 2mm, standard strongly recurved, obovate, blade 6×5mm, wings and keel falcate or upcurved, 4−6×3mm. Stamens 10 in 2 groups of 5. Pods oblong-elliptic, 7−10×2.5−3cm, thickened over the solitary seed.

Bhutan: S—Samchi, Phuntsholing, Sarbhang and Gaylegphug districts; **Sikkim.** Subtropical forest, 400−900m. April−May.

Bark and roots used as fish poison (146).

5. D. pinnata (Loureiro) Prain; *D. tamarindifolia* Roxb. Dz: *Olla Sema;* Nep: *Siris Lahara*

Similar to *D. stipulacea* but leaves 9−15cm, leaflets 25−41, oblong,

1.5−2.5×0.5−0.8cm, obtuse or emarginate, base oblique, rounded, appressed pubescent especially beneath; stipules lanceolate, 0.5mm, deciduous; panicles 2−3cm, brown pubescent, branches bearing numerous ovate bracts c 1.5mm; calyx including teeth c 4mm, appressed pubescent; petals creamy white; claws c 3mm, standard reflexed, blade ovate c 3×2mm, wings and keel oblong 3×1mm; stamens 9, filaments united into sheath divided by a slit on upper side; pods oblong 4−7×1−1.3cm, stipitate, 1−3-seeded, uniformly veiny throughout.

Bhutan: S—Samchi, Sarbhang and Gaylegphug districts, **C**—Punakha and Tongsa districts; **Sikkim**. On subtropical and warm forest slopes, 300−1400m. March−April.

6. D. mimosoides Franchet

Similar to *D. stipulacea* and *D. pinnata* but twigs surrounded at base persistent ovate bracts 2−3mm; leaflets 25−35, oblong, 10−15×3−6mm, obtuse, base rounded, sparsely appressed pubescent beneath; stipules ovate, deciduous; panicles 2−3cm, bracts and bracteoles deciduous; calyx including teeth 3−4mm; petals yellowish white c 5mm; stamens 9 united into sheath split along upper side; pods oblong, 3−7×1−1.5cm, thickened and veiny over 1−2 seeds.

Bhutan: C—Thimphu district (Nahi); **Sikkim:** Tista valley below Choong-Thang. 2000m. April−July.

7. D. sericea G. Don; *D. hircina* Bentham, *D. stenocarpa* Kurz. Dz: *Pchang;* Nep: *Bandre Siris*

Tree 15−20m. Leaves 18−20cm, leaflets (9−)13−19(−21), ovate-elliptic, 2.5−4×1.7−2.5cm, acute, obtuse or emarginate, base rounded, appressed pubescent on both surfaces; stipules lanceolate 8−10mm, deciduous. Panicles 3−4cm. Calyx 3−4mm. Petals white or pale mauve, claws c1.5mm, standard obovate, blade 4×3mm, wings oblique 4.5×2mm, keel strongly curved c 3×2mm. Stamens 10 in 2 bundles of 5. Pods linear-elliptic, 2−6×0.6−0.8cm, prominently veiny over the 1−5 seeds.

Bhutan: C—Punakha, Tongsa and Mongar districts; **Sikkim**. In Warm broad-leaved forests, 500−1500m. April−May.

8. D. assamica Bentham; *D. bhutanica* Thothathri. Nep: *Laha Siris* (34)

Similar to *D. sericea* but leaves 25−35cm, leaflets 13−21, oblong-elliptic 3−5(−8)×1.8−3cm obtuse or emarginate, base rounded, glabrous above, appressed pubescent beneath; panicles 10−15cm, bracts and bracteoles deciduous; calyx including teeth 3−4mm; petals white 6−7mm; pods oblong elliptic, 3.5−7.5×1.2−1.5cm, 1−4-seeded.

Sikkim: Kalimpong.

9. D. volubilis Roxb.
Similar to *D. sericea* but a large woody twiner; leaves 10−17cm, leaflets 11−13, obovate or ovate-oblong 2.5−5×1.75−3cm, obtuse or emarginate, base rounded, glabrous; panicles terminal, 20−30cm; calyx including teeth 3mm; petals purplish, claws c 2mm, standard suborbicular 4×6mm, wings and keel elliptic 4.5×3mm; pods oblong-elliptic, 5−8×1.5−2cm, 1−2-seeded.
Sikkim: Terai. January−February.

23. ABRUS Adanson

Slender shrubby twiners. Leaves even-pinnate, leaflets numerous, stipels present; stipules small, caducous. Flowers in axillary or terminal racemes borne in clusters on swellings in upper part of rachis. Calyx campanulate, teeth short or obsolete. Petals clawed, wings shorter and narrower than upcurved keel. Stamens 9, connate into a split sheath. Pods oblong, compressed thinly septate, seeds hard and glossy, arillate or not.

1. A. pulchellus Thwaites
Leaves 5−12cm, rachis ending in a fine bristle; leaflets 6−10 pairs, oblong, 1−3.5×0.6−1.5cm, obtuse and rounded at apex and base, sparsely appressed white pubescent beneath, glabrous above, stipels subulate c1mm; stipules lanceolate 4−5mm. Peduncles 5−12cm, accrescent. Calyx 3−4mm including short broad teeth. Petals purplish 10−12mm, standard obovate c 8mm broad, wings oblong, blade 8×2.5mm, keel curved and sharply pointed, 10×4mm. Pods 6−12×1−1.3cm, sparsely appressed white pubescent, seeds 8−12, oblong, ± compressed, 5×3mm, brown, with a pale coloured aril c 0.5mm around subterminal hilum.
Bhutan: S—Samchi, Phuntsholing, Gaylegphug and Deothang districts; **Sikkim.** Subtropical forest, 510−800m. August−October.

The record (71) of *A. fruticulosus* Wight & Arnott, a peninsular species, from Bhutan, probably refers to this species. They were once considered to be identical but have since been shown to be distinct.

2. A. precatorius L. Nep: *Lalgeri* (34)
Similar to *A. pulchellus* but leaflets 10−16 pairs, oblong, 12−20×4−6mm, rounded at both ends or sometimes emarginate at apex; racemes 2−10cm; calyx 2−3mm; petals white or purplish, 10−12mm; pods 2−4×1−1.5cm, densely pubescent with short pale hairs; seeds 3−7, ovoid, not compressed, 6−7×5−6mm, scarlet with a black area around subapical hilum, exarillate.
Bhutan: C—Tashigang district (Gamri Chu); **Sikkim.** In dry valley, 1100−1250m, September.

24. DERRIS Loureiro

Woody twiners. Leaves odd-pinnate, leaflets entire; stipels absent; stipules persistent. Flowers in axillary or terminal racemose panicles, bracts deciduous, bracteoles sometimes persisting. Calyx cup-shaped, truncate or teeth short and broad. Petals clawed, standard reflexed and sometimes with a pair of callose thickenings near base, wings slightly adherent to incurved keel. Stamens monadelphous, all connate or the posterior one free at base. Pods compressed, leathery, oblong, winged along one or both sutures, indehiscent, seeds 1–3.

1. Standard with a pair of callose thickenings at base .... **Species 5 & 6**
\+ Standard without callose thickenings at base 2

2. Leaves beneath, leaf rachises and young shoots brown pubescent
   4. D. ferruginea
\+ Leaves beneath, leaf rachises and young shoots minutely puberulous or glabrous .... **Species 1–3**

1. D. acuminata Bentham

Branchlets shortly brown puberulous at first. Leaves 20–35cm, leaflets 7, thinly coriaceous, elliptic or obovate, 9–15×4–8cm, shortly acuminate to a blunt tip, base rounded or cuneate, glabrous above, sparsely puberulous along veins beneath, prominently reticulate on lower surface; stipules broadly triangular, c 2.5mm, coriaceous. Flowers in clusters of 3–5 along inflorescence branches, pedicels 2–3mm, bracteoles ovate, c 0.4mm, borne at calyx base. Calyx c 3.5mm, obscurely toothed, brown puberulous. Petals white or purplish, claws c 3mm, standard suborbicular 5–6mm long and broad, wings and keel oblong 5×1.5mm. Pods (immature) oblong, narrowly winged along both sutures.

Bhutan: C—Punakha, Tongsa and Mongar districts. Warm broad-leaved forests, 1370–1830m. August–September.

2. D. polystachya Bentham

Similar to *D. acuminata* but leaflets 7–9, not prominently reticulate beneath; flowers red or pink, borne in short racemes 1–1.5cm along inflorescence branches; standard suborbicular or broadly elliptic, 6–8×6mm, wings and keel 6–7×2mm; pods asymmetrically oblong-elliptic, 11×3cm, acuminate, base rounded, upper wing 4–5mm broad, lower wing 2–3mm, 1–2-seeded.

Bhutan: S—Samchi and Chukka districts; **Sikkim**. Subtropical forest, 300–1200m. August–September.

3. D. monticola (Kurz) Prain; *D. acuminata* Bentham var. *sikkimensis* Thothathri

Similar to *D. acuminata* but leaflets 9−11, oblong, 6.5−9.5×2−3cm, thin, glabrous; flowers borne in clusters of 3−5, pedicels 5−7mm; calyx 5−6mm; petals pink or purple, claws 5−6mm, standard broadly ovate 11−12×9−10mm, wings and keel oblong, 10×3mm; pods oblong-elliptic, 7−10×1.5−2.5cm, winged on both sutures, the upper one broader than the lower.

Sikkim: Darjeeling. August.

4. D. ferruginea (Roxb.) Bentham

Branchlets and young parts densely brown pubescent. Leaves 18−25cm, rachis brown pubescent, leaflets 7−9, obovate-elliptic, 7−10×3−4cm, bluntly acuminate, base cuneate, glabrous above, brown pubescent especially on veins beneath; stipules ovate, 2−2.5mm. Panicle branches elongate, 15−20cm, brown pubescent, flowers borne on short racemes. Calyx 4−5mm. Petals reddish, claws 3−4mm, standard broadly elliptic, 10×8mm, wings oblong, 9−1.5mm, keel 9×3.5mm. Pods oblong-elliptic, 4.5−7×2−3.5cm, brown pubescent at first, upper suture wing 5−6mm broad, lower suture wing c 1.5mm broad.

Sikkim: Terai. February−May.

5. D. cuneifolia Bentham; *D. discolor* Bentham. Nep: *Baiari* (34)

Branchlets puberulous at first. Leaves 20−30cm, leaflets 5−9, membranous, ovate-elliptic, 7−10×3−5cm, shortly and bluntly acuminate, base cuneate, minutely puberulous beneath at first. Panicles 7−15cm, flowers borne on short branchlets, pedicels 3−4mm. Calyx c 3 mm, subglabrous. Petals red, 1−1.3cm, standard ovate or suborbicular with 2 callose thickenings near base. Pods oblong-elliptic, 3−5×2.5−3cm, upper suture with wing c 5mm broad, lower suture obscurely winged, 1−2-seeded, puberulous at first.

Sikkim: Terai. May.

Some specimens from Sikkim with leaflets up to 13×9cm and pedicels 1−1.2cm have been described as var. **longipedicellata** Thothathri.

6. D. microptera Bentham

Similar to *D. cuneifolia* but leaflets oblong-elliptic, 6−10×3.5−4.5cm, acuminate, base rounded, glabrous; inflorescence branches 20−35cm, flowers on pedicels 4−6mm; calyx 3−4mm; petals purple, 1−1.3cm, standard ovate or suborbicular bearing 2 callose thickenings near base; pods oblong-elliptic, 4−7×2.5−3cm, glabrous, upper suture narrowly (1−2mm) winged, 1−3-seeded.

Sikkim: Terai. April−June.

25. MILLETTIA Wight & Arnott

Trees or woody climbers. Leaves odd-pinnate, usually evergreen; stipels commonly present; stipules small, caducous or persistent. Flowers in axillary or terminal panicles, sometimes racemiform with short lateral branches. Calyx campanulate, teeth 5, sometimes very short. Standard spreading or reflexed, wings oblong, free from keel, auricled and clawed at base, keel upcurved, clawed at base. Stamens monadelphous. Pods oblong, flat or constricted between seeds, leathery or woody, few-seeded.

1. Standard pubescent or sericeous dorsally; pods tomentose; leaves 5–9-foliate .... **Species 1 & 2**
+ Standard glabrous dorsally; pods glabrous (pubescent when young in *M. pachycarpa* but then leaves 11–17-foliate) .... **Species 3 & 4**

1. M. extensa (Bentham) Baker; *M. auriculata* Brandis. Nep: *Kurku*

Large woody climber. Leaves 7–9-foliate, 20–40cm, leaflets obovate or oblong, 7–15(–18)×4–7(–11)cm, subacute or shortly acuminate, base rounded, glabrous above pubescent beneath, sometimes only on veins; stipels subulate, 1–2mm, deciduous. Panicles racemiform, 12–15cm, flowers borne in small clusters. Calyx c 4mm, obscurely toothed, sericeous. Petals yellowish-green, standard suborbicular, c 12×12mm, bilobed at apex, base shortly clawed and with 2 rounded auricles, sericeous dorsally, wings and keel ± as long as standard, obovate, narrowed into a long claw at base. Pods compressed, 10–18×2–2.7cm, woody, densely brown tomentose, seeds 4–6(–8), broadly elliptic, c15×10mm, dark brown, glossy.

Bhutan: S—Samchi, Phuntsholing and Sarbhang districts; **Sikkim.** Subtropical forests, climbing often on tall trees, 400–600m. May–June.

The leaves and twigs are lopped for cattle fodder (48).

2. M. cinerea Bentham

Similar to *M. extensa* but leaves 5-foliate, 15–25cm, leaflets elliptic-obovate, 6–15×3.5–6cm, shortly acuminate, base rounded, sparsely pubescent on veins beneath; stipels subulate 3–5mm, ± persistent; flowers in terminal branched panicles; calyx c 6mm, teeth a third as long; petals crimson or purplish, 12–15mm, standard broadly ovate; pods up to 13×2cm, 1.5cm thick, constricted between the seeds, velvety.

Bhutan: S—Samchi and Deothang districts, **C**—Tongsa, Mongar and Tashigang districts; **Sikkim.** In warm moist broad-leaved forests, 1050–1700m. May–June.

Leaves are used as cattle fodder (146).

3. M. pachycarpa Bentham. Nep: *Kurkus* (34), *Kakushbish* (34)

Tall climber. Leaves 11–17-foliate, 30–50cm, leaflets obovate or oblanceolate, 7−15×2.5−5cm, acuminate, base cuneate, glabrous above, densely appressed brownish pubescent beneath; stipules broadly ovate-triangular 4×5mm, persistent. Panicles 12−30cm, branches racemiform. Calyx 6−7mm, densely brown pubescent, teeth short and broad. Petals purplish, standard broadly elliptic, c 3×2cm, bilobed at apex, glabrous, wings and keel oblong, 2.5−2.75×0.7−0.8cm. Pods 6−23× 3−5cm, 2−3cm thick, constricted between seeds, unwinged, pubescent at first later glabrous; seeds reniform c 3×2.5×2cm.

Bhutan: C—Punakha district (near Chusom), Tongsa district (Birti) and Tashigang district (between Chazam and Duksum); **Sikkim.** Amongst scrub in arid valleys, 900−1350m. April−May.

The bark, roots and pods are apparently used as fish poison.

4. M. glaucescens Kurz; *M. prainii* Dunn

Similar to *M. pachycarpa* but a large tree; leaves 7−9-foliate, 15−30cm, leaflets oblanceolate, 8−13×2−3.5cm, acuminate, base cuneate, glabrous, greyish glaucous beneath; panicles axillary, 16−20cm, racemiform; calyx c 3mm, teeth very short and broad, minutely pubescent; petals purplish, 10−12mm; pods 8−13×3cm, with a wing 4−5mm broad along both sutures, glabrous, glaucous.

Sikkim: Kukna, Sivoke, etc. 600m. March−April.

The record of *M. piscidia* (Roxb.) Wight from Sikkim (80) is based on misidentified specimens of *Derris microptera* Bentham.

26. TEPHROSIA Persoon

Shrubs. Leaves odd-pinnate, leaflets numerous, entire, lateral veins numerous, parallel, uniting in a marginal vein; stipels absent; stipules deciduous. Flowers in axillary racemes or panicles. Calyx campanulate, teeth 5, short. Standard suborbicular, wings obovate or oblong, slightly adherent to blunt upcurved keel. Stamens diadelphous. Pods linear, compressed, coriaceous, continuous within, c10-seeded.

1. T. candida DC. Sha: *Kumchumo Shing;* Nep: *Bun Mara*

Stems erect c 2m, densely brown tomentose. Leaves 10−25cm, leaflets 11−23, elliptic-oblong, 3−7×0.8−1.8cm, acute, mucronate or acuminate, base cuneate, margins narrowly revolute, sparsely pubescent or glabrous above, pale sericeous beneath; stipules lanceolate, 6−7mm. Calyx 5−7mm, divided to middle into ovate-triangular teeth, sericeous. Standard c 2×2cm, base abruptly narrowed to a short claw, wings and keel c 20×8mm. Style shortly bearded on inside near apex. Pods 5−9×0.7−0.8cm, sericeous; seeds ellipsoid, c 5×3mm, brown with white fleshy aril around hilum.

76. LEGUMINOSAE

Bhutan: S—Chukka, Gaylegphug and Deothang districts, **C**—Tongsa district; **Sikkim.** Warm broad-leaved forests, 1050−1200m. June−November.

The leaves and bark are used to poison fish (48).

27. GLIRICIDIA Humboldt, Bonpland & Kunth

Small trees. Leaves odd-pinnate, deciduous, leaflets 7−21, entire; stipels absent; stipules minute. Flowers showy, often appearing before leaves, in axillary racemes. Calyx campanulate, truncate, teeth obsolete. Standard suborbicular, reflexed, wings oblong slightly longer than keel, keel upcurved almost at right angles at distal end. Stamens diadelphous. Pods linear-oblong or oblanceolate, compressed, leathery, shortly stalked, somewhat thickened along both sutures, not septate within, valves coiling on dehiscence.

1. G. sepium (Jacquin) Walpers
Tree 3−10m. Leaves 15−25cm, leaflets oblong-ovate, 4.5−7×2−3cm, bluntly acute, base rounded, often oblique, pubescent on veins beneath, otherwise glabrous; petiolules 3−5mm. Racemes 7−15cm. Calyx 5−6mm, glabrous. Petals pinkish-white, standard c 2×2cm, bilobed at apex, wings 20×4mm, deeper pink, keel c 1.75×4mm. Pods 10−14×1.5cm, on basal stalks c 7.5mm.
Bhutan: S—Phuntsholing. 250m. February.
Native of S. America, cultivated as an ornamental roadside tree. Timber is used for firewood (130).

28. SESBANIA Scopoli

Shrubs or small trees. Leaves even-pinnate, rachis ending in a short point, leaflets numerous, narrow; stipels absent; stipules small. Flowers in

---

FIG. 42. **Leguminosae.** a & b, *Butea monosperma:* a, portion of inflorescence; b, pod. c−e, *Zornia gibbosa:* c, portion of flowering shoot; d, part of inflorescence, bract overlying flower removed; e, pod with one bract removed. f, *Vigna vexillata:* ovary and style. g & h, *Vicia sativa:* g, portion of flowering and fruiting shoot; h, detail of stipule. i−k, *Christia vespertilionis:* i, portion of flowering shoot; j, flower; k, pod within part of opened calyx. l−n, *Cicer arietinum:* l, portion of flowering shoot; m, pod; n, seed. Scale: a, b × ½; g, i × ⅔; c, l, m × 1; f, n × 1¼; d, k × 2; j × 2½; e, h × 3.

axillary racemes. Calyx campanulate, shortly 5-toothed. Petals all clawed, standard suborbicular, spreading or reflexed with two ascending spurs at top of claw, wings oblong, longer than upcurved keel. Stamens diadelphous. Pods linear, terete, constricted and septate between the numerous seeds.

1. S. sesban (L.) Merrill; *S. aegyptiaca* Persoon

Stems 2–3m, sparsely pubescent at first. Leaves 7–15cm, leaflets 9–15(–20) pairs, oblong or oblanceolate, 1–2.5×0.3–0.6cm, obtuse, base rounded or cuneate, rather oblique, glabrous; stipules lanceolate 2–3mm. Racemes 3–10-flowered, 3–7cm. Calyx 5–6mm, teeth triangular. Petals yellow or purplish; standard c 1.8×1.8cm, wings 18×6mm, yellow, keel 14×5mm. Pods 9–23×0.3–0.4cm, usually darkly coloured on constrictions; seeds oblong, c 3.5×2mm, brown or black.

Sikkim: Cultivated in the Terai (34).

The flowers may be pure yellow or the standard purple spotted (var. *picta* Prain) or deep purple (var. *bicolor* Wight & Arnott). It is often planted as a nurse tree for crops (16); branches of this rapidly growing plant are lopped for fodder (125).

29. INDIGOFERA L.

The name *Re Sheng* (Tashigang dialect) applies to several unspecified *Indigofera* species.

Annual herbs or perennial shrubs, sometimes trees, usually bearing appressed medifixed, branched hairs. Leaves odd-pinnate, sometimes digitately 3-foliate or simple; stipels sometimes present; stipules small, linear-lanceolate or subulate. Flowers in lax or dense axillary racemes or clusters, bracts mostly small, deciduous. Calyx tube campanulate, teeth 5, usually subequal. Standard broadly elliptic, ovate or suborbicular, wings oblong or obovate, slightly adherent to keel, keel straight or curved, lanceolate or oblanceolate, obtuse at apex with a short spur or pouch on each side near base. Stamens diadelphous. Pods linear or oblong, straight or curved, terete, rarely subglobose, transversely septate between seeds.

1. Leaves simple .... **1. I. linifolia**
+ Leaves 3-foliate or pinnate 2

2. Leaves 3-foliate; leaflets gland-dotted beneath .... **2. I. trifoliata**
+ Leaves pinnate; leaflets not gland-dotted 3

3. Annual herb .... **3. I. astragalina**
+ Shrubs 4

4. Racemes often borne on old wood with a cluster of scales at base of peduncles .... **4. I. cassioides**
+ Racemes borne on young wood without a cluster of scales at base 5

5. Young shoots ± densely spreading hirsute .... **5. I. dosua**
+ Young shoots glabrous or bearing appressed hairs 6

6. Petals 3−5mm .... **Species 6−8**
+ Petals 7−10mm 7

7. Bracts minute, 1−2mm; leaflets small, 5−20×3−9mm ...... **Species 9−12**
+ Bracts larger, 3−7mm; leaflets larger, 15−60×10−25mm (but as little as 10×5mm in *I. bracteata*) .... **Species 13 & 14**

1. I. linifolia (L.f.) Retzius
Annual herb, sometimes quite woody at base, stems prostrate or diffuse, up to 60cm. Leaves simple, elliptic or oblanceolate, 7−25×2−4mm, obtuse or acute, mucronate, base rounded or attenuate, appressed silvery pubescent on both surfaces; stipules linear-lanceolate, 1.5−2.5mm. Racemes up to 12-flowered, at first crowded, later up to 3cm. Calyx c 3mm, teeth tapering to fine subulate points. Petals c 3.5mm, bright red, standard obovate, c 3mm broad. Pods subglobose, c 2.5mm, densely appressed silvery pubescent, 1-seeded.
Bhutan: S—Phuntsholing district (near Phuntsholing), **C**—Mongar district (Lingmethang). Weed of cultivated ground and on riverside shingle, 200−950m. February−July.

2. I. trifoliata L.
Trailing or suberect herb, somewhat woody at base, stems up to 60cm, at first appressed pubescent, later glabrous. Leaves 3-foliate, leaflets elliptic to oblanceolate, 7−20×3−7mm, obtuse or acute, base cuneate, appressed pubescent on both surfaces, gland-dotted beneath; petioles 0.5−1cm; stipules subulate c 1mm. Flowers 6−7, clustered in leaf axils, scarcely racemose. Calyx c 2mm, teeth subulate. Petals c 4mm, standard and keel greenish outside pink within, wings pink. Pods linear, 10−15×1.5mm, appressed pubescent, 6−8-seeded.
Bhutan: S—Deothang district (Kheri), **C**—Punakha, Mongar and Tashigang districts. On dry banks and gravel, 950−1525m. June−August.

3. I. astragalina DC.; *I. hirsuta* sensu F.B.I. non L.
Annual. Stems erect, 0.25−1.5m, young growth ± densely spreading brownish-white hirsute. Leaves pinnate, 8−15cm, leaflets 7−11, elliptic-obovate, 2−6×1−2.5cm, obtuse or subacute, base rounded or cuneate, appressed pubescent on both surfaces; stipules subulate c 1cm.

Calyx c 5mm, teeth subulate, densely brown hirsute. Petals scarcely exserted beyond calyx, reddish. Pods linear, deflexed, 10−20×4mm, densely whitish hirsute, 5−7-seeded.

Sikkim: locality unknown. 300−900m.

4. I. cassioides DC.; *I. pulchella* sensu F.B.I. non Roxb.

Shrub 1.5−4(−10)m. Leaves pinnate, 7−15cm, leaflets 11−21 ovate-elliptic 1−2.5×0.7−1.5cm, obtuse or subacute,mucronate, base rounded, appressed pubescent on both surfaces; stipules linear 2−4mm, deciduous. Racemes showy, 5−12cm, often borne on old wood and unfolding before leaves, rachis bearing a cluster of small (1−2mm) ovate scales at base, bracts ovate, 2−10×1−4mm. Calyx 2−3mm, teeth subequal. Petals deep pink, standard obovate, 12−15×7−9mm, wings oblong 10−12×3−4mm, keel 12−15×2−3mm. Pods linear, 2.5−4.5×0.3−0.4mm, crimson when immature, 8−12-seeded.

Bhutan: S—Chukka district, **C**—Punakha and Tashigang districts; **Sikkim.** Warm broad-leaved forest, 1200−1500m. March−May.

5. I. dosua D. Don. Sha: *Kumchingma Shing;* Nep: *Chiringi Jhar* (34)

Shrub 0.3−4m, young growth spreading pale or brownish hirsute. Leaves pinnate, leaflets 15−53, oblong or broadly elliptic, 5−35×3−8mm, obtuse or acute, mucronate, base rounded or cuneate, appressed pubescent on both surfaces; stipules narrowly lanceolate, 6−12mm. Calyx c 2mm, teeth broadly triangular. Petals pink to purple, standard broadly oblong, c 9×6mm, wings oblanceolate, c 8×2mm, keel ± as long as wings, acute. Pods linear, ± deflexed, 20−35×2.5−3mm, pubescent, 10−12-seeded.

Bhutan: S—Chukka, Gaylegphug and Deothang districts, **C**—Thimphu to Tashigang districts; **Sikkim.** Open hillsides and roadside banks, 1000−2000m. April−August.

There are two more or less distinct varieties: var. **dosua,** generally a low growing shrub less than 1m, shortly pale hirsute, leaves 3−8cm with 15−31 oblong-elliptic leaflets 5−10×3−5mm, racemes 10−30-flowered, 5−10cm, and var. **tomentosa** Baker, generally a taller shrub up to 4m, densely brown hirsute, leaves 12−18cm with 37−53 narrowly oblong leaflets 10−35×3−8mm, racemes always many-flowered, 10−20cm. The latter appears to be the commoner variety in Bhutan and is sometimes considered as a distinct species: *I. stachyodes* Lindley.

6. I. caerulea Roxb.; *I. argentea* L. non Burman f.

Shrub 1−2m, stems appressed whitish pubescent. Leaves 6−10cm, leaflets 7−11, obovate, 1.5−3.5×1−2cm, obtuse or retuse, base cuneate, glabrous above, appressed pubescent beneath. Racemes erect, 3−10cm, densely flowered to base. Calyx c1.5mm, teeth triangular. Petals 3−4mm, reddish yellow. Pods terete, 10−15×15×4mm, deflexed but upcurved,

appressed whitish pubescent when young, 3−4-seeded, somewhat torulose.
Sikkim: cultivated as a source of Indigo (34).

7. I. tinctoria L.
Similar to *I. caerulea* but stems sparsely appressed pubescent, not whitish; racemes 3−7cm, lax-flowered to base; pods 15−30×2−2.5mm, deflexed but scarcely upcurved, sparsely pubescent when young, 8−12-seeded, hardly torulose.
Sikkim: cultivated as a source of Indigo (17).

8. I. zollingeriana Miquel
Shrub or small tree 2−12m. Leaves 15−25cm, leaflets 11−17(−23), ovate-elliptic, 3−8×1−2.5cm, acuminate to a fine mucro c 1mm, base rounded, minutely appressed pubescent on both surfaces; petioles 2−3mm. Racemes 8−20cm, pedicels 3−4mm. Calyx c 2.5mm, appressed pubescent. Standard ovate, 4−5×3−4mm, red or purplish, pubescent on the back, wings 3−3.5×1.5mm, pink, keel 4−4.5×1.5mm, white tinged pink. Pods spreading, straight or slightly curved, 2.5−4cm, 5−6mm thick, surface becoming transversely cracked and fissured, brown, indehiscent.
Bhutan: S—Phuntsholing district (Phuntsholing). Cultivated, 250m. June−July.

Grown for its decorative flowers and foliage; native of China, Indochina and Malesia.

9. I. heterantha Brandis; *I. gerardiana* Baker
Slender erect shrub up to 2(−3)m, young growth appressed greyish pubescent. Leaves 1−3cm, leaflets 5−7, obovate or oblanceolate, 5−10×3−4mm, obtuse, mucronate, base cuneate, appressed pubescent on both surfaces; stipules subulate, 1−2mm. Racemes up to 3cm bearing flowers to base of rachis, bracts narrowly lanceolate c 1.5mm. Calyx 2−3mm,, divided to middle into narrowly lanceolate teeth. Petals pink or purple, standard obovate, 5−8×3−5mm, wings and keel 4−5×1.5mm. Pods linear, 20−30×2.5mm, ± deflexed, appressed pubescent; fruiting pedicels 2−3mm, stout.
Bhutan: S—Chukka district (Chimakothi), **C**—Thimphu, Punakha, Tongsa and Tashigang districts, **N**—Upper Kuru Chu district (Shabling). Open hillsides and banks, 1400−2550m. May−July.

10. I. exilis Grierson & Long; *I. leptostachya* sensu F.B.I. non DC.
Similar to *I. heterantha* but young growth less densely appressed whitish pubescent; leaves 7−14cm, leaflets 15−23, oblanceolate or obovate, 8−20×4−9mm, minutely appressed pubescent on both surfaces but more sparsely so and paler beneath; racemes lax, 6−10cm, flowers borne in upper half of rachis; calyx teeth broadly triangular; petals slightly larger, standard 8.5×5.5mm, wings and keel c 8×2mm; pods glabrous, spreading.

76. LEGUMINOSAE

Bhutan: S—Chukka district (near Chukka); **Sikkim:** Lachen and Lachung. 1400m. July.

11. I. cylindracea Baker; ?*I. heterantha* Brandis var. *longipedicellata* Thothathri

Similar to *I. heterantha* but smaller, stems up to 1m, sparsely appressed brownish pubescent; leaflets 11−15, elliptic-oblong, 9−16×4−8mm, appressed pubescent on both surfaces but hairs minute on upper one; racemes 5−8cm, flowers borne in upper half of rachis, bracts subulate c 2mm; calyx teeth narrowly lanceolate; petals larger, standard 7−8×5.5mm, strongly reflexed, wings and keel c 7×2mm; pods glabrous, ± erect, fruiting pedicels 8−10mm, slender.

Bhutan: C—Thimphu district (Taba and Chapcha); **Sikkim.** On grassy banks among scattered pines, 2440−2700m. July−August.

12. I. pseudoreticulata Grierson & Long

Similar to *I. heterantha* but stems trailing or prostrate, sparsely appressed brown pubescent when young; leaflets 9−15, narrow to broadly elliptic, 6−13×4−7mm, appressed pubescent on both surfaces, veins prominently reticulate; racemes 4−9cm, flowers borne in upper half of rachis; petals larger, standard c 8×4mm, wings and keel 7−8×2mm.

Bhutan: C—Thimphu district (Dotena) and Tongsa district (Rukubji), **N**—Upper Mo Chu district (Tamji) and Upper Kuru Chu district (Julu). On rocks and grassy banks, 2300−2800m. July−August.

13. I. hebepetala Baker

Erect shrub 1−2m. Leaves 6−18cm, leaflets 5−13, elliptic, 1.5−5×0.5−2.5cm, obtuse or emarginate, mucronate, base rounded or cuneate, appressed pubescent beneath, glabrous or sometimes appressed pubescent on upper surface; stipules narrowly lanceolate, 2−3mm. Racemes 7−18cm bearing flowers almost to base of rachis, bracts boat-shaped, 4−5×2−3mm, hooded and enclosing bud, abruptly narrowed to a subulate point 1.5−2mm. Calyx c 2.5mm, teeth short and broad. Standard obovate, c 10×7mm, standing at right angles to keel, dark crimson, wing and keel c 10×3mm, pink. Pods terete, straight, 30−50×3mm, glabrous, 8−10-seeded.

Bhutan: C—Thimphu and Punakha districts, **N**—Upper Mo Chu district; **Sikkim.** Banks of streams, 2100−2500m. May−July.

The typical var. **hebepetala** with leaflets appressed pubescent on the upper surface was described from Sikkim. In all the material seen so far from Bhutan the leaflets are glabrous on the upper surface and belong to var. **glabra** Ali.

14. I. bracteata Baker

Similar to *I. hebepetala* but stems prostrate or trailing; leaves 4−9cm, leaflets 5−7, elliptic or obovate, 1−3×0.5−1.2cm, obtuse, base cuneate, glabrous or weakly appressed pubescent on upper surface, more strongly appressed pubescent beneath; stipules subulate 2−3mm, ± persistent; racemes 7−20cm bearing flowers in upper half of rachis, naked below, bracts elliptic 4−6×2mm; calyx c 2mm, teeth short and broad; petals purple, standard broadly elliptic, c 10×6mm, wings and keel c 10×2.5mm; pods (immature) linear, glabrous, deflexed.

Bhutan/Chumbi frontier: Sharna to Choidiponkay; **Sikkim.** 2800m. July.

15. I. atropurpurea Hornemann

Similar to *I. hebepetala* but larger, erect; leaves 10−30cm, leaflets 7−15, elliptic or obovate, 3−6×1.5−2.5cm, acute or obtuse, 3−6×1.5−2.5cm, acute or obtuse, base rounded or cuneate, appressed pubescent on both surfaces, petiolules 3−4mm; stipules narrowly lanceolate c3mm; racemes 10−25cm, bearing flowers nearly to base of rachis, bracts narrowly lanceolate c 3mm; calyx c 2mm, broadly toothed; standard broadly elliptic c 8×7mm, blackish purple, wings oblong c 8×2mm, pink, keel oblanceolate, c 8×2.5mm, purple; pods ± terete, straight, 35−40×3−4mm, deflexed.

Bhutan: C—Tongsa district (Dakpai) (117). 1650m. August.

30. DESMODIUM Desveaux

Perennial herbs, shrubs or trees. Leaves pinnately 3-foliate, sometimes 1-foliate, margins entire or undulate; stipels usually conspicuous; stipules persistent or deciduous. Flowers in axillary or terminal racemes, panicles, corymbs or umbels; bracts large or small, persistent or deciduous. Calyx tube short, teeth 5, upper 2 connate into a bidentate lip, lower lip 3-dentate. Petals clawed, wings adherent to keel. Stamens monadelphous or diadelphous. Pods compressed, ± constricted between seeds, dehiscent along lower suture or indehiscent and breaking up into 1-seeded segments.

1. Petioles broadly (up to 1cm) winged; leaves all 1-foliate
 1. D. triquetrum
+ Petioles unwinged; leaves 3-foliate or sometimes 1-foliate 2

2. Inflorescence racemiform; flowers concealed by pairs of leafy orbicular persistent bracts .... **2. D. pulchellum**
+ Inflorescence umbellate, corymbose, racemose or paniculate; bracts conspicuous only in bud, ovate-lanceolate, caducous, not leafy 3

3. Stems triangular in section; inflorescence of axillary shortly stalked umbels or corymbs .... **3. D. triangulare**
+ Stems terete; inflorescence of racemes or panicles 4

4. Leaves 1-foliate or, if 3-foliate, lateral leaflets less than half size of terminal leaflet (see also *D. styracifolium*); keel petals with short lateral appendages; pods dehiscing along lower suture; seeds with a cup-shaped aril .... **Species 4 & 5**
+ Leaves 1-foliate or 3-foliate but lateral leaflets more than half size of terminal leaflet; keel petals unappendaged; pods breaking up into 1-seeded segments (dehiscing along lower suture only in *D. microphyllum*); seeds exarillate (or bearing a rim-like aril in *D. microphyllum*) 5

5. Leaves 1-foliate 6
+ Leaves 3-foliate (sometimes 1-foliate in *D. microphyllum* and *D. styracifolium*) 7

6. Herbs, usually prostrate, or weak shrubs; leaves broader than long
9. D. reniforme
+ Shrubs; leaves longer than broad .... **Species 6–8**

7. Flowers borne on wood of previous season .... **10. D. oojeinense**
+ Flowers borne on growth of current season (inflorescences sometimes on wood of previous season in *D. elegans*) 8

8. Leaflets 1.5cm or less, broadly oblong, orbicular or obovate
Species 11 & 12
+ Leaflets more than 2cm, usually ovate or elliptic (rarely orbicular in *D. styracifolium*) 9

9. Pods indented along lower suture only 10
+ Pods indented along both sutures 13

10. Indentations between pod segments shallow, less than half width of pod
11
+ Indentations between pod segments deep, almost reaching upper suture
12

11. Leaflets acute or acuminate .... **Species 13–15**
+ Leaflets subacute, obtuse or rounded at apex .... **Species 16 & 17**

12. Leaflets ovate or ovate-elliptic, acuminate or acute .... **Species 18–21**
+ Leaflets elliptic or obovate, obtuse or acute .... **22. D. concinnum**

13. Segments of pod ± isodiametric .... **Species 23 & 24**
+ Segments of pod longer than broad .... **Species 25 – 27**

1. D. triquetrum (L.) DC.; *Tadehagi triquetrum* (L.) Ohashi

Subshrub 0.5 – 2m, branches triquetrous. Leaves 1-foliate, coriaceous, ovate-oblong, 5 – 11×2 – 5.5cm, acute or acuminate, base rounded, glabrous or sparsely pubescent above, pilose along veins beneath; petiole winged, 2 – 4×0.5 – 1cm; stipules ovate, 1.3 – 2×0.4 – 0.6cm; stipels lanceolate, 2 – 3mm, at apex of petiole wing. Racemes or panicles axillary or terminal, elongate, bracts linear-lanceolate c 7×1mm, persistent. Calyx including teeth 5 – 7mm. Petals purplish, 7 – 8mm, standard suborbicular, 7 – 8mm broad, wings obovate, blade 6×3mm, with a short narrow auricle at base, keel ovate, 5×3mm, upwardly curved, acute. Pods oblong, 2.5 – 3.5cm, shallowly indented along lower or both sutures and dividing into 5 – 7 squarish segments c 6×7mm, pubescent only along sutures.

Bhutan: S—Gaylegphug district (near Birti and Tama) (117). 600 – 1450m. April.

Because of their marginally pubescent pods and broader leaves (less than 3 times longer than broad), the E Himalayan plants have been referred to subsp. **pseudotriquetrum** (DC.) Prain in contrast to pods that are pubescent on the lateral surfaces and more lanceolate leaves (more than 3 times longer than broad) in the typical subsp. *triquetrum*.

2. D. pulchellum (L.) Baker; *Phyllodium pulchellum* (L.) Desveaux. Nep: *Sarkinu* (13)

Shrub 0.5 – 2.5m. Leaves 3-foliate, coriaceous, leaflets elliptic or ovate 4.5 – 12×2.5 – 6cm, terminal leaflet at least twice as large as lateral ones, acute or obtuse, base rounded, margins ± undulate, pubescent especially beneath; petioles up to 1cm; stipules lanceolate 6 – 8×2mm acuminate; stipels subulate 3mm. Inflorescence racemiform, axillary or terminal, consisting of an elongate series of paired leafy elliptic or suborbicular bracts 1 – 1.5×1 – 1.5cm, each concealing 3 – 6 flowers in fascicles. Calyx 2 – 3mm. Petals white or pale yellow, 6 – 8mm, standard elliptic c 4mm broad, wings oblong, blade 3.5×1mm, shortly auriculate at base, keel spathulate 6×2.5mm. Pods 1 – 2-seeded, indented along both sutures into suborbicular segments 2 – 3mm diameter, pubescent along sutures, otherwise glabrous.

Sikkim: Terai. August – October.

The bark and flowers are used medicinally (13).

3. D. triangulare (Retzius) Merrill; *D. cephalotes* (Roxb.) Wight & Arnott, *Dendrolobium triangulare* (Retzius) Schindler

Shrub 0.5 – 2m. Stems triquetrous, appressed white sericeous when young. Leaves 3-foliate, leaflets oblong-elliptic, 4 – 15×1.5 – 6cm, terminal leaflet 1.5 – 2 times as large as lateral ones, acuminate, base rounded or cuneate, appressed pubescent on veins beneath; stipels narrowly lanceolate, 6−8×1mm, acuminate; stipules narrowly ovate, 8−18×3−5mm acuminate, deciduous. Flowers in short dense axillary umbels or corymbs, peduncles less than 10mm. Calyx whitish sericeous, 6−10mm including teeth. Petals white or yellowish, 8−10mm, standard broadly elliptic c 6mm broad, wings oblong, blade 6−7×1.5mm, slightly auricled at base, keel spathulate 5−6×3mm. Pods linear, 1−2cm, undulate along lower suture, sparsely appressed pubescent, breaking up into 2−5 squarish segments.

Sikkim: Terai. 300−1200m. August−October.

4. D. motorium (Houttuyn) Merrill; *D. gyrans* (L.f.) DC., including var. *roylei* (Wight & Arnott) Baker, *Codariocalyx motorius* (Houttuyn) Ohashi. Fig. 43 a&b

Shrub 0.5−2m. Leaves 1 or 3-foliate, terminal leaflet ovate-oblong 4−13×1−4cm, acute, subacute or obtuse, base rounded, appressed pubescent beneath, lateral leaflets when present linear-oblanceolate, 1−1.5×0.2−0.3cm; stipels linear-lanceolate, 2−5mm; stipules ovate-acuminate, 3−15mm; petioles 1−4cm. Flowers in terminal and axillary racemes enclosed at first by overlapping ovate deciduous scales 8−10×4−5mm; peduncles densely pubescent with spreading or reflexed hooked hairs, pedicels 1−4mm in flower, 3−7mm in fruit. Calyx 2−2.5mm. Petals 6−10mm, orange, red or purplish, standard suborbicular, 7−10mm broad, wings 7−9×4−5mm, keel 4−5×3mm. Pods linear, 3−4× 0.5−0.6cm, curved, sparsely pubescent with short hooked hairs, undulating along lower margin and divided within into 5−11 compartments, dehiscing along lower suture. Seeds black, elliptic, 3×2mm, hilum surrounded by a whitish cup-shaped aril 0.5−0.75mm tall.

Bhutan: S—Samchi and Gaylegphug districts, **C**—Punakha and Mongar districts; **Sikkim.** On dry banks, 300−1650m. August−October.

Sometimes cultivated as an ornamental.

5. D. gyroides (Link) DC.; *Codariocalyx gyroides* (Link) Hasskarl. Sha: *Mardum Kumchimo Shing*

Similar to *D. motorium* but terminal leaflet broadly elliptic or obovate, 3−6×2−4cm, broadly ovate, mucronate at apex, lateral leaflets elliptic

FIG. 43. **Leguminosae.** a & b, *Desmodium motorium:* a, mature pod; b, seed. c−e, *Desmodium oblongum:* c, pod; d, segment of pod; e, seed. f & g, *Desmodium podocarpum:* f, pod; g, seed. h, *Dalbergia sissoo:* portion of shoot with leaf and pods. i, *Sophora wightii* subsp. *bhutanica:* pod. j, *Alysicarpus vaginalis:* pod. k−m, *Mucuna macrocarpa:* k, flower; l, pod (5 segments omitted); m, seed. n−p, *Mucuna imbricata:* n, pod; o, diagrammatic section of pod; p, seed. q−s, *Mucuna pruriens:* q, pod; r, section of pod showing attachment of seeds; s, seed. t−w, *Cajanus cajan:* t, portion of shoot with leaf and inflorescence; u, pods, one partly dissected to show attachment of seed; v, seed, lateral view; w, seed, hylar view. Scale: l, o × ⅓; n × ⅔; s, u × ½; i, k, p, t × ⅔; h, q, r × ¾; m × ⅘; a × 1⅓; f, j, v, w × 2; c × 2½; d × 4; g × 5; b × 6; e × 8.

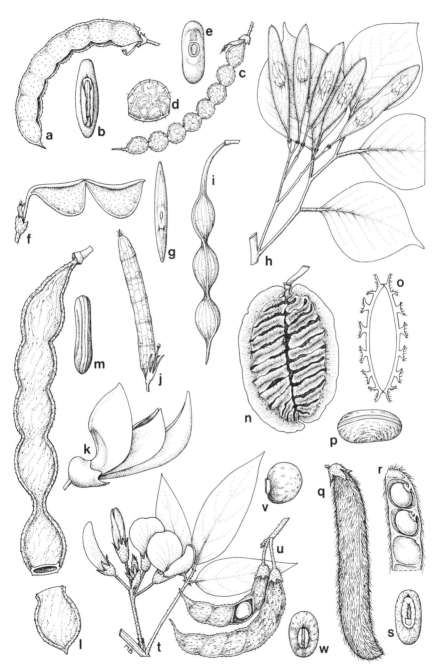

1−2×0.7−1.3cm; racemes shorter and denser, peduncles appressed whitish pubescent, pedicels 4−9mm in flower, 6−12mm in fruit; pods with 8−13 compartments, densely spreading brownish pubescent; aril less than 0.25mm tall.

Bhutan: S—Deothang district (Gomchu); **Sikkim:** Sivoke Terai. August−October.

6. D. oblongum Bentham. Fig. 43 c−e

Shrub 1−3m. Leaves 1-foliate, coriaceous, narrowly ovate or elliptic, 5−14×2−3.5cm, acute or obtuse, base rounded, appressed pubescent especially beneath; stipels narrowly triangular, 2−4mm; stipules lanceolate, 8−12×2−2.5mm, deciduous; petioles 0.5−1cm. Flowers in axillary or terminal racemes or panicles, buds concealed at first by ovate acuminate bracts 7−10mm, pedicels 1.5−2cm. Calyx 2.5−3mm. Petals blue-purple, 8−12mm, standard elliptic or suborbicular, 6−12mm broad, wings oblong, blade 6−10×4mm, keel incurved 7−9×3.5−4mm. Pods linear, curved, 2−3×0.3cm, deeply indented along both sutures into 6−10 suborbicular segments, ± glabrous.

Bhutan: C—Punakha, Mongar and Tashigang districts. Open rocky hillsides and by streams, 1525−1830m. August−October.

7. D. gangeticum (L.) DC.

Similar to *D. oblongum* but leaves herbaceous, generally larger 3.5−20×2.5−7cm, acute or acuminate, densely appressed pubescent beneath; inflorescence branches very slender, bracts subulate, 3−7mm, pedicels 3−4mm; calyx c 2 mm; petals white, yellow or purplish, 3−4mm, standard obovate, 3−4mm broad; pods 1−2.5×0.25cm, pubescent with minute hooked hairs, indented along lower suture and dividing into 4−8 oblong rounded segments.

Bhutan: S—Gaylegphug district, **C**—Punakha and Tongsa districts; **Sikkim.** Margin of cultivated land, 750−1400m. March−April.

8. D. velutinum (Willdenow) DC.; *D. latifolium* DC.

Similar to *D. oblongum* but leaves broadly ovate, 4−13×3−9cm, densely pale pubescent especially beneath; stipules triangular 2−7×1−3.5mm, abruptly narrowed from broad base to subulate apex; flowers in fascicles of 2−5 forming ± dense racemes; calyx 2−3mm; petals purplish, 4−6mm, standard obovate, 4−5mm broad; pods narrowly oblong, 1.5−2.5cm, indented along both sutures and dividing into 5−7 oblong segments c 3×2mm, densley pubescent.

Bhutan: S—Gaylegphug district (near Gaylegphug) (117); **Sikkim.** 270m.

9. D. renifolium (L.) Schindler; *D. reniforme* (L.) DC.
Weak prostrate herb with stems c 30cm, or thin shrub up to 2m. Leaves transversely elliptic or reniform, 0.5−2(−3)×1−4(−5)cm, broadly acute, obtuse or emarginate, base rounded, cuneate or cordate, ± glabrous; petioles 1−2cm; stipels subulate 0.5−1mm; stipules narrowly triangular, 3−6×0.5−1mm. Racemes terminal or axillary, 5−15cm. Calyx c 2 mm. Petals white or bluish, 4−5mm, standard obovate, 3−4mm broad. Pods oblong, 1.5−2.5cm, indented along lower suture and dividing into 3−5 segments 3−5×2.5−3mm, glabrous.
Bhutan: C—Thimphu and Tongsa districts. On grassy slopes, 1300−2300m. August.

10. D. oojeinense (Roxb.) Ohashi; *Ougeinia dalbergioides* Bentham. Nep: *Sandan Pipli* (34)
Tree 5−12m. Leaves coriaceous, pinnately 3-foliate, lateral leaflets ovate, 5.5−9×3.5−7.5cm, terminal leaflets broadly rhombic-elliptic, 9−16×5−12cm, acute or obtuse, base rounded or cuneate, glabrous; petioles 4−10cm; stipels subulate 3−4mm; stipules triangular 3−3.5× 0.5mm. Flowers in dense clusters of short racemes on branches of previous year. Calyx 3−4.5mm. Petals pink or purplish, 9−11mm, standard elliptic 7−8mm broad, wings and keel oblong, blade 6.5−7×2.5−3mm. Pods oblong, 2.5−6×0.7−1.2cm, undulate along both sutures and dividing into 2−4 rounded segments 1−3×0.75−1cm, glabrous, leathery.
Sikkim. March.
The wood is tough and strong and is used for making carriages and furniture (48).

11. D. microphyllum (Thunberg) DC.; *D. parvifolium* DC.
Slender ± prostrate subshrub; stems 15−60(−150)cm. Leaves 3-foliate or sometimes 1-foliate, leaflets oblong or suborbicular, 2−6 (−17)×1−6(−10)mm subacute, obtuse or emarginate, mucronate, rounded or cordate at base, sparsely and weakly hirsute beneath; petioles 3−8mm; stipels minute; stipules ovate, 2.5−5×0.5−1mm. Racemes few (c 5)-flowered. Calyx teeth lanceolate, 3−3.5×0.5mm, densely brown pilose. Petals purplish, 4−5mm, standard obovate 3.5−4mm broad. Pods oblong, 5−13(−20)×2−3mm, undulate along both sutures to form 2−4 compartments, dehiscing along lower suture.
Bhutan: C—Thimphu, Punakha, Tongsa, Mongar and Tashigang districts. On sand and gravel in open grassland or by roadsides, 1675−2500m. April−October.

12. D. triflorum (L.) DC.
Similar to *D. microphyllum* but leaflets obovate, 3−10(−12)mm long and broad, apex emarginate, base broadly cuneate; flowers 2−5 in whorls at

leaf axils; calyx 2.5−3mm; petals purplish 3.5−4.5mm; pods oblong, 6−17×2−3mm, undulate along lower suture and dividing into 2−5 segments.

Bhutan: S—Phuntsholing and Gaylegphug district, **C**—Punakha, Tongsa and Tashigang districts; **Sikkim**. Damp sandy river banks, 300−1650m. March−October.

13. D. kulhaitense Prain

Shrub 50−150cm, branches greyish pubescent at first. Leaves coriaceous, 3-foliate, leaflets ovate-elliptic, 3.5−10×1.5−3.5cm, acuminate, base rounded, glabrous above, whitish tomentose beneath; petioles 3−5cm; stipels ovate, 2−3×0.5mm; stipules ovate-elliptic, 7−15×1.5−2.5mm. Racemes or panicles axillary or terminal. Calyx 4−5mm. Petals rose-purple, 8−12mm, standard obovate, 5−6mm broad. Pods undulate along lower suture and dividing into 8−9 squarish segments 4.5×3.5mm, glabrous.

Sikkim: Penlong La near Gangtok. 2100m. September−October.

14. D. khasianum Prain

Similar to *D. kulhaitense* but leaves herbaceous, elliptic or ovate, 2.5−6×2−3cm, appressed pubescent beneath; stipules triangular 4−5×2−2.5mm, deciduous; pods dividing into 5−8 oblong segments 5−7×4mm, glabrous.

Bhutan: S—Sarbhang district (near Noonpani), **C**—Tongsa district (below Tama); **Sikkim; Chumbi**. Subtropical forest slopes, 800−1050m.

15. D. elegans DC.; *D. tiliifolium* (D. Don) Wall., *D. oxyphyllum* sensu F.B.I. Dz: *Tatur Shi;* Sha: *Beymangrobu, Neptang Shing*; Nep: *Sarkinu* (34)

Shrub or small tree 1.5−4m. Leaves 3-foliate, leaflets ovate or broadly elliptic, lateral ones somewhat smaller than the terminal, 3.5−8×2−6cm, acute, base rounded or cuneate, margins entire or undulate, finely sericeous especially beneath; petioles 2−9cm; stipels linear 1−4mm; stipules lanceolate, 4−10×1−3mm, deciduous. Racemes axillary or terminal sometimes from the axils of fallen leaves. Calyx 3−4mm. Petals mauve or bright purple, 10−15mm, standard broadly elliptic, 9−12mm broad, wings and keel oblong, blade 10×4mm. Pods oblong, undulate along both sutures and dividing into 3−7 elliptic segments 7−10×4−6mm, appressed pubescent.

Bhutan: S—Deothang district, **C**—Thimphu, Punakha, Tongsa and Tashigang districts, **N**—Upper Mo Chu district; **Sikkim**. Dry rocky hillsides, 750−2450m. March−July.

16. D. heterocarpon (L.) DC.; *D. polycarpum* (Poiret) DC.

Prostrate or ascending herb or shrub up to 3m. Leaves usually 3-foliate but sometimes also 1-foliate leaves present, leaflets elliptic or obovate

(1−)2−7.5×(0.7−)1−3.5cm, obtuse or emarginate, base rounded, appressed greyish pubescent beneath; petioles 1−3cm; stipels subulate 1−7mm; stipules lanceolate, 7−18×2−3mm. Racemes elongate, axillary or terminal, bracts ovate, 7−8×2.5−4mm, acuminate. Calyx 2.5−3mm. Petals purplish 4.5−6.5mm, standard elliptic to suborbicular 4−6mm broad, wings and keel oblong or obovate, blades 3.5−5×1.5−2.5mm. Pods undulate along lower suture and dividing into 3−7 squarish segments 3−4×3mm, sparsely pubescent with hooked hairs.

Bhutan: S—Gaylegphug district (Tama), C—Punakha district (Tinlegang) and Tongsa district (near Tongsa); **Sikkim.** On dry soil, 1850−1950m. September−November.

There are two varieties: var. *heterocarpon* in which the inflorescence rachis bears spreading hooked hairs and var. *strigosum* van Meeuwen in which the rachis is densely covered in appressed straight hairs. Both varieties are known from Sikkim but only var. *heterocarpon* has so far been found in Bhutan.

17. D. styracifolium (Osbeck) Merrill; *D. retroflexum* (L.) DC., *D. capitatum* (Burman f.) DC.

Prostrate, ascending or erect shrub 0.5−2m. Leaves 3-foliate or sometimes 1-foliate, terminal leaflets twice as large as laterals, broadly elliptic to orbicular, 2−6×2−5.5cm, obtuse, mucronate, base rounded or cordate, densely greyish sericeous beneath, lateral leaflets 7−25×5−18mm; petioles 1.5−3cm; stipels subulate c 5mm; stipules lanceolate, 10−15×2mm. Racemes short (1−3cm) axillary and terminal, bracts ovate, 3−4× 2.5−3mm, white pilose. Calyx 3−3.5mm, white pilose. Petals purplish 4−5mm, standard suborbicular, 5mm broad. Pods 1.5−2cm borne on sharply deflexed pedicels, segments 3−5, oblong, 3.5×2.5mm, folded back on each other first, becoming straight later, minutely pubescent with hooked hairs.

Sikkim: Terai.

18. D. podocarpum DC.; *D. oxyphyllum* DC. Fig. 43 f&g

Erect herb 30−150cm. Leaves 3-foliate, leaflets ovate-elliptic, 4−9.5×1.5−3.5cm, acuminate, base rounded or cuneate, margins entire, principal lateral veins reaching margins, ± glabrous; petioles 5−10cm; stipels subulate, 1−4mm; stipules linear, 5−10×0.5−1mm. Racemes slender, axillary and terminal, pedicels 4−5mm. Calyx 1.5mm. Petals white, pink or purple, 3.5−4mm, standard broadly elliptic, 2.5−3mm broad. Pods with basal stalk 3−4mm, deeply indented along lower suture into 2−3 semicircular segments 7−9×3−4mm, minutely pubescent with hooked hairs.

Bhutan: S—Chukka district (near Chukka), **C**—Punakha district (Tinglegang and Rinchu) and Tongsa district (Shamgong); **Sikkim.** 1400–2500m. July–August.

Because of their leaflets, which are narrower than in the type, the E Himalayan plants have been placed in subsp. **oxyphyllum** (DC.) Ohashi.

19. D. laxum DC.; *D. podocarpum* DC. var. *laxum* (DC.) Baker

Similar to *D. podocarpum* but leaflets subcoriaceous, sparsely pubescent, lateral veins forming a network before reaching margin; stipules ovate-lanceolate 8–10×2–3mm; pedicels 7–9mm; petals 5–7mm; pod segments 7–18×4–6mm.

Bhutan: S—Gaylegphug district (near Gaylegphug), **C**—Punakha district (between Rinchu and Mishichen); **Sikkim.** 270–1500m. May.

20. D. williamsii Ohashi

Similar to *D. podocarpum* but leaflets broadly ovate, 3.5–8×2.5–6cm, lateral veins reaching margins, sparsely appressed pubescent on upper surface; stipules triangular, 5–6×1mm; pedicels 1.5–2cm; petals 12–16mm; mature fruit unknown.

Bhutan: C—Punakha and Tongsa districts, **N**—Upper Kuru Chu district. On grassy banks at forest margins, 1350–2450m. July–August.

21. D. duclouxii Pampanini

Similar to *D. podocarpum* but leaflets ovate, 5–7×4–6cm, acuminate, margins somewhat undulate; stipels subulate 1–2mm; stipules membranous, broadly ovate, 6–10(–15)×4–6(–9)mm; bracts similar to stipules; calyx 5–6mm; petals pink or yellow, 12–15mm; pods segments 10–15×6–7mm.

Bhutan: C—Mongar district (near Lhuntse and Denchung). Damp areas in forest, 2100m. July–August.

22. D. concinnum DC.

Shrub up to 1.5m with slender drooping branches. Leaves 3-foliate, leaflets elliptic or narrowly obovate, 2–7×1–3.5cm, obtuse or subacute, base rounded, softly appressed pubescent beneath, lateral veins parallel, prominent beneath; petioles 1–2cm; stipels linear c 5mm; stipules lanceolate, 10–20×2–5mm. Racemes elongate, axillary or terminal. Calyx 2.5mm. Petals purple, 5–7mm, standard broadly obovate, 5–6mm broad. Pods deeply indented along lower suture into 3–6 rounded segments 3–4.5×2–3mm, minutely pubescent with hooked hairs.

Bhutan: locality unknown.

23. D. multiflorum DC.; *D. floribundum* (D. Don.) G. Don

Shrub 0.5–2m. Leaves ± coriaceous, 3-foliate, elliptic or obovate,

2−6(−9)×1.5−3.5(−4)cm, acute or obtuse, base rounded, appressed pubescent especially beneath; petioles 2−5cm; stipels subulate 2−3mm; stipules lanceolate, 6−10×1.5−2mm. Calyx 3−4mm. Petals purplish, 7−8mm, standard obovate, 6mm broad. Pods undulate along both sutures but more deeply so along lower one, and dividing into 4−7 elliptic segments 3−4×3mm; appressed pubescent.

Bhutan: S—Gaylegphug and Deothang districts, C—Thimphu to Tashigang districts, N—Upper Mo Chu district; **Sikkim.** On banks in Evergreen oak forests, 1200−2750m. July−September.

24. D. sequax Wall.; *D. sinuatum* (Miquel) Baker

Similar to *D. multiflorum* but leaves rhombic or ovate, 2.5−9.5× 1.5−7cm, acute or acuminate, rounded at base, margins undulate, appressed pubescent on both surfaces; petioles 1.5−5cm; stipels filiform 1−4mm; stipules linear, 4−5×1mm, deciduous; pods dividing into 9−14 ellipsoid segments, c 3×2.5mm, densely covered with short hooked hairs.

Bhutan: S—Phuntsholing district (Kamji), C—Tongsa district (Shamgong); **Sikkim.** 300−1200m. September−October.

25. D. confertum DC.

Shrub up to 2m. Leaves coriaceous, 3-foliate, leaflets elliptic or obovate 4−12×2−6.5cm, obtuse or acute, base rounded, appressed pubescent especially beneath; petioles 3−6.5cm; stipels linear 2−4mm; stipules ovate-lanceolate, 5−7×2−4mm, early deciduous. Racemes axillary and terminal, enclosed at first within deciduous overlapping ovate bracts 8−12×4−5mm forming a cone-like inflorescence. Calyx 3−5mm. Petals purplish, 10−14mm, standard obovate 6−9mm broad. Pods densely pale spreading hirsute, dividing into 3−4 elliptic segments 7−10×3−4mm.

Bhutan: S—Samchi district, C—Mongar and Tashigang districts; **Sikkim.** In Chir Pine forest, 800−1500m. August−November.

26. D. caudatum (Thunberg) DC.; *D. laburnifolium* (Poiret) DC.

Shrub 30−100cm. Leaves 3-foliate, leaflets elliptic, 3−11×1−3cm, acuminate, base cuneate, sparsely appressed pubescent beneath; petioles 3−5cm; stipels subulate 3−5mm; stipules narrowly triangular at base, subulate above, 6−10×1mm. Racemes axillary and terminal, 5−15cm or more, flowers each with 2 linear bracteoles at base of calyx. Calyx 4−5mm. Petals white, 8−10mm, standard elliptic 5−6mm broad. Pods undulate along both sutures and dividing into 4−6 elliptic segments 12−15×4−5mm, pubescent with minute hooked hairs.

Bhutan: S—Gaylegphug district (near Gaylegphug), C—Mongar district (Ngasamp); **Sikkim:** Dikchu. 270−1200m. August−September.

76. LEGUMINOSAE

27. D. laxiflorum DC.

Shrub 30–150cm, ± erect. Leaves 3-foliate or some 1-foliate, leaflets ovate-elliptic 4–13×1.7–7.5cm, acute or acuminate, rounded or cuneate at base, appressed pubescent especially beneath; petioles 2–6cm; stipels narrowly lanceolate, 10×1mm; stipules narrowly triangular, 10–15× 2–3mm. Calyx 3–4mm. Petals white or blue, 5–6mm, standard obovate 3–4mm broad. Pods 2–5cm, densely pubescent with minute hooked hairs, shallowly indented along both sutures and dividing into 6–12 segments 4–6×1–2mm.

Bhutan: S—Phuntsholing district (Phuntsholing), **C**—Punakha district (between Punakha and Sinchu La); **Sikkim:** Darogadara. Subtropical forest slopes, 250–1370m. September–October.

31. URARIA Desveaux

Shrubs or perennial herbs. Leaves odd-pinnate, leaflets 1–9, margins entire or undulate; stipels present; stipules persistent. Flowers in terminal racemes or panicles, bracts persistent or deciduous, pedicels becoming hooked distally. Calyx deeply divided into 5 teeth, the two upper ones shorter, not increasing in size. Petals shortly clawed, wings adhering to longer upcurved keel. Stamens diadelphous. Pods compressed, divided transversely into rounded segments each folded over the other or segments ± triangular and becoming coiled into a flat circle.

1 Upper leaves 5–9-foliate; leaflets linear-lanceolate, 1(–2)cm broad
 .... **1. U. picta**
+ Upper leaves 1–3-foliate; leaflets ovate, elliptic or obovate, mostly 1.5–6cm broad, seldom less than 1cm 2

2. Bracts persistent; calyx longer than petals .... **2. U. lagopodioides**
+   Bracts deciduous; calyx shorter than petals 3

3. Herbs, stems up to 30cm; leaflets always solitary .... **3. U. prunellifolia**
+   Shrubs, stems 0.5–4m; leaflets usually 3 (sometimes solitary in *U. rufescens*) .... **Species 4–6**

1. U. picta (Jacquin) Desveaux

Herb, stems 1–2m, covered with minute, pale, hooked hairs. Leaflets (3–)5–7(–9), linear-lanceolate, 4–20×0.5–1(–2)cm, obtuse or acuminate, base rounded, usually with a pale median band on upper surface, glabrous above, pubescent beneath; stipules lanceolate 0.5–1.5cm; stipels subulate 2–8mm. Flowers in elongate racemes 9–24(–40)cm, bracts lanceolate 1–1.5cm, deciduous, pedicels 5mm in flower, 10mm in fruit, becoming hooked. Calyx teeth subulate c 5mm, pilose. Petals purple, 7–8mm, standard obovate c 4mm broad, wings oblong, blade 4×2.5mm, keel spathulate c 7×3mm. Pod segments usually 4, ovate, c 4mm, black with whitish reticulations, folded back on each other.

Bhutan: C—Tashigang district (Dangme Chu); **Sikkim:** Darjeeling. Dry grassy banks, 900m. June–July.

2. U. lagopodioides (L.) Desveaux; *U. lagopoides* (Burman f.) DC.

Herb, stems prostrate or ascending, 20–75cm, soft pale brown pubescent. Leaves 3-foliate or sometimes 1-foliate, leaflets ovate to obovate 1.5–6.5×0.7–4cm, obtuse or emarginate, mucronate, base rounded; stipules lanceolate 4–8mm, acuminate; stipels subulate 2–4mm. Racemes dense 2.5–8cm, bracts ovate 8–10×5mm, acuminate, hirsute, persistent. Flowers bluish, scarcely exserted from bracts and calyces, pedicels c 4mm. Calyx 8–10mm teeth subulate, pilose. Pod segments 2–3, elliptic, c 2.5mm, dark brown.

Bhutan: C—Thimphu district (near Thimphu); **Sikkim:** Darjeeling. On dry sandy soil, 2300m. August.

3. U. prunellifolia Baker

Herb, stems 15–30cm, densely pubescent with minute hooked hairs. Leaves 1-foliate, oblong or ovate, 4–12×1.5–3.5cm, acute or obtuse, base rounded or cordate, pubescent especially on veins beneath; petioles up to 5mm; stipules lanceolate 6–9mm, acuminate; stipels subulate 2.5–3mm. Racemes 3–6cm, bracts ovate, 4–5×2.5mm, pedicels c10mm, pilose, becoming hooked at maturity. Calyx 3–4mm. Petals 5–6mm. Pod segments 6–7, somewhat triangular, becoming coiled into a flat circle.

Assam Duars: S of Deothang; **W Bengal Duars:** Siliguri. April–May.

4. U. lagopus DC.

Shrub 1–4m, branches erect or subscandent, spreading brown hirsute. Leaves 3-foliate, leaflets elliptic or ovate, 5–10×2.5–6cm, acute or obtuse base rounded, margin entire or undulate, brown pubescent especially on veins beneath; petioles 7–10cm; stipules ovate, 10–12mm, acuminate to a subulate point; stipels similar but half as long as stipules. Flowers in dense, elongate, brown hirsute racemes 8–22cm, pedicels c10mm, bracts ovate 8–9×4–5mm, acuminate. Calyx 4–5mm. Petals 8–9mm, purplish, standard suborbicular 7–8mm broad, wings oblong, blade 7–8×3mm, keel obtuse c 6×3mm. Pod segments 2–6, ovate, 3.5×3mm, pale brown, pubescent, folded back upon each other.

Bhutan: S—Chukka district, **C**—Punakha, Tongsa and Tashigang districts; **Sikkim.** Gravelly soil and among boulders, 1300–2200m. September–October.

5. U. rufescens (DC.) Schindler; *U. hamosa* (Roxb.) Wight & Arnott

Similar to *U. lagopus* but stems up to 2m; leaves 1−3-foliate, leaflets ovate-elliptic, 4−12×2.5−6cm; inflorescence lax, pedicels 3−4mm, bracts ovate 8−12mm, long acuminate, deciduous; calyx 3−4mm; petals 6−7mm purplish; pod segments 5−6, rounded, c 2mm diameter, black, tightly folded on each other.

Sikkim: Mahanadi Terai. August−October.

6. U. sinensis Franchet

Similar to *U. lagopus* and *U. rufescens* but more slender, stems 60−100cm; leaves smaller, elliptic or obovate, 2−6×1.5−3.5cm, obtuse or emarginate, base cuneate or rounded, sparsely pubescent especially beneath; petioles 1−3cm; stipules lanceolate c 8mm; stipels linear c 2mm; flowers in loose racemes or panicles, bracts ovate 6−7mm, acute, pedicels 1−1.5cm, sparsely pubescent; calyx 3−4mm, teeth triangular; petals 8−12mm, purple; pods segments 6−8, rounded 3−3.5mm diameter, brown, glabrous or sparsely pubescent, surface reticulate.

Bhutan: C—Mongar district (near Lhuntse). On open dry hillsides, 1830−2550m. July−August.

32. CHRISTIA Moench

Annual herbs. Leaves 3-foliate but often reduced to terminal leaflet only; stipels present; stipules minute. Flowers in axillary or terminal racemes. Calyx campanulate, divided to middle into 5 subequal teeth, becoming enlarged in fruit. Petals clawed, scarcely longer than calyx, wings adherent to keel. Stamens diadelphous. Pods compressed, divided transversely into 1-seeded segments, folded on top of each other within persistent calyx, indehiscent.

1. C. vespertilionis (L.f.) van Meeuwen; *Lourea vespertilionis* (L.f.) Desveaux. Fig. 42 i−k

Stems erect, 30−70cm, sparsely pubescent. Terminal leaflet transversely elliptic or bat-wing-like, 0.7−2×4−8cm, apex emarginate, base rounded or emarginate, finely pubescent, whitish along veins, lateral leaflets, if present, obcordate or obdeltoid, 1−2×1−2cm; stipules linear lanceolate 1−2mm; petioles 1.5−3cm with a pair of subulate stipels c1mm near apex. Racemes 10−20-flowered. Calyx in flower 5−6mm, membranous, prominently reticulate-veined, pilose. Petals white, 5−6mm, standard obovate, 2.5−3mm broad, wings and keel blades c 3.5mm. Pod segments 4−5, elliptic c 3×2mm, within persistent calyx 9−12mm.

Bhutan: C—Punakha district (Chusom). 1500m. April−September.

Also cultivated at Phuntsholing for its decorative foliage. Native of Indo-China and Malaysia.

33. ALYSICARPUS Desveaux

Erect or spreading annual herbs. Leaves 1−3-foliate; stipels present; stipules persistent. Flowers in terminal or axillary racemes, bracts scarious, deciduous. Calyx stiff, papery, deeply divided, teeth sometimes imbricate at base, upper 2 often connate to near apex. Petals scarcely longer than calyx, shortly clawed, wings adherent on keel. Stamens diadelphous, stamens alternately long and short. Pods compressed or terete, transversely constricted or dividing into 1-seeded segments, dehiscent or indehiscent.

1. A. rugosus (Willdenow) DC.

Stems 15−60cm, glabrous except for a line of hair below each leaf. Leaves 1−3-foliate, leaflets elliptic or lanceolate 1−7×0.5−3.5cm, acute or obtuse, base rounded or cuneate, often with a pale median band on upper surface, minutely pubescent beneath with longer stiff hairs on veins, sessile or on petioles up to 5mm; stipules lanceolate 0.7−1.5cm, acuminate, brownish, scarious; stipels linear-lanceolate c 2mm. Racemes dense, bracts broadly ovate 6mm. Calyx 6−8mm, accrescent, teeth lanceolate, imbricate at base, parallel veined, ciliate. Petals reddish, 5−6mm, standard obovate, 3−4mm broad. Pods linear, constricted into 3−5 elliptic segments 2−3mm, compressed, brown, transversely rugose, concealed within accrescent calyx.

Sikkim: Terai. 300−900m. September−November.

2. A. vaginalis (L.) DC. Fig 43j

Similar to *A. rugosus* but stems pubescent with minute hooked hairs, leaflet always solitary, broadly elliptic to narrow lanceolate, 0.5−8.5×0.5−3cm, obtuse, acute or acuminate, base rounded or cordate, pubescent beneath; petioles 2−12mm; racemes lax, slender; calyx teeth not imbricate at base 3−4mm; petals whitish or pink, c 5mm; pods terete, 1−2.5×0.25cm, faintly reticulate, rugose, distinctly jointed and dividing into 3−7 open-ended segments 2.5−3mm, dehiscent.

Bhutan: C—Punakha district (Punakha); **Sikkim:** Terai. On margin of paddy fields, 1300m. August−September.

34. CAMPYLOTROPIS Bunge

Erect shrubs. Leaves pinnately 3-foliate. Racemes axillary, bracteate. Calyx campanulate, divided to below middle into lanceolate teeth, upper pair connate to near apex. Petals twice as long as calyx, standard ovate, wings adherent on upwardly curved keel. Stamens diadelphous. Pods compressed, 1-seeded, indehiscent.

1. C. speciosa (Schindler) Schindler; *C. eriocarpa* (Maximowicz) Schindler, *Lespedeza eriocarpa* sensu F.B.I. non DC.

Shrub 1−3m. Leaves subcoriaceous, leaflets oblanceolate or obovate,

1.5−3×0.5−1.75cm, acute, obtuse or emarginate, mucronate, base cuneate, veins ± prominent on upper surface, glabrous above, appressed pubescent beneath; petioles 0.5−3cm; stipules lanceolate 5−6mm. Racemes many flowered, 5−12cm, bracts lanceolate 3−4mm, deciduous. Calyx 4−5mm, appressed brownish pubescent. Petals purplish, 8−10mm, standard c 10×6mm, folded forward over wings and keel, wings oblong c 10×3mm, keel c 10×2mm. Pods elliptic, 7−8×3−4mm, acuminate, style c6mm deciduous, but mucronate base 2−3mm persisting, appressed brownish pubescent.

Bhutan: C—Thimphu, Punakha, Tongsa and Tashigang districts; **Sikkim.** On open banks at margins of oak forests, 2400−2500m. September−October.

2. C. griffithii Schindler; *C. macrostyla* (D. Don) Miquel var. *griffithii* (Schindler) Ohashi

Similar to *C. speciosa* but leaflets usually obtuse, emarginate or retuse; stipules 2−3mm; racemes 5−10-flowered, 2−3cm, bracts narrowly ovate c1mm, deciduous; pods 15−18×6−7mm, acuminate, sparsely pubescent and finely reticulate, style c 6mm, deciduous.

Bhutan: C—Punakha district (near Wangdu Phodrang) and Tongsa district (near Banjormani). Warm broad-leaved forest slopes, 1150−1400m. April−May.

35. LESPEDEZA Michaux

Subshrubs. Leaves digitately 3-foliate, leaflets entire, subsessile; stipels absent; stipules persistent. Flowers solitary or crowded in axillary subumbellate racemes, bracts linear, small. Calyx deeply divided into 5 subequal teeth. Petals clawed, wings adherent to keel. Stamens diadelphous. Pods ellipsoid, compressed, 1-seeded, indehiscent.

1. L. juncea (L.f.) Persoon; *L. sericea* Miquel, *L. cuneata* (Dumont de Courset) G. Don

Stems 60−100cm, often simple, unbranched. Leaflets spathulate or oblanceolate, 6−15×2−4mm, obtuse, truncate or emarginate, mucronate (mucro c 1mm), base cuneate, glabrous or sparsely pubescent above, more densely and often whitish sericeous beneath; petioles 1−2mm; stipules linear-lanceolate, 3−4mm, brownish. Flowers solitary or up to 6, ± sessile in leaf axils. Calyx c 5mm, teeth linear-lanceolate, brown, appressed pubescent. Petals white tinged purple, 6−7mm, standard obovate c 5mm broad, wings oblong, blade c 4×1.5mm, keel spathulate c 4.5×2mm. Pods c 3×2mm coriaceous, appressed pubescent.

Bhutan: S—Chukka and Deothang districts, **C**—Thimphu, Punakha and Tashigang districts. On dry sandy banks, 1200−2550m. July−September.

The species is variable in the length of the peduncle and density of indumentum on the leaves. In typical var. *juncea* from Western Himalaya the inflorescence is borne on a short (5−10mm) peduncle. In East Himalayan plants, however, the flowers are always tightly clustered around the leaf axils and hence belong either to var. **variegata** (Cambessedes) Ali, in which the leaves are moderately pubescent, or to var. **sericea** (Thunberg) Lace & Hemsley in which the leaves are more densely pubescent.

2. L. gerardiana Maximowicz

Similar to *L. juncea* but leaflets broader, commonly up to 5mm; flowers 3−7 in crowded subumbellate racemes, on peduncles up to 2.5cm; calyx 8−9mm; petals 10−12mm, yellow, keel purplish at tip.

Bhutan: C—Thimphu, Tongsa and Bumthang districts; **Chumbi; Sikkim.** On open banks, 2450−3650m. August−September.

36. ERYTHRINA L.

The name *Khelsho* (Med) applies to several unspecified species of *Erythrina*.

Shrubs or trees, branchlets often armed with thorns or conical spines and sometimes stellately pubescent. Leaves pinnately 3-foliate, leaflets entire or margins sometimes undulate in *E. suberosa*; stipels fleshy, gland-like, usually one at base of each lateral leaflet and 2 at base of terminal leaflet; stipules small, deciduous or persistent. Flowers often appearing before or with new leaves, borne in axillary or terminal racemes. Calyx segments connate to form an ovoid or ellipsoid body in bud through which expanding corolla bursts to leave a spathaceous calyx, deeply divided dorsally, or a bilabiate or truncate calyx, individual teeth absent or, as in *E. variegata,* appearing as weak subulate points. Petals shortly clawed, standard much larger than and folded over wings and keel. Stamens monadelphous, ± as long as standard. Pods ± stalked, oblong-ellipsoid, scarcely or deeply constricted between seeds.

1. Calyx spathaceous, deeply divided dorsally .... **Species 1 & 2**
+ Calyx otherwise 2

2. Calyx truncate or weakly bilabiate .... **3. E. arborescens**
+ Calyx bilabiate to middle .... **4. E. suberosa**

1. E. stricta Roxb. Nep: *Phaledo* (34)

Tree 6−20m, branchlets armed with whitish conical spines. Leaflets ovate, (6−)15−20×(4−)8−20cm, acute or acuminate, base rounded, brownish pubescent at first. Flowers appearing before leaves. Calyx c 15mm deeply divided dorsally, densely brown pubescent. Standard narrowly ovate, 4.5−5.5×1.8−2cm, scarlet, wings oblong c 7×2.5mm, keel ovate-elliptic 2−2.5×0.7cm pale. Pods 10−15×1.5cm, thinly coriaceous, on stalks 1−1.5cm, 1−3-seeded, scarcely constricted; seeds oblong c 1.5× 0.6cm.

Bhutan: S—Phuntsholing and Sarbhang districts, **C**—Tongsa and Tashigang districts; **Sikkim**. Subtropical and Warm broad-leaved forests, 200−1450m. February−April.

2. E. variegata L.; *E. indica* Lamarck. Nep: *Phaledo* (34)

Similar to *E. stricta* but trees up to 30 or 40m, spines brown or blackish; calyx 2−2.5cm, brown stellately pubescent at first, glabrescent, subulate teeth projecting c 2mm at apex; standard 4.5−5×1.5−1.7cm, wings and keel c 1.5cm; pods coriaceous 15−30cm, ± constricted between seeds, somewhat reticulately veiny, spongy inner wall of pod separating from outer; seeds oblong, 17×8×8mm, reddish-brown.

Sikkim: locality unknown.

Cultivated in hedgerows and in tea-gardens (34).

3. E. arborescens Roxb. Dz: *Chassee;* Nep: *Phaledo* (34), *Roringa* (34). Fig. 41 s−u

Tree 5−15m, branchlets with brown spines. Leaflets ovate, 12−20×9−16cm, acute or shortly acuminate, base rounded or shallowly cordate, sparsely or densely pubescent beneath; stipels c1mm; petioles 12−25cm sometimes bearing a few spines; stipules lanceolate c3mm deciduous. Flowers appearing after leaf development, racemes axillary on peduncles up to 30cm. Calyx 8−10mm, truncate or very shallowly bilabiate, sparsely brown pubescent. Standard ovate-elliptic, 3.5−4.7×1.5−2.2cm, orange-scarlet, wings oblong, 12×5mm, pale coloured, keel triangular, 15−20×10mm, whitish. Pods oblong-ellipsoid, 15−25×1.5−2.5cm, scarcely constricted between the 4−6 reniform black seeds 1.5−2×1×1.3cm.

Bhutan: S—Chukka and Gaylegphug districts, **C**—Punakha to Tashigang districts; **Sikkim**. Warm broad-leaved forests, 1525−2440m. July−August.

4. E. suberosa Roxb.

Similar to *E. arborescens* but branchlets without spines; leaflets ovate, 9−20×7−17cm, tapering to a blunt point at apex, base rounded or cuneate, margin entire or undulate, densely pubescent beneath at first; flowers appearing with young leaves; calyx c1cm bilabiate to middle or below; standard ovate-elliptic 3−3.5×1.5cm, scarlet, keel greenish, ovate 12×8mm, wings oblong or ovate 6−7×3−4mm; pods thinly coriaceous

7−9×1cm, scarcely constricted between seeds, seeds dark brown or black c 9×5mm.

Bhutan: locality unknown; **Sikkim.** March−May.

The wood, though light, is durable and is used for making jars and boxes for household use (48).

37. MUCUNA Adanson

Herbaceous or woody twiners. Leaves pinnately 3-foliate; leaflets entire, lateral ones very asymmetric, stipels usually present; stipules small, caducous. Flowers in racemes, axillary or borne on old leafless wood. Calyx broadly campanulate, bilabiate, upper lip entire or 2-lobed, lower one 3-lobed. Standard rounded, auricled at base, usually smaller than other petals and folded forward over them, wings oblong, keel linear-oblong, incurved, acute and hardened at tip, as long as or longer than the adherent wings. Stamens 10, diadelphous. Pods oblong or S-shaped, terete or ± compressed, coriaceous or woody, sometimes covered with irritant bristles; seeds oblong with a short hilum or discoid, ± compressed, with hilum usually extending around more than half the circumference, arillate or exarillate.

1. Large woody twiners; inflorescences borne on old wood; pods linear, ± straight, valves without lamellae, ± covered with short brown hair but without irritant bristles 2
+ Woody or semi-woody twiners; inflorescences borne in axils of leaves on upper branches; pods oblong, elliptic or S-shaped, valves with or without lamellae, ± densely covered with brown irritant bristles (but lacking them in *M. pruriens* var. *utilis*) 3

2. Standard ciliate apically, greenish white; sutures of pod unthickened
 1. M. macrocarpa
+ Standard eciliate apically, dark purple; sutures of pod thickened
 2. M. sempervirens

3. Leaves ± densely appressed white pubescent beneath; keel rarely more than 4cm; pods ± S-shaped, rounded in section, densely covered with brown irritant bristles (except var. *utilis*) .... **5. M. pruriens**
+ Leaves glabrous or shortly pubescent on veins beneath; keel (4.2−) 4.8−5.7cm; pods oblong or broadly elliptic, ± compressed, valves covered with lamellae bearing sparse brown irritant bristles 4

4. Bracts 2.5−3cm broad; calyx tube c 2cm broad; lamellae T or Y-shaped in section, interrupted at midline leaving a narrow gap that none crosses
 3. M. imbricata
+ Bracts up to 2cm broad; calyx tube 1−1.5cm broad; lamellae simple, ± continuous across pod .... **4. M. nigricans**

76. LEGUMINOSAE

1. M. macrocarpa Wall. Nep: *Baldengra*. Fig. 43 k—m

Woody twiner. Leaves subcoriaceous, leaflets ovate or elliptic, 11–18×6–10cm, obtuse or subacute, base rounded, appressed pubescent beneath at first; stipels not persisting; petioles 10–15cm. Racemes (5–)10–15(–20)cm borne on old leafless wood, pedicels c10mm. Calyx tube c 10mm, ± covered with appressed brown bristles, teeth 2–5mm. Standard c 3cm, greenish white, ciliate near apex, wings 4.5–6cm, dark purple, ciliate near apex, keel 6–7cm, greenish purple. Pods (25–)30–45×4–5cm, compressed, woody, constricted between the 8–15 seeds, finely appressed brown pubescent, glabrescent in parts, sutures not thickened; seeds discoid 2.5–3cm diameter, black, hilum extending to ¾ of circumference.

Bhutan: S—Phuntsholing, Chukka and Sarbhang districts; **Sikkim**. Subtropical forests, 300–2100m. March–April.

2. M. sempervirens Hemsley

Similar to *M. macrocarpa* but leaflets acuminate; petals all purple, standard and wings eciliate; sutures of pod thickened and rounded.

Bhutan: S—Chukka district (E of Jumudag). Evergreen oak forest, 2050m.

3. M. imbricata Baker; *M. imbricata* sensu F.B.I. p.p., *M. interrupta* Gagnepain. Fig. 43 n–p

Large woody twiner. Leaflets ovate or rhombic, 10–15×5–10cm, acute, mucronate, base rounded or cuneate, glabrous or with a few fine brown appressed hairs. Inflorescence 5–20cm, bracts ± persistent, broadly ovate, 2.5–3×2–2.5cm. Calyx tube broadly campanulate, c 1×2cm, teeth triangular, 5–14×4–6mm, ± densely appressed brown pubescent. Petals white (or possibly red), standard 3–3.5cm, wings and keel 5–5.5cm. Pods thinly woody, oblong-elliptic 5–12×3.5–6cm, ± compressed, valves covered with oblique lamellae, T or Y-shaped in section, all interrupted near midline to leave a longitudinal gap crossed by none, each suture with 2 marginal wings c 10mm broad, wings and lamellae sparsely covered with brown irritant bristles. Seeds 2–3, discoid, c 2.5cm diameter, purplish streaked with black.

Bhutan: S—Sarbhang district (Kami Khola), **C**—Tongsa district (Shamgong). Subtropical forest on river bank, 400m.

4. M. nigricans (Loureiro) Steudel; *M. imbricata* sensu F.B.I. p.p. Med: *Chhinpashosha;* Nep: *Baldengra, Kaoso* (34)

Very similar to *M. imbricata* but bracts up to 2cm broad; calyx tube

1–1.5cm broad, finely greyish brown appressed pubescent intermixed with a few reddish brown irritant bristles; pods up to 15×6cm, valves bearing simple oblique lamellae, some at least traversing the whole width, sparsely covered with reddish brown irritant bristles.

Sikkim: Badamtam. 600m. August.

Because of past confusion between this species and *M. imbricata*, the record (117) of *M. nigricans* from Phuntsholing requires confirmation.

A paste made from the seeds is used medicinally (146).

5. M. pruriens (L.) DC. Sha: *Zalibi*; Nep: *Kauso, Kautcho, Kaochir* (34),*Kuach* (34). Fig. 43 q−s

Annual or short-lived perennial, stems silvery pubescent at first. Leaflets elliptic or ovate, 6−15×5−10(−15)cm, obtuse or acute, rounded at base, silvery appressed pubescent especially beneath; stipels subulate c5mm; petioles 15−25cm. Racemes 5−30cm, pendent, pedicels 1−5mm. Calyx tube 5−7mm, appressed silvery pubescent intermixed with some longer fine brown bristles, teeth 5−7mm. Petals dark purple or white, standard 1.5−2.5cm, wings 3.5−4cm, keel c 4cm. Pods S-shaped, 6−8×1.5cm, terete, somewhat fleshy, covered with irritant brownish hairs 2−3mm long, divided internally into 4−6 compartments; seeds oblong-ellipsoid, 8−15×5−9×4−7mm, blackish, glossy, hilum 4−6mm surrounded by a whitish aril.

Bhutan: S—Samchi, Phuntsholing and Gaylegphug districts; **Sikkim.** Subtropical forest slopes, 200−450m. September−November.

Var. **utilis** (Wight) Burck (*M. capitata* Wight & Arnott) differs from typical var. *pruriens* in lacking the coarse brown hairs on inflorescence and calyx; petals purple or whitish, somewhat smaller, standard c1.5cm, wings 2−3cm, keel 3−3.5cm; pods up to 2cm broad, lacking the brown irritant hairs; seeds whitish brown or black, sometimes marbled.

Sikkim: Terai

This variety is sometimes cultivated as a fodder crop; the young pods and seeds are edible.

38. BUTEA Roxburgh

Erect perennial herbs, twining shrubs or tall trees. Leaves pinnately 3-foliate, leaflets entire, stipels present, stipules deciduous. Flowers in dense terminal and axillary racemes or panicles. Calyx 5-toothed, upper 2 teeth connate. Petals clawed, standard sometimes recurved, wings adherent on upcurved keel. Stamens diadelphous. Pods oblong, compressed, wing-like containing one orbicular distal seed, indehiscent.

1. Perennial herb, 2−3m; stems deeply ridged and grooved
 1. B. buteiformis
+ Twining shrub or tree 15−20m; stems not deeply ridged and grooved ... 2

76. LEGUMINOSAE

2. Twining shrub; petals 8−9mm .... **2. B. parviflora**
+ Tree; petals 4−6cm .... **3. B. monosperma**

1. B. buteiformis (Voigt) Grierson & Long; *B. minor* Baker. Sha: *Phrogpa Laga*

Erect perennial herb, stems 2−3m, deeply ridged and grooved, brownish pubescent. Leaflets ovate-elliptic, 15−45×12−35cm, acute, base rounded, appressed pubescent above, brownish sericeous beneath, stipels linear 0.7cm; petioles 10−20cm; stipules ovate 1.5×1cm. Panicle branches 15−50cm. Calyx 6−7mm very shortly toothed, brown pubescent. Petals orange-red, standard broadly elliptic, 1.5×1cm, reflexed, with 2 lines of pubescence in lower half, wings narrowly ovate, curved, blade 1.3×0.5cm, keel ovate 1.7×0.7cm; pubescent. Pods 9−10×2−3cm, erect, subacute, densely brownish tomentose.

Bhutan: C—Mongar district (near Lhuntse) and Tashigang district (between Rongtung and Tashigang); **Sikkim.** On dry open hillsides, 1220−1440m. June−September.

2. B. parviflora Roxb.; *Spatholobus roxburghii* Bentham. Nep: *Birali Lahara, Gaunji Lahara, Debri Lahara, Baldengra*

Large woody twiner. Leaflets broadly elliptic, 11−18(−30)×6.5−15(−20)cm, acute or obtuse, base rounded or cuneate, lateral ones strongly asymmetric, sparsely appressed pubescent beneath at first, stipels subulate 3−4mm; petioles 7−25cm; stipules broadly triangular 4×6mm, deciduous. Flowers in large terminal panicles, elongating in fruit. Calyx 6−10mm divided to middle into lanceolate teeth, densely pale pubescent. Petals creamy-white or pink, 7−12mm, standard broadly ovate 9−11×6−8mm, wings oblong, blade 5−7×3mm, keel oblanceolate 3−5×2−3mm. Pods 10−12×3−4cm, coriaceous, densely brown velvety.

Bhutan: S—Samchi and Sarbhang districts, **C**—Tongsa district (Dakpai); **Sikkim.** Subtropical forest slopes, 400−650m. July−October.

3. B. monosperma (Lamarck) O. Kuntze; *B. frondosa* Roxb. Nep: *Palas* (34), *Mauwa*. Fig. 42 a&b

Tree to 20m. Leaves coriaceous, leaflets ovate or rhombic, 12−20×10−18cm, obtuse or emarginate, base cuneate, brownish pubescent especially beneath; stipels linear c 2mm; petioles 10−20cm; stipules ovate 5−6mm. Flowers in dense racemes forming terminal panicles 20−40cm; appearing while almost leafless. Calyx 1.2−1.8cm divided almost to middle into broad teeth, densely brown pubescent. Petals orange red, standard ovate, 4−5×2cm, reflexed, wings lanceolate, blade 4−4.5×1cm, curved, keel 4−5×1.5cm, strongly upcurved and almost beaked. Pods

9−15×3−4cm, densely brownish pubescent, borne on basal stalks c 2.5cm.
Bhutan: S—Deothang district (Samdrup Jongkhar); **Sikkim**. Subtropical and terai forests, 500m. February−April.

Valuable for reclaiming alkaline salt-lands (126). It is also an excellent host plant for the lac insect.

39. APIOS Fabricius

Twining perennial herbs. Leaves odd-pinnate, leaflets (3−)5−7, distinctly petiolulate; stipels present; stipules deciduous. Flowers in lax axillary racemes, borne in pairs on thickened nodes. Calyx campanulate, 5-toothed, upper 2 ± connate. Petals shortly clawed, standard broadly elliptic to suborbicular, shortly auricled at base, wings obovate, adherent to narrow upcurved beak. Stamens diadelphous. Style coiled through 360°. Pods linear, compressed. Seeds 10−15.

1. A. carnea (Wall.) Baker. Fig. 41m

Leaves 15−30cm; leaflets ovate or elliptic, 5−12×2.5−7cm, acute or acuminate, base rounded or cuneate, appressed pubescent; petiolules 2−3mm; stipels subulate 2−4mm; stipules lanceolate 3−4mm. Racemes 12−40cm. Calyx tube 4−5mm, upper and lateral teeth triangular 4mm, lowest tooth lanceolate c6mm. Petals pink, blue or purple, 1.7−2.5cm, standard 1.75−2cm broad, wings 1−1.2×0.8−0.9cm, keel 2.5×0.5cm. Pods 10−15×0.5−0.6cm; seeds oblong 5×3mm, brown, hilum surrounded by a low aril.

Bhutan: C—Thimphu, Punakha, Tongsa and Tashigang districts, **N**—Upper Mo Chu and Upper Kulong Chu districts. Margins of moist broad-leaved forests, 1830−2750m. July−September.

40. COCHLIANTHUS Bentham

Slender twining perennials, leaves and flowers becoming blackish on drying. Leaves pinnately 3-foliate, leaflets entire, stipels and stipules minute or absent. Flowers in axillary racemes. Calyx 5-toothed, upper 2 ± connate. Petals shortly clawed, standard with 2 minute inflexed auricles at base, wings oblanceolate slightly longer than standard and bearing a linear appendage at base, keel linear, coiled. Stamens diadelphous. Stigma capitate. Pods linear, incurved.

1. C. gracilis Bentham

Leaflets ovate, 4−10×2.5−5.5cm, acuminate, base rounded or cuneate, appressed pubescent; petioles 4−10cm. Racemes 5−20-flowered on peduncles 2−8.5cm. Calyx 7.5mm, divided to middle into triangular teeth, appressed pubescent. Petals reddish-purple, standard elliptic 1.7×1.2cm, wings 2×0.4cm, basal appendage ± as long as claw, keel ± as long as wings, c 2mm broad, coiled through 360°. Pods not seen.

Bhutan: N—Upper Bumthang Chu district (Shazo). 3050m. September.

41. CANAVALIA DC.

Annual or perennial herbaceous twiners. Leaves pinnately 3-foliate, leaflets entire, stipels present, caducous; stipules deciduous. Racemes axillary, flowers borne in pairs on swollen nodes. Calyx campanulate, bilabiate, upper lip large, 2-lobed, lower lip smaller 3-toothed. Petals shortly clawed, standard reflexed, wings minutely auricled near base, keel upcurved, bluntly beaked. Stamens monadelphous. Pods oblong, flattened, each valve with prominent rib along ventral suture and an extra rib parallel and close to it, dehiscence by twisting of valves; seeds 10−15, ellipsoid, funicle often persisting, hilum linear.

1. C. gladiata (Jacquin) DC. Eng: *Sword Bean*

Leaflets ovate, 10−15(−20)×5.5−12cm, acuminate, base rounded, sparsely whitish pubescent at first; stipels subulate 2−3mm; petioles 3−8cm; stipules lanceolate c 5mm. Peduncles 20−25cm, 12−20-flowered. Calyx 15−20mm, upper lip with 2 rounded lobes, lower lip with acute teeth 2−3mm. Petals purplish, standard broadly elliptic, 3.5−4×2.5cm, wings oblong, blade 2.5×0.7cm, keel 2.5×0.8−0.9cm. Pods thickly coriaceous, 20−25(−40)×3.5−4.5(−5)cm, with extra rib 3−6mm from sutural rib; seeds brown or reddish, 25−35×12−20×10−14mm, hilum 15−20mm with a small flag-like aril 2−3mm at micropylar end.

Sikkim: Gareedhura Terai.

The young pods are eaten as a vegetable.

2. C. gladiolata Sauer

Similar to *C. gladiata* but pods not more than 3.5cm wide, and with extra rib 1−3mm from sutural rib, seeds not more than 2cm long.

Sikkim: locality unknown.

Probably not specifically distinct from *C. gladiata*.

42. PACHYRHIZUS DC.

Tuberous-rooted twining perennial herb. Leaves pinnately 3-foliate, leaflets often angular, stipels and stipules present. Racemes elongate, flowers borne in small clusters on short peg-like projections, bracts and bracteoles caducous. Calyx tube campanulate, two upper teeth united into a bidentate lip. Petals shortly clawed, standard auricled at base, wings with a slender basal appendage almost as long as the claw, keel incurved. Stamens diadelphous. Style thickened and strongly recurved or hooked at apex, bearded along inner side. Pods linear, ± compressed, transversely depressed between seeds, weakly septate within.

1. P. erosus (L.) Urban; *P. angulosus* DC.
Leaf rachis 6−15cm, terminal leaflet very broadly elliptic or fan-shaped, 5.5−13×8−17cm, acute, base cuneate, margin with several broad-pointed teeth, lateral leaflets broadly ovate, 5−12×4.5−12cm, acute, base oblique, margin angular or with a few broad teeth, appressed pubescent; stipels subulate 5−7mm; stipules lanceolate 7−8mm. Calyx c 10mm, lower teeth lanceolate, half as long, appressed brown pubescent. Petals bluish, standard broadly elliptic or suborbicular 1.5cm, wings elliptic, blade 15×5mm, keel narrowly ovate 12×6mm. Pods subcoriaceous, 10×1.5cm, 9−10-seeded, appressed pubescent.

Bhutan: S—Chukka district (Chasilakha); **Sikkim.** 1200m. August−November.

Native of Central America cultivated for its edible tubers. The young pods may also be eaten.

43. PUERARIA DC.

Herbaceous perennial or woody twiners. Leaves pinnately 3-foliate, leaflets entire or 2−3-lobed, stipels present, stipules usually persistent. Racemes or panicles axillary, flowers often clustered on node-like swellings, bracts deciduous. Calyx campanulate, 5-toothed, upper 2 teeth ± connate. Petals shortly clawed, standard with 2 inflexed auricles at base, wings auricled and adherent on keel. Stamens monadelphous or diadelphous. Pods linear, compressed, continuous within or septate between seeds.

1. Plants ± leafless at flowering time .... **1.P. sikkimensis**
+ Plants bearing leaves at flowering time 2

2. Calyx 5−7mm; stipules basifixed .... **Species 2−4**
+ Calyx (8−)10−15mm; stipules medifixed .... **Species 5 & 6**

1. P. sikkimensis Prain. Nep: *Birali Lahara*
Plants strongly woody, almost entirely leafless at flowering time. Leaflets broadly ovate, c 15×12cm, acute or shortly acuminate, base rounded or truncate, sparsely pubescent. Inflorescence branches up to 30cm, brownish pubescent. Calyx 8−9mm, divided to middle into ovate teeth, brown

pubescent. Petals white or pale blue, wings always blue or dark blue, 15—16mm, standard suborbicular, 15mm broad, wings oblong c 3mm broad, keel upcurved 4—5mm broad. Immature pods brown pubescent, mature pods unknown.

Bhutan: S—Samchi district (Daina Khola and W bank of Torsa River) and Sarbhang district (Singi Khola); **Sikkim.** Terai and subtropical forest slopes, 200—400m. February—March.

2. **P. lobata** (Willdenow) Ohwi; *P. thunbergiana* (Siebold & Zuccarini) Bentham, *P. thomsonii* Bentham

Herbaceous twiner. Leaflets ovate, 10—17×7—15cm, shortly acuminate, base rounded, simple or 3-lobed, appressed brownish pubescent; stipels linear 10—12mm; petioles 10—15cm; stipules medifixed, portions above and below attachment lanceolate 10—12×3—5mm. Calyx 12—20mm, divided to lower third into lanceolate teeth, brown pubescent. Petals blue or purple 2—2.5cm, standard broadly elliptic 1.5—2.5cm, wings obliquely oblong 5—7mm broad, keel 7—9mm broad. Pods 8—9×1cm, densely spreading brown hirsute, 8—12-seeded.

Bhutan: C—Punakha district (Neptengka) and Tashigang district (Shali); **Sikkim.** Margins of forests, 300—1800m. August—September.

The above description relates to var. **thomsonii** (Bentham) van der Maesen.

3. **P. edulis** Pampanini; *P. quadristipellata* W. W. Smith

Similar to *P. lobata* but leaflets smaller, 10—15×6—9cm, simple or 2—3-lobed; stipels lanceolate 6—10mm, those of lateral leaflets deeply bifid; calyx 8—10mm ± glabrous; petals 14—16mm.

Bhutan: C—Thimphu, Punakha, Mongar and Tashigang districts; **Sikkim.** Amongst shrubs on hillsides, 1525—2400m. July—October.

4. **P. peduncularis** (Bentham) Bentham

Herbaceous twiner, stems softly brown pubescent. Leaflets broadly ovate-elliptic 8—17×4—11cm, acuminate, base cuneate or rounded, often oblique, unlobed, appressed brownish pubescent; stipels linear-lanceolate 3—6mm; petioles 5—25cm; stipules lanceolate, 7—8×1.5—2.5mm, basifixed. Racemes (7—)25—40cm. Calyx 5—6mm, shortly toothed. Petals 1.3—1.5cm, blue or purple but white near base, standard obovate 7mm, wings oblong, blade 9×2.5mm, keel oblanceolate c 10×4mm. Pods 5—8×0.7cm, glabrous, 4—8-seeded.

Bhutan: S—Chukka and Deothang districts, **C**—Thimphu to Tashigang districts; **Sikkim.** Cool broad-leaved forest, 2100—3050m. June—August.

5. **P. wallichii** DC. Nep: *Birali*

Similar to *P. peduncularis* but more woody; leaflets minutely white pubescent beneath, stipels subulate 2−3mm; stipules linear 5−7× 0.5−1mm; calyx obscurely toothed; petals white or greenish yellow; pods linear, 10×0.8cm, conspicuously thickened along both sutures.

Sikkim: Chumthang and Lamteng.

6. P. phaseoloides (Roxb.) Bentham

Similar to *P. peduncularis* but stems and petioles spreading brown hirsute; leaflets with sinuate or lobed margins; stipels linear-lanceolate 4−6mm; stipules ovate 5−8×3−4mm; calyx 6−8mm, divided to middle into elongate fine pointed teeth, hirsute; petals 1.2−2cm, purplish; pods 5−12×0.4−0.5cm, appressed pubescent.

Bhutan: S—Gaylegphug district (near Gaylegphug) and Deothang district (Kheri); **Sikkim:** Selim. 300m. August−October.

The above description refers to var. **subspicata** (Bentham) van der Maesen which has the leaves more deeply lobed, and flowers and pods larger than in the typical variety.

44. GLYCINE Willdenow

Annual herbs. Leaves pinnately 3-foliate, leaflets entire, stipellate; stipules small, free. Flowers in short racemes, sometimes solitary, axillary; bracteoles 2, small, at base of calyx. Calyx campanulate, 5-toothed, upper 2 connate to above middle. Petals clawed, standard suborbicular, weakly auricled, wings small, adherent on blunt keel. Stamens usually monadelphous. Pods oblong, weakly septate between seeds, dehiscent by twisting of valves; seeds oblong, hilum lateral, surrounded by low papery aril.

1. G. max (L.) Merrill; *G. soja* sensu F.B.I. non Siebold and Zuccarini, *G. hispida* (Moench) Maximowicz. Eng: *Soybean*

Stems 30−200cm, erect, spreading brownish hirsute. Leaflets ovate, 5−9(−14)×3−6(−10)cm, acute, base cuneate or rounded, appressed pubescent, stipels narrowly lanceolate 1−3.5mm; petioles 7−20cm; stipules ovate 3−7mm. Racemes usually 5−8-flowered; bracteoles narrowly lanceolate 2−3mm. Calyx 5−7mm, brown hirsute. Petals purplish or white, 6−8mm, wings narrowly oblong, longer than keel. Pods oblong 4−5× 0.75−1.2cm, hirsute; seeds 2−3, 6−10×5−7mm, creamy-white or blackish.

Bhutan: C—Tongsa district (Shamgong); **Sikkim.** 1950m. July−September.

Cultivated for its nutritious seeds.

45. TERAMNUS P. Browne

Twining perennial herbs. Leaves pinnately 3-foliate, leaflets entire, stipels present; stipules small. Flowers in subsessile racemes, axillary, bracteoles 2 at calyx base. Calyx campanulate, teeth 5, upper 2 ± connate. Petals clawed, standard obovate, wings narrowly oblong, adherent on the straight blunt keel. Stamens monadelphous, 5 longer with fertile anthers alternating with 5 shorter staminodes. Pods linear, bearing thick persistent hooked style distally, thinly septate between 8–10 seeds.

1. T. flexilis Bentham

Leaflets ovate-elliptic 7–12×2.5–8cm, acute or acuminate, base rounded, sparsely appressed pubescent; stipels subulate c 2mm; petioles 3–10cm; stipules narrowly lanceolate c4mm. Racemes up to 3.5cm. Calyx c 4mm, appressed pubescent. Petals 5–6mm, white or reddish. Pods linear 5–7×0.4–0.5cm, compressed, sparsely appressed pubescent.

Bhutan: S—Deothang district (near Wamrung, 117). 2300m. September.

46. MASTERSIA Bentham

Woody twiner. Leaves pinnately 3-foliate, leaflets entire, stipels present; stipules caducous. Racemes axillary and terminal, flowers borne in clusters of 2–3 on small cushion-like outgrowths, bracts deciduous, bracteoles ± persistent, obovate, sheathing base of calyx, striate. Calyx tube turbinate, teeth lanceolate, the upper two united into a broad entire lip. Petals shortly clawed, standard suborbicular, wings oblique, oblong, keel broad, incurved. Stamens diadelphous. Style not bearded, capitate. Pods oblong, compressed, narrowly winged along upper suture, indehiscent, seeds oblong, transversely arranged.

1. M. assamica Bentham; *M. cleistocarpa* Baker

Leaf rachis 6–15cm; leaflets elliptic or ovate 8–14×6–12.5cm, acuminate, base rounded, glabrous above, appressed pubescent beneath, stipels linear 7–8mm. Racemes 30cm or more, bracts ovate-lanceolate, 7–8mm, ciliate, bracteoles 5–7×3–4mm. Calyx tube 4–5mm, teeth 7–12mm. Petals purplish 10–18mm. Pods thinly coriaceous, 10–12× 2.5–3cm, glabrous, blackish; seeds 5×2.5mm.

Bhutan: S—Sarbhang district (near Sarbhang and Phipsoo) and Gaylegphug district (above Sher Camp). Subtropical forest slopes, 300–1000m. June.

47. SHUTERIA Wight & Arnott

Perennial twining herbs. Leaves pinnately 3-foliate, leaflets entire, stipels present; stipules striate, persistent. Racemes axillary, bracts and bracteoles persistent or deciduous. Calyx campanulate, upper 2 teeth connate. Petals clawed, wings adherent to keel. Stamens diadelphous. Pods linear, compressed, non-septate, seeds 4−6, ellipsoid.

1. **S. involucrata** (Wall.) Wight & Arnott; *S. vestita* Wight & Arnott
Stems hirsute with spreading or retrorse hairs. Leaflets ovate, 3−6×2.5−4cm, subacute or rounded at base and apex, pinnately veined, sparsely appressed pubescent especially beneath; petioles 5−7cm; stipels linear 4−6×0.5mm; stipules lanceolate 7−8×1.5−2mm. Racemes many-flowered, short and crowded, bracts narrowly lanceolate 4−5mm, bracteoles similar, c 2.5mm, appressed to base of calyx. Calyx 4−6mm, densely pubescent. Petals 8−12mm, purplish, standard broadly elliptic 5.5−6.5mm broad, wings oblong, blade 5×1.5mm, keel upcurved distally 5×2mm. Pods 2.5−3×0.5cm, brownish pubescent; seeds dark brown or mottled, 3×1.75mm.

Bhutan: S—Samchi to Gaylegphug districts, **C**—Tongsa and Tashigang districts; **Sikkim.** Subtropical and warm forests, 200−1600m. November−December.

The above description refers to var. **glabrata** (Wight & Arnott) Ohashi which lacks the involucre of 2−3 rounded leaf-like bracts at the base of the inflorescence which characterises the West Himalayan typical variety.

2. **S. ferruginea** (Bentham) Baker; *Amphicarpaea ferruginea* Bentham
Similar to *S. involucrata* but stems appressed pubescent; leaflets ovate 4−9×3−5cm, acute or acuminate, apiculate, base rounded, asymmetric in lateral leaflets, 3-veined at base, appressed brownish pubescent; stipules ovate-lanceolate 4−7×1.5−2mm; stipels subulate 2−3mm; racemes 7−15-flowered, bracts ovate 6−7×2mm, deciduous; calyx 7−8mm; petals bluish, 1.5−1.7cm; pods unknown.

Bhutan: C—Thimphu, Tongsa, Bumthang, Mongar and Tashigang districts. Dense forest, 2100−2900m. July−August.

3. **S. hirsuta** Baker
Similar to *S. involucrata* but stems more strongly brown hirsute, leaflets up to 15×8.5cm, acuminate, 3-veined at base; stipules linear-lanceolate 7−10×1mm; racemes elongate up to 50-flowered, bracts linear-lanceolate 10−12×1mm, persistent; calyx 7−8mm; petals 12−17mm, blue; pods 6−7×0.3cm slightly constricted laterally between the 12−14 seeds.

Bhutan: C—Tongsa district (Shamgong); **W. Bengal Duars:** Buxa; **Sikkim:** Rungeet Valley. Amongst scrub, 250−300m. October−December.

48. DUMASIA DC.

Twining perennial herb. Leaves pinnately 3-foliate, leaflets entire, stipels present, stipules persistent. Flowers in slender axillary racemes, bracts persistent. Calyx campanulate, teeth obsolete. Petals clawed, standard with 2 small auricles at base, wings adherent to keel by longitudinal folds. Stamens diadelphous. Pods oblong, constricted between ellipsoid seeds.

1. D. villosa DC.

Leaflets ovate, 2−6×1−4cm, acute or obtuse, mucronate, base cuneate or rounded, ± appressed pubescent, stipels subulate c 2mm; petioles 1.5−6.5cm; stipules linear-lanceolate 4−5×1mm. Racemes 5−30-flowered, flowers widely spaced, usually in pairs, bracts subulate 4−5mm. Calyx 7−9mm, obliquely truncate. Petals yellow 1.5−1.7cm, standard oblong 6−7mm broad, wings oblong, blade 5−6×2−3mm, keel oblique 5×3mm. Pods 2−3×0.7−0.8cm, ± appressed pubescent; seeds 2−4, blue-black, glaucous 5×4mm.

Bhutan: S—Phuntsholing, Chukka, Gaylegphug and Deothang districts, **C**—Punakha and Tashigang districts; **Sikkim.** Broad-leaved forests, 1500−2000m. August−September.

Both var. **villosa** and var. **leiocarpa** (Bentham) Baker, the latter with subglabrous pods and leaflets, have been recorded from Bhutan.

49. CLITORIA L.

Twining perennial herbs. Leaves odd-pinnate or 3-foliate, leaflets entire; stipels and stipules present. Flowers large, resupinate, solitary axillary or paired in short racemes. Bracts persistent at base of calyx. Calyx tubular, 5-toothed. Standard much larger than other petals, wings oblong, clawed, adherent to keel, keel shorter than wings, incurved, acute. Stamens diadelphous, anthers alternately dorsifixed and basifixed. Pods linear, compressed, beaked by persistent style, usually membranously septate between the seeds.

1. C. ternatea L.

Leaves odd-pinnate; leaflets 5−9, ovate 2−6×1−4cm, obtuse or emarginate, base rounded, minutely pubescent, stipels subulate c 2mm; stipules narrowly lanceolate c4mm. Flowers solitary, axillary, bracts ovate 6×2mm in flower, 10×9mm in fruit. Calyx tube 10−12mm, teeth ovate 7−8mm. Petals dark blue fading to greenish yellow at base, standard broadly ovate, 4.5×2cm, wing petals blade 10×6mm, keel 8×4mm. Pods coriaceous, 8.5−9.5×1−1.2cm, seeds 8−10, black.

Bhutan: S—Phuntsholing district (Phuntsholing). In garden, 250m. February.

Cultivated for its decorative flowers.

2. C. mariana L.

Similar to *C. ternatea* but leaves pinnately 3-foliate, leaflets ovate-elliptic, 6.5−9×3−4.5cm, acute or acuminate, base rounded, glabrous above, pubescent beneath, stipels linear 7−8mm; stipules ovate c 10×4mm, striate; flowers borne in pairs in racemes 6−15cm, bracts ovate 8×3mm; calyx tube 12−15mm, teeth ovate 8×3.5mm; petals mauve, standard c 4×2.5cm, wings 10×7mm, keel 8×3.5mm.

Bhutan: S—Deothang district (Chungkar), N−Upper Kuru Chu district (Shawa). In dense scrub, 1050m. June−August.

50. DYSOLOBIUM (Bentham) Prain

Herbaceous twiners. Leaves pinnately 3-foliate, stipellate, stipules rounded or truncate at base. Flowers borne in 2's or 3's on glandular swellings in racemes. Calyx 4-toothed, upper tooth 2-lobed, lowest tooth longer than others and upturned. Petals clawed, standard without appendages, wings with a recurved upper auricle and a small tooth on lower margin, keel strongly incurved. Style bearded below apex. Pods terete, septate, ridged along both sutures, densely tomentose, dehiscence by twisting valves; seeds with a rim-like aril around hilum.

1. D. grande (Bentham) Prain; *Phaseolus velutinus* Baker

Woody twiner, stems densely brown pubescent. Leaflets ovate or broadly elliptic, 12−18×9−18cm, acute or shortly acuminate, base rounded or truncate, brownish pubescent, stipels linear-lanceolate 5−7mm; petioles 12−15cm; stipules lanceolate 8−10×3mm. Racemes up to 60cm bearing 30−40 flowers. Calyx c1.5cm, lowest tooth narrower and longer than others. Petals purplish, 3.5−4cm, standard broadly orbicular, c 4.5cm broad, keel strongly incurved through 270°. Pods oblong, 15−22×2−2.5cm, ± terete, woody, densely brown tomentose.

Sikkim: locality unknown. July.

51. LABLAB Adanson

Twining perennial herbs. Leaves pinnately 3-foliate, leaflets entire, stipels and stipules small. Flowers in axillary racemes, borne in clusters of 3−6 on glandular swellings, bracts deciduous, bracteoles present at flowering time. Calyx campanulate, teeth short, upper 2 united. Standard reflexed, auricled at base and with 2 longitudinal thickenings on either side of mid-line, wings laterally pouched, adherent to standard, keel hook-shaped. Style abruptly upturned, laterally compressed, bearded in upper part. Pods straight or curved, compressed, sinuous with minute warts along both sutures, shallowly septate between seeds within; seeds arillate.

1. L. purpureus (L.) Sweet. Dz: *Semchu;* Sha: *Orey;* Nep: *Shimi;* Eng: *Hyacinth Bean* or *Bonavist Bean.* Fig. 44a

Leaf rachis 5−10cm; leaflets ovate, 4−13×3−11cm, acuminate, base truncate or cuneate, appressed pubescent, stipels triangular 5−6mm; stipules ovate 3−5mm. Racemes 15−35cm, bracteoles elliptic 5−6×3mm. Calyx 4−5mm pubescent at mouth. Petals white or purple, standard suborbicular 12mm, wing oblong-obovate, blade 10×8mm, keel 3mm broad, hooked at middle. Pods oblong 7−13×2−3cm, bearing persistent style, 5−6-seeded, seeds ellipsoid compressed, 9−12×6−9mm, black, brown or white, hilum covered by white aril which extends around a third of margin of seed.

Bhutan: S—Samchi, Chukka and Sarbhang districts. In fields and gardens, 460−1250m. February−March.

Cultivated for its edible young pods and seeds.

52. DOLICHOS L.

Twining perennial herbs. Leaves pinnately 3-foliate, leaflets entire or lobed, stipels and stipules present. Flowers in short axillary racemes. Calyx campanulate, teeth short, broad, upper 2 united. Petals shortly clawed, standard bearing 2 fleshy hooked appendages near centre and 2 inflexed auricles at base; wings oblong, adhering to keel; keel smaller, incurved. Stamens diadelphous. Style bearded at apex on inner side. Pods oblong, compressed, sutures somewhat thickened.

1. D. tenuicaulis (Baker) Craib; *Phaseolus tenuicaulis* Baker, *D. falcatus* sensu F.B.I. p.p. non Willdenow

Leaf rachis 3−5cm, leaflets ovate-lanceolate, 3−5(−10)×1.5−2(−7)cm, acute or acuminate, base rounded or cuneate, margin entire or weakly 3-lobed, appressed pubescent; stipels narrowly ovate 2−3mm; stipules triangular 2.5−5×1−2mm. Racemes 2−8(−14)-flowered, 1−3(−5)cm. Calyx 5−6mm. Petals blue or purplish, standard suborbicular 10−12mm, wing petals blade 8×5mm, keel 7×3mm. Pods glabrous 5−7×0.8−1cm, 6−8-seeded, seeds brown, ellipsoid or reniform, 4−6.5×3−4mm, ± compressed, hilum on margin c 2mm, central, rim aril prominent 1−1.5mm.

Bhutan: C—Punakha and Mongar districts; **Sikkim.** Warm broad-leaved forests, 1400−1800m. August.

53. MACROTYLOMA (Wight & Arnott) Verdcourt

Slender annual or perennial twining herbs. Leaves pinnately 3-foliate, leaflets entire, stipels and stipules present. Flowers in small axillary clusters.

Calyx campanulate, teeth 4–5, upper 2 ± connate. Standard reflexed, bearing a lamelliform appendage on either side of mid-line in lower half, wings very narrow, keel almost straight. Stamens diadelphous. Style filiform, not bearded. Pods straight or curved, linear-oblong, compressed, not septate; seeds oblong, rounded, hilum small, central on lateral margin, minutely arillate.

1. M. uniflorum (Lamarck) Verdcourt; *Dolichos biflorus* sensu F.B.I. non L.

Petals 1–7cm, leaflets elliptic or obovate, 1–5(–8)×0.7–3(–7.5)cm, subacute or obtuse, base rounded, pubescent; stipels filiform 2mm; stipules lanceolate 4–8mm, veiny. Flowers subsessile, in clusters of (1–)2–3(–5), peduncles up to 1.5cm. Calyx tube 2–3mm, teeth triangular 3–8mm, acuminate. Petals yellow or greenish, standard obovate-oblong, 6–12×4–7mm, usually with a central purple spot, lamelliform appendages c 5mm, wings 4–10×1mm, keel slightly curved 5–10×1.5mm. Pods 3–5.5×0.5–0.8cm, pubescent; seeds brownish 3–6.5×3–5mm.

Bhutan: S—Deothang district (Kheri); **Sikkim:** Selim. 600m. September–October.

54. VIGNA Savi

Annual or perennial, erect or twining herbs. Leaves pinnately 3-foliate, leaflets entire or lobed, stipels present; stipules auriculate or extending as a lobe below point of attachment. Racemes axillary, flowers borne on glandular swellings. Calyx campanulate, teeth lanceolate, upper 2 ± connate. Petals clawed, standard with inflexed auricles, wings shorter than standard, auricled, keel incurved making almost a complete turn. Style bearded on inner side near tip and usually ending beyond stigma in a distinct beak. Pods linear, ± cylindric, incompletely septate, valves thin, subcoriaceous, becoming strongly twisted on dehiscence.

1. Flowers blue, purple or white 2
+ Flowers yellow 3

2. Stipules auriculate or bilobed at base .... **1. V. vexillata**
+ Stipules rounded at base .... **2. V. pilosa**

3. Stipules more than 3mm broad; stems, petioles and peduncles stiffly pilose 4
+ Stipules less than 2mm broad; stems, petioles and peduncles softly pubescent or glabrous 5

4. Stems twining; leaflets obtuse or broadly acute, usually 2–3-lobed
   3. V. trilobata
+ Stems erect; leaflets acute or acuminate, usually unlobed ... **4. V. radiata**

5. Bracteoles subulate or narrowly lanceolate, longer than calyx; pods ± glabrous .... **5. V. umbellata**
+ Bracteoles minute, shorter than calyx; pods densely appressed pubescent
6. V. clarkei

1. V. vexillata (L.) A. Richard. Fig. 42f

Slender twiner, stems and petioles deflexed hirsute at first. Leaflets ovate-lanceolate, 3–12×1–2cm, acuminate, base rounded., appressed brownish pubescent, stipels linear-lanceolate, 2–5mm; petioles 3–6cm; stipules ovate-lanceolate, 5–6×2mm, auriculate or bilobed at base. Racemes 1–3-flowered, 5–30cm. Calyx 10–12mm, divided to middle into slender teeth. Petals white or purplish, standard obovate 18×15mm, wing blade 15×10mm, keel 4–5mm broad. Pods 5–10×0.5cm, brown hirsute at first. Seeds 10–20, reniform, 3–4×2mm, black.

Bhutan: C—Punakha, Mongar and Tashigang districts. On grassy hillsides, 1370–2000m. July–September.

2. V. pilosa (Willdenow) Baker; *Dysolobium pilosum* (Willdenow) Marechal

Similar to *V. vexillata* but leaflets usually broader, 2.5–5(–7)cm, sparsely hirsute; petioles 4–10cm; stipules 5×2mm rounded at base; racemes 3–13cm bearing 10–20 flowers; calyx 4–5mm; petals c15mm; pods 7–8(–13)×0.7cm, densely brown hirsute with spreading hairs.

Bhutan: S—Gaylegphug district (near Gaylegphug, 117); **Sikkim.** 270m.

3. V. trilobata (L.) Verdcourt; *Phaseolus trilobus* Aiton

Straggling annual, stems and petioles deflexed hirsute. Leaflets unequally 3-lobed (entire, broadly elliptic in seedlings), 1.3–4.5×1–3.5cm, lobes oblong-spathulate, middle one longer and broader, obtuse; stipels narrowly elliptic 1–2mm; petioles 2–10cm; stipules ovate 4–20×3–12mm, cordate at base or extending as a rounded lobe below point of insertion. Racemes 2–10cm, 2–5(–8)-flowered. Calyx 2–2.5mm. Petals yellow, 6–7mm, standard suborbicular, 7–8mm broad. Pods 3–5×0.3cm, sparsely appressed pubescent; seeds 6–12, reniform c 2×1.5mm, black.

Bhutan: C—Thimphu district (Paro, 117). Margins of cultivated ground, 2550m. August.

4. V. radiata (L.) Wilczek; *Phaseolus mungo* sensu F.B.I. non L., *P. aureus* Roxb. Eng: *Mung, Green Gram*

Similar to *V. trilobata* but stems erect; leaflets ovate, 4–10×2.5–7cm, shortly acuminate, base rounded or cuneate, margin entire or undulate, stipels linear-lanceolate 3–6mm; petioles 6–14cm; stipules medifixed ovate-elliptic 7—12×3—4mm; flowers 5—15 terminally clustered on racemes 2—9cm; calyx 2—3mm; petals 7—12mm; pods 4—7×0.5—0.6cm, shortly brown pubescent; seeds 10—15, green or blackish, hilum not protuberant, ungrooved.

Bhutan: C—Thimphu district (Paro, 117). 2500m. Cultivated for its edible seeds.

The above record may refer to *V. mungo* (L.) Hepper (*Phaseolus mungo* L. Eng: *Urd* or *Black Gram*) which has been confused with *V. radiata* and may only be a variety of it. It is distinguished by its more densely hirsute pods and seeds with raised centrally grooved hilum.

5. V. umbellata (Thunberg) Ohwi & Ohashi; *Phaseolus calcaratus* Roxb.

Stems twining, ± glabrous. Leaflets ovate, 4—8×2.5—5cm, acuminate, base cuneate or rounded, margin entire or shallowly lobed, stipels linear-lanceolate 3—4mm; petioles 5—9cm; stipules medifixed, elliptic 5—7×1.5—2mm. Racemes elongate, 10—12-flowered, on peduncles 4—6cm. Calyx 3—4mm. Petals 10—12cm, yellow. Pods 7—8×0.5cm, glabrous.

Bhutan: C—Punakha district (near Punakha and Rinchu); **Sikkim**. Margin of paddy fields, 1270—1525m. August—September.

6. V. clarkei Prain

Similar to *V. umbellata* but leaflets narrowly lanceolate, 8—12×1—2cm, rounded at base; stipules minute; racemes few-flowered, c1.5cm, on peduncles up to 12cm; calyx c6mm; petals 10—12mm; pods 6.5—7cm densely brown pubescent.

Sikkim: Terai (Dalkajar). 150m.

55. PHASEOLUS L.

Twining annual or perennial herbs. Leaves pinnately 3-foliate, leaflets simple or shallowly lobed, stipellate; stipules striate, not extending below point of attachment, persistent. Racemes axillary, flowers borne on swollen nodes in clusters of 2—4. Calyx campanulate, bilabiate, upper pair of teeth ± connate, lower lip 3-toothed. Petals clawed, standard auricled at base, reflexed, wings adherent on keel, keel linear, coiled spirally through 1—2 turns. Stamens diadelphous. Pods oblong or linear, ± compressed, dehiscent; seeds several to many.

1. P. coccineus L. Eng: *Scarlet Runner Bean*

Stems pubescent with minute hooked hairs. Leaflets ovate-rhombic, 6—10×5—9cm, acuminate, base cuneate or truncate, lateral leaflets asymmetric at base, pubescent, stipels subulate c 5mm; petioles 8—11cm; stipules triangular, 3—4×1.5—2mm. Peduncles 10—17cm, pubescent, bracts and bracteoles inconspicuous. Calyx tube c 3mm, teeth of lower lip triangular c1mm. Petals red, standard suborbicular 16−18×16−19mm, wings obovate 24−27×16−18mm, keel c 10mm. Pods 9−20×1.8−2.5cm, obliquely ridged and slightly tuberculate, ± glabrous; seeds oblong, 15−25×10−15mm, glossy black with purple mosaic.

Sikkim: Darjeeling.

Native of Central America widely cultivated for its young pods which are cooked and eaten.

2. P. lunatus L. Eng: *Lima Bean*

Similar to *P. coccineus* but peduncles 2−10cm, standard suborbicular 10−12×10−12mm, erect but hooded enclosing top of coiled keel, yellowish-green, wings obovate c 12×6mm, white, keel 5−7mm, yellowish green; pods 5−12×1.5−3cm, pod wall striate slightly pubescent; seeds reniform, 12−24×9−12mm, white with or without red-purple spots or mosaic, sometimes dark purplish-brown.

Bhutan: cultivated in subtropical regions; **Sikkim.**

Native of Central America cultivated for its young pods and mature seeds which are cooked and eaten.

3. P. vulgaris L. Fig. 41 n−r

Similar to *P. coccineus* but peduncles 3−8(−15)cm, bracts conspicuous, ovate 7−8×4−5mm, subtending base of calyx; petals white or purplish, standard suborbicular 10−15×10−15mm, wings obovate 15−20× 5−11mm, keel c 10mm; pods 9−18×1−1.5cm, smooth or finely striated; seeds oblong or reniform, 8−17×5−7mm white, purplish brown to black or variously spotted and mottled.

Bhutan: C—Thimphu district (Thimphu); **Sikkim.** 2370m. September.

Native of Central America, cultivated for its young pods which are cooked and eaten.

56. CAJANUS DC.

Erect shrubs or herbaceous or woody twiners. Leaves pinnately or subdigitately 3-foliate, leaflets entire, dotted beneath with subsessile red or yellowish glands, stipels and stipules usually present. Racemes axillary, flowers sometimes solitary or few, bracts caducous or persisting until flowering time. Calyx campanulate, teeth 5, upper 2 united into an entire or bifid lip. Standard reflexed, bearing 2 inflexed auricles at base, wings as long as incurved obtuse keel. Stamens diadelphous. Style somewhat thickened below stigma. Pods linear-oblong, ± compressed, grooved diagonally or almost at right angles between seeds, ± septate within. Seeds oblong or rounded with a small basal oblong hilum, arillate.

1. Erect shrub .... **1. C. cajan**
+ Twiners, herbaceous or woody 2

2. Leaves pinnately trifoliate 3
+ Leaves subdigitately trifoliate 4

3. Leaflets small 2−4.5cm, elliptic or obovate, acute or subacute
 2. C. scarabaeoides
+ Leaflets larger 5−10cm, obovate, acuminate .... **3. C. grandiflorus**

4. Racemes up to 15-flowered .... **4. C. mollis**
+ Flowers solitary or up to 5 5

5. Slender herbaceous twiner; pods 1.7−2.5×0.5−0.8cm, 3−4-seeded
 5. C. elongatus
+ More robust twiner; pods larger 2−3.5×1cm, 5−6-seeded .. **6. C. villosus**

1. C. cajan (L.) Millspaugh; *C. indicus* Sprengel. Eng: *Pigeon Pea.* Fig. 43 t−w

Erect grey silky shrub 1−4m, young stems angular. Leaves pinnately trifoliate; petioles 1−4.5cm; leaflets ovate-elliptic 4−9×1−3cm, acuminate, base cuneate, greyish pubescent especially beneath; stipels subulate 1−2mm; stipules narrowly triangular 4−5mm. Racemes 10−20-flowered, 5−15cm, forming leafy panicles, bracts deciduous. Calyx tube 5mm, teeth triangular 3−5mm. Petals yellow or tinged red, 1.7−2cm, standard suborbicular 1.3cm, wings ovate, blade 12×6mm, keel lunate 12×5mm. Pods 4−7×0.7−1cm acuminate, grooved obliquely between seeds, 3−6-seeded. Seeds subglobose, somewhat compressed, 6−7mm, white, red, brown or blackish, almost glossy, hilum surrounded by a whitish rim aril.

Bhutan: S—Gaylegphug district (near Gaylegphug), **C**—Tashigang district (Dangme Chu Valley); **Sikkim**. In fields, 270−900m. October−December.

Cultivated for the young pods and seeds which may be eaten raw or boiled. Mature seed is usually split, boiled and eaten as dhal. It is also cultivated as a host plant for the lac insect (126). The thick main stems are used for firewood (130).

2. C. scarabaeoides (L.) Wall.; *Atylosia scarabaeoides* (L.) Bentham

Slender twiner, stems finely pubescent. Leaves pinnately trifoliate; petioles 1−2cm; leaflets ovate-elliptic or obovate 2−4.5×0.7−2cm, acute or subacute, base cuneate, pale pubescent especially beneath; stipels and stipules minute, deciduous. Flowers solitary or 5, peduncles 0.5−1.5cm. Calyx tube 2mm, teeth narrowly triangular 5−8mm. Petals yellow, standard obovate 8×7mm, wing oblong, blade 7×2mm, keel lunate 7×3mm. Pods 1.5−2.5×0.6−0.8cm, densely and softly pubescent, grooved almost at right angles to sutures, 5−6-seeded. Seeds oblong c 3.5×2.5mm, dark brown, aril pale c 0.5mm.

Bhutan: C—Punakha, Tongsa and Tashigang districts; **Sikkim**. On dry rocks and sand, 1370−1650m. September−November.

3. C. grandiflorus (Baker) van der Maesen; *Atylosia grandiflora* Baker, *Dunbaria pulchra* Baker

Woody twiner. Leaves pinnately trifoliate; petioles 5−9cm; leaflets broadly ovate or rhombic, 5−10×4−8cm, acuminate, base rounded, pubescent and gland dotted especially beneath; stipels subulate 3−4mm; stipules caducous. Racemes c 10-flowered, 12−18cm, buds enclosed at first by deciduous boat-shaped bracts 3×2.5cm shortly 3-toothed at apex. Calyx tube 5mm, teeth triangular 7−12mm, acuminate, bearing scattered bulbous-based hairs especially at margins. Petals yellow, standard suborbicular 2.5−3×2−2.5cm, wings oblong-obovate, blade 1.5−2.5× 1cm, keel lunate 1.5−2.5×1cm. Pods 4−4.5×1cm, densely brown hirsute, grooved almost at right angles to sutures, 6−7-seeded. Seeds oblong 4−5×3.5mm, blackish, aril yellowish c 1mm.

Bhutan: C—Mongar district (Kuru Chu Valley); **Sikkim**. 1200m. August.

4. C. mollis (Bentham) van der Maesen; *Atylosia mollis* Bentham. Dz: *Semchung Robjay*.

Twining perennial herb. Leaves subdigitately 3-foliate; petioles 3−8cm; leaflets ovate-elliptic or rhombic 4−7×2.5−6cm, acute, base cuneate, pubescent especially beneath, glandular; stipels subulate 2−3mm; stipules minute, deciduous. Racemes 6−15-flowered, 5−7cm, bracts ovate 8−14×6−10mm, deciduous. Calyx tube 5mm, teeth 3−5mm. Petals yellow 1.3−2.5cm, standard broadly oblong or obovate, 13−22×12−15mm, wings elliptic, blade 10−17×4−5mm with 2 auricles at base, the lower one short, the upper almost as long as basal claw, keel oblong 10−16×5−6mm, upcurved and rounded distally. Pods 4−5×0.7−0.9cm, rounded or acute at apex, often bearing persistent style, densely pubescent, grooved almost at right angles to sutures, 6−9-seeded; seeds oblong, brown or blackish 3.5−4×2.5−3mm, aril 0.5mm, yellowish.

Bhutan: C—Punakha district (near Wangdu Phodrang) and Tashigang district (Dangme Chu Valley), **N**—Upper Mo Chu district (Gasa). Climbing on trees, 1500−2130m. August−September.

5. C. elongatus (Bentham) van der Maesen; *Atylosia elongata* Bentham

Stems sparsely pubescent with long fine hairs. Leaves digitately trifoliate, leaflets ovate or broadly elliptic 2.5−4×2−3.5cm, acute, base rounded, sparsely pubescent with long fine hairs and glands especially beneath; stipels absent; stipules ovate c 4×2mm. Flowers 1−4, peduncles slender 3−4cm, bracts elliptic c 4mm, 3-toothed at apex. Calyx tube 2−4mm, teeth narrowly triangular 5−8mm. Petals 10−12mm. Pods 1.7−2.5×0.5−0.8cm, 3−4-seeded.

Bhutan: C—Tongsa district (Nimshong) and Tashigang district (Damoitsi). On dry ground, 2000m. August.

6. C. villosus (Baker) van der Maesen; *Atylosia villosa* Baker

Stems shortly pubescent. Leaves subdigitately trifoliate, leaflets narrowly obovate 2.5−4×1.2−2cm, acute, base cuneate, pubescent especially beneath. Racemes up to 5-flowered, 2.5−5cm. Calyx tube 3−5mm, teeth lanceolate 3−5mm. Standard obovate c1.8×1.5cm, wings oblong-obovate c 15×5mm, keel oblong c 10×5mm. Pods oblong, 2−3.5×1cm, densely covered with spreading brown hairs, transverely grooved almost at right angles to sutures, 5−6-seeded.

Sikkim. 150−1200m.

57. DUNBARIA Wight & Arnott

Twining herbs. Leaves pinnately 3-foliate, leaflets entire bearing reddish gland dots especially beneath; stipels minute; stipules small ± persistent. Flowers solitary axillary or in axillary racemes. Calyx 5-toothed, upper 2 ± connate into a lip. Standard suborbicular, erect or spreading, auricled at base, wings obliquely obovate or oblong, keel blunt, incurved. Stamens diadelphous. Pods linear-oblong, straight or curved, compressed, valves thin, ± septate between 4−11 seeds, funicles broadened, hilum surrounded by rim-like aril.

1. Flowers in racemes .... **1. D. circinalis**
+ Flowers solitary, axillary .... **Species 2 & 3**

1. D. circinalis (Bentham) Baker

Stems finely pubescent. Leaflets ovate-rhombic, 2−4.5(−6)×1.3−4.5(−7)cm, acute or acuminate, base rounded or cuneate, finely pubescent on both surfaces, terminal leaflet larger than oblique laterals; petioles 1.5−3.5cm; stipules lanceolate 3−4×1mm. Flowers few or many, racemes 4−20cm, pedicels 2−3mm. Calyx tube 3.5−4mm, teeth lanceolate, 3−7mm, gland dotted and pubescent with simple and glandular hairs. Petals yellow 1−1.5cm. Pods 5−6×1cm dotted with reddish glands and pubescent with yellowish gland-based hairs; seeds rounded or ± reniform, 5×4.5mm, black.

Sikkim: Sookna. 150m. July.

76. LEGUMINOSAE

2. D. rotundifolia (Loureiro) Merrill; *D. conspersa* Bentham

Similar to *D. circinalis* but leaves 1.2−3×1−3.5cm, apex subacute; petioles 1−2.5cm; flowers solitary, axillary on peduncles 3−5mm; calyx tube c 4mm, teeth 4−6mm, dotted with reddish glands and pale pubescent; petals 1.25−1.5cm; pods 3.5−5×0.6−0.8cm, subglabrous; seeds c 3mm, brown, aril c 0.75mm.

Sikkim: Darjeeling Terai. September.

3. D. debilis Baker

Similar to *D. circinalis* and more especially to *D. rotundifolia* but leaves oblong, 10−30×3−10mm, 3−4 times as long as broad; petioles acute, 5−12mm; flowers solitary axillary; calyx c 6mm, greyish pubescent; petals yellow c1.2cm; pods 3.5−5×0.6cm.

Sikkim: Terai. 150m.

58. FLEMINGIA Aiton

Erect shrubs. Leaves simple or digitately 3-foliate, usually bearing yellow or reddish glands beneath; stipels absent; stipules present, striate. Flowers in axillary and terminal racemes, panicles or globose heads, bracts often conspicuous, deciduous or persistent, bracteoles absent. Calyx campanulate, teeth 5, acute. Petals clawed, standard with 2 basal inflexed auricles, wings usually adherent to incurved keel. Stamens diadelphous, filament of free stamen often thickened or kneed near base. Style slightly thickened above middle. Pods short, ± inflated, not septate, 1−2-seeded.

1. Leaves simple 2
+ Leaves 3-foliate 3

2. Bracts ovate, deciduous, flowers visible .... **3. F. paniculata**
+ Bracts suborbicular, persistent, enfolding and obscuring flowers
                                                                          Species 1 & 2

3. Flowers in rounded heads surrounded by brown persistent bracts
                                                                          4. F. involucrata
+ Flowers in short or long racemes or panicles, bracts deciduous 4

4. Shrub almost leafless at flowering time; petals yellow ..... **7. F. bhutanica**
+ Shrubs, leafy at flowering time; petals pink or blue-purple .. **Species 5 & 6**

1. F. strobilifera (L.) Aiton; *Moghania strobilifera* (L.) Kuntze, *F. bracteata* (Roxb.) Wight

Shrub 1−5m. Leaves ovate-lanceolate 7−15×4−8cm, acute or

acuminate, base rounded or truncate, minutely glandular and pubescent along veins beneath; stipules lanceolate 6−9mm; petioles 1−2cm. Flowers 1−2 in axils of reniform or suborbicular membranous bracts 1.7−2.5×3−4cm. Calyx tube 1.5−2mm, teeth lanceolate 3−5mm, gland dotted. Petals white, standard suborbicular 7−8mm long and broad, often bearing purplish lines near middle, wings oblong, blade 3−4mm, keel strongly incurved 7−8mm. Pods ellipsoid c 10×7mm, finely pubescent.

Bhutan: locality unknown; **Sikkim.** 1250−1400m. February−March.

2. **F. fruticulosa** Bentham; *Moghania fruticulosa* (Bentham) Mukerjee

Similar to *F. strobilifera* but shrub 20−40(−100)cm; leaves ovate-oblong 3−6×1.5−4cm, obtuse or subacute, base cordate; stipules 5−8mm; petals pink.

Bhutan: S—Chukka district, **C**—Thimphu, Punakha, Tongsa and Mongar districts. 1500−2100m. July−September.

3. **F. paniculata** Bentham; *Moghania paniculata* (Bentham) Li

Shrub 1−2m. Leaves ovate, 9−15×4−8cm, acute or acuminate, base rounded or cordate, pubescent and glandular especially beneath; stipules lanceolate, 1−1.5cm, deciduous; petioles 1−3cm. Flowers in axillary or terminal racemes or panicles, bracts ovate c 5×2mm, deciduous. Calyx tube 2.5mm, teeth lanceolate 4−5mm. Petals pink, standard obovate 6−8mm long and broad, wings oblong, blade 5×2mm, keel lunate 6×3mm, greenish. Pods oblong c 12×7mm.

Sikkim: foothills. 600m. February−April.

4. **F. involucrata** Bentham; *Moghania involucrata* (Bentham) Kuntze

Shrub c1m. Leaflets elliptic or narrowly obovate, 6−9×1.5−3.5cm, acute, base rounded or cuneate, pale beneath, pubescent and glandular especially on underside; petioles 1−2cm; stipules lanceolate 6−8mm, deciduous. Flowers in axillary or terminal globose clusters 2−3cm diameter, surrounded by 10−15 brown ovate bracts 1.5−2×0.5−0.7cm. Calyx tube 3−4mm, teeth lanceolate 10−12mm, densely white hirsute. Petals pink, standard elliptic 9−10×4−5mm, wings oblanceolate, blade 6×1.5mm, keel obliquely elliptic 6×2.5mm. Pods ovoid 6×3mm, blackish.

Sikkim: Terai. October−February.

5. **F. macrophylla** (Willdenow) Merrill; *F. congesta* Aiton, *F. semialata* Roxb., *F. wightiana* Wight & Arnott, *Moghania macrophylla* (Willdenow) Kuntze. Nep: *Batwasi* (13)

Shrub 1−2m, branches squarish or triangular in section, pale appressed sericeous when young. Leaflets elliptic-obovate, 6−9×1.5−3.5cm, acuminate, base rounded or cuneate, sparsely or densely appressed pubescent and gland dotted especially beneath; petioles 3−9cm, narrowly (c1mm) winged; stipules lanceolate 8−9mm, deciduous. Flowers in axillary racemes 5−10cm or forming terminal panicles, bracts ovate, 8−10× 2.5−4mm, caducous, appressed brown sericeous. Calyx tube 2−3mm, teeth lanceolate 5−10mm, appressed greyish brown pubescent. Petals pink, standard ovate or obovate, 10−12×5−8mm, wings obovate-elliptic, blade 6−8×2−3mm, keel falcate 7−10×3−4mm. Pods ovoid-ellipsoid, 10−12×7−8mm, finely greyish brown pubescent.

Bhutan: S—Samchi, Chukka, Gaylegphug and Deothang districts, **C**—Punakha, Tongsa and Mongar districts; **Sikkim.** Warm broad-leaved forest slopes, 300−2000m. April−May.

The species is variable in respect to length of inflorescence, development of the petiole wings and density of indumentum; extreme examples of these variables have become the basis for the names here placed in synonymy.

The leaves and bark are used medicinally (13).

6. F. stricta Roxb.; *Moghania stricta* (Roxb.) Kuntze

Similar to *F. macrophylla* but shrub up to 3m; leaflets ovate-lanceolate 17−30×5−15cm, acuminate, base rounded or cuneate; petioles 8−10cm, narrowly winged; stipules 5−6.5cm; racemes 7−11cm, bracts lanceolate 1.5−2cm, deciduous; calyx tube 2−3mm, teeth 3−7mm; petals purple, standard obovate or suborbicular 7−8×6−7mm, wings oblanceolate, blade 6×2mm, keel obliquely elliptic 6.5×3mm; pods ellipsoid 15×8mm.

Sikkim: Singleghora. 360−600m. January−March.

7. F. bhutanica Grierson

Similar to *F. macrophylla* but branches cylindrical, striate, pubescent at first, almost leafless at flowering time; leaflets ovate-elliptic, 5.5−7×1.5−4cm; petioles 2.5−4cm, scarcely winged; racemes 2.5−3cm; calyx tube 2−3mm, teeth lanceolate 4−8mm, pubescent and bearing reddish glands; petals yellow, standard elliptic 9−10×5−6mm, wings oblong, blade 6−7×1.5−2mm, keel lunate 8.5−10×2.5−3mm; pods ellipsoid 10−12×6−7mm, finely pubescent and densely reddish glandular.

Bhutan: C—Punakha (near Lobesa and Wangdu Phodrang). River banks and Warm broad-leaved forests, 1300−1600m. April.

<u>Endemic to Bhutan.</u>

59. RHYNCHOSIA Loureiro

Twining perennial herbs often with woody rootstocks. Leaves pinnately 3-foliate, leaflets entire, gland-dotted beneath, stipels and stipules present. Flowers in axillary racemes. Calyx campanulate, 5-toothed. Petals clawed, standard spreading or reflexed, auricled at base, wings shorter than the keel.

Stamens diadelphous, filament of free stamen kneed at base. Pods oblong or narrowly obovate, compressed, 2-seeded.

1. R. minima (L.) DC.
Petioles 0.7−2cm; leaflets rhombic-ovate, 1−2×0.7−2cm, acute, base rounded or broadly cuneate, subglabrous or puberulous above, densely glandular beneath; stipels subulate; stipules narrowly ovate 3−4mm. Racemes 3−10-flowered, 2−12cm. Calyx tube 1.5mm, teeth narrowly lanceolate 3−3.5mm, glandular. Petals yellow with purple veins 4.5−6mm, standard obovate 5×2.5cm, wings oblong, blade 3−3.5×0.5mm, keel upcurved, rounded distally, 4×2mm. Pods 1.3−1.7×0.4−0.5cm minutely pubescent. Seeds reniform, 3×2mm, dark brown.

Bhutan: C—Punakha district (Wangdu Phodrang and Chusom). On dry sandy rocks, 1300−1400m. April−September.

2. R. harae Ohashi & Tateishi
Similar to *R. minima* but much larger, stems densely pubescent with long hairs; petioles 10−18cm; leaflets ovate-elliptic 18−23×10−17cm; racemes many-flowered, (5−)11−21cm; calyx tube 3.5−4mm, teeth 7−7.5mm; petals 15−18mm brownish, standard 14−15×11−13mm, wings 10×4mm, keel 10−11×4.5−5mm; pods 2×0.7cm densely pubescent with long hairs; seeds oblong-globose 5−6×4mm.

Bhutan: locality unknown; **Sikkim:** Darjeeling. 300−900m. September.

60. ERIOSEMA (DC.) G. Don

Tuberous-rooted perennial herb or subshrub, stems simple or little branched. Leaves simple, entire, gland-dotted beneath, stipels absent, stipules inconspicuous. Flowers solitary, axillary. Calyx campanulate, 5-toothed. Petals clawed, standard auricled at base, pubescent on outer surface, wings and keel shorter. Stamens diadelphous. Pods oblong-elliptic, ± compressed, 1−2-seeded, densely villous.

1. E. himalaicum Ohashi; *E. chinense* sensu F.B.I. non Vogel, *E. tuberosum* (D. Don) Wang & Tang non Richter
Stems erect, 8−70cm, densely pubescent. Leaves oblong-elliptic, 2−4.5×0.25−0.75(−1)cm, acute or acuminate, base rounded or cuneate, pubescent especially beneath, ± sessile; stipules narrowly lanceolate c 3mm. Flowers on peduncles up to 5mm. Calyx 5−7mm divided to middle into lanceolate teeth. Petals yellow, 8mm, standard obovate 7×6mm, wings oblong, blade 6×2mm, keel narrowly obovate 5×2.5mm. Pods 7−8×5−6mm; seeds oblong c 3mm, black with a low whitish aril extending almost along its length.

76. LEGUMINOSAE

Bhutan: C—Punakha district (Rinchu) and Tashigang district (Tashiyangsi), N—Upper Mo Chu district (Tamji). 1525–2130m. June–August.

61. AESCHYNOMENE L.

Perennial herbs or subshrubs. Leaves even-pinnate, rachis ending in a short point, leaflets entire, sensitive, stipels absent; stipules medifixed, deciduous. Flowers in short axillary racemes. Calyx bilabiate, lips obscurely toothed. Petals shortly clawed. Stamens in two bundles of 5. Pods linear, compressed, transversely divided into 1-seeded segments, indehiscent.

1. A. indica L.

Stems 15–100(–250)cm, thick (sometimes up to 5cm) and pithy near base. Leaves 4–12cm, leaflets gradually decreasing in size towards apex of leaf, 20–40 pairs, oblong 2–9×0.75–2mm, obtuse, mucronate, base rounded, glabrous, rachis glandular near base; stipules lanceolate 8–10×3mm, auricled below point of attachment 2–3mm. Peduncles bearing sessile glands, glabrous or finely pubescent. Calyx 4–6mm, glabrous. Petals yellow or purplish, glabrous, 8–9(–12)mm, minutely glandular at margin, standard broadly ovate 7–8mm broad, wings elliptic, blade 6–7×3–4mm, keel upcurved, c 7×3mm. Pods 3–5.5×0.3–0.5cm borne on stalks 5–10mm, transversely divided into 7–10 squarish segments, sparsely gland-dotted.

Bhutan: S—Deothang district (near Deothang); **Sikkim:** Terai. Marshy ground, 200–300m. August.

2. A. aspera L.

Similar to *A. indica* but more robust, leaflets 30–50 pairs; peduncles and flowers coarsely pubescent; calyx c 9mm; petals 15–20mm, pubescent outside; pods 5–7×0.7–0.8cm on basal stalks 1.5–2cm, ± coarsely papillate over each seed.

Sikkim: locality unknown (73).

The light pith from this species was formerly used in the manufacture of sola topis (125).

62. SMITHIA Aiton

Annual or perennial herbs or subshrubs. Leaves even-pinnate, rachis ending in a bristle, sensitive, leaflets entire, stipels absent, stipules medifixed, persistent. Flowers in axillary or terminal racemes or panicles, bracts small, caducous, bracteoles paired, sheathing base of calyx, persistent. Calyx deeply bilabiate, lips ± entire. Petals clawed. Stamens in two bundles of 5. Pods compressed, consisting of several rounded 1-seeded segments folded on top of one another within persistent, accrescent calyx, indehiscent.

1. Annuals with slender prostrate or spreading stems; flowers less than 1cm
Species 1 & 2
+ Perennial herbs or subshrubs with robust erect stems; flowers c 2.5cm
3. S. grandis

1. S. sensitiva Aiton
Annual. Stems prostrate or spreading, 15−50(−100)cm. Leaves 1−3cm, rachis ending in a bristle, leaflets 3−8 pairs, 3−8×1.5−3mm, obtuse, mucronate, base obliquely rounded, finely ciliate and with a few whitish bristles on midrib beneath; stipules membranuous, lanceolate 3−5mm, auricle attenuate ± as long, extending below point of attachment, with a rounded lobe near base. Racemes 2−7-flowered, axillary, peduncles slender 1−2cm, bracteoles ovate 2−3mm. Calyx 4−5mm, lips ovate, acute, closely parallel-veined, bearing a few bristles on midrib. Petals yellow, 7−9mm, standard broadly elliptic or suborbicular, 5−6mm broad, wings oblong 5−6×1mm, keel spathulate 6×2mm, upcurved and rounded distally. Pods consisting of 4−6 segments, c1.5mm diameter, within persistent calyx 8−9mm.
Bhutan: S—Phuntsholing district (near Phuntsholing); **Sikkim:** Mongpu. Roadsides and marshy ground, 200m. August−November.

2. S. ciliata Royle
Similar to *S. sensitiva* but stipules unlobed below point of attachment; flowers secund in small racemes on short (up to 1.5cm) peduncles; calyx upper lip truncate or shortly toothed, margins stiffly ciliate, veins anastomosing, not closely parallel; petals white or pale blue, 5−6mm.
Bhutan: C—Thimphu district (near Thimphu and Paro); **Sikkim.** Roadsides and riverside shingle, 2350−2500m. September−October.

3. S. grandis Baker
Perennial herb or subshrub. Stems erect, robust 1−1.3m, glabrous. Leaves 5−7cm, leaflets 10−12 pairs, oblong 15−20×3−4mm, obtuse, mucronate, base obliquely rounded, glabrous; stipules lanceolate 1.5−2cm, auricles 3−7mm. Flowers in terminal, lax corymbose panicles. Calyx at first 8−10mm, lips obovate rounded, sparsely bristly. Petals c 2.5cm. Pod segments 4−5mm diameter, concealed in calyx 18−20mm.
Sikkim: Terai.

76. LEGUMINOSAE

63. ZORNIA Gmelin

Annual herbs. Leaves with a pair of opposing entire leaflets, gland-dotted, stipels absent, stipules medifixed, persistent. Racemes axillary, elongate, each flower concealed between a pair of medifixed bracts resembling the stipules. Calyx membranous, irregularly toothed in upper half. Petals shortly clawed. Stamens monadelphous. Pods compressed, lower margin especially deeply indented, dividing into several 1-seeded segments, each covered with retrorsely barbed bristles, indehiscent.

1. Z. gibbosa Spanoghe; *Z. diphylla* sensu F.B.I. p.p. non Persoon. Fig. 42c−e

Stems spreading 15−35cm. Leaflets ovate-lanceolate 10−25×3−9mm, acuminate, base rounded or cuneate, glabrous; petioles 5−15mm; stipules lanceolate 5−10×1−1.5mm, with an oblong lobe extending 2−5mm below point of insertion, gland-dotted. Racemes 5−15cm, bracts ovate-lanceolate 8−15×3−4mm, gland-dotted, margins ciliate. Calyx 2.5−3mm. Petals yellow, 3−5mm, standard broadly ovate 2.5−3mm broad, wings oblong, blade 2−3.5×1−1.5mm, keel narrowly ovate 2.5−3.5×1−2mm. Pods separating into 2−3(−4) broadly elliptic segments 2−3mm.

Bhutan: C—Punakha district (near Punakha and Wangdu Phodrang) and Tashigang district (Dangme Chu). Roadsides and on dry soil by fields, 900−1350m. August−October.

64. ARACHIS L.

Annual, stems ± prostrate, rooting at lower nodes. Leaves even-pinnate, leaflets in 2 pairs, entire, stipels absent; stipules elongated, adnate to petiole. Flowers 1−3(−6) sessile, axillary. Receptacular (calycine) tube elongate-filiform, resembling peduncle, calyx bilabiate, upper lip 4-toothed, lower lip linear. Standard shortly clawed, suborbicular, wings auricled at base, keel incurved, beaked. Stamens monadelphous, anthers alternately long and short. Ovary ± sessile at base of calyx tube borne on a gynophore which elongates and hardens after flowering forcing the developing fruit below ground. Pods cylindrical, constricted but not septate between the 2(−6) seeds, sometimes seed solitary.

1. A. hypogaea L. Eng: *Ground Nut, Pea Nut, Monkey Nut*

Stems 20−70cm, sparsely white villous. Leaf rachis 5−7cm ending blindly; leaflets oblong-obovate, 2.5−3.5×1.5−2.5cm, subacute, obtuse or emarginate, base rounded, ciliate otherwise ± glabrous; petioles 3−4cm, hirsute; stipules lanceolate 2−2.5cm united with petiole in lower half. Calyx tube 2.5−4cm, teeth 10−12mm, upper lip ovate. Petals 10−12mm, yellow.

Pods oblong 2−4(−6)cm finely reticulate, borne on gynophore 6−7cm; seeds oblong-ovoid 10−15mm.

Bhutan: C—Punakha district (Punakha). In garden, 1270m. September.

Native of tropical America, cultivated for its edible seeds from which oil may also be extracted.

65. CARAGANA Fabricius

Spiny shrubs. Leaves even-pinnate, leaflets small, entire, deciduous, rachis often persistent becoming woody and spine-tipped; stipules lanceolate sometimes spinescent, stipels absent. Flowers solitary or few, axillary. Calyx campanulate, teeth subequal, upper pair usually smaller. Petals all clawed, standard ovate or obovate, wings oblong with or without a linear appendage at base, keel elliptic, rounded. Stamens diadelphous. Pods ellipsoid, inflated, dehiscing by twisting of valves, 4−6-seeded.

1. Wing petals with a linear appendage at base almost as long as the claw
 Species 1 & 2
+ Wing petals with rounded auricles at base .... **3. C. spinifera**

1. C. jubata (Pallas) Poiret; *C. chumbica* Prain. Fig. 44 t−x

Stems ± erect, 0.5−1.5(−4)m, covered with numerous spiny leaf rachises. Leaves 2−4cm borne in tufts, leaflets 4−5 pairs, ovate-elliptic 5−10×2−3mm, acute, base rounded, margin inrolled, pubescent with spreading hairs on both surfaces, but very dense beneath; stipules adnate to rachis, lanceolate or subulate, 8−9mm. Calyx 1.3−1.5cm divided to middle into lanceolate teeth; reddish green. Petals c 2.5cm, creamy white streaked or blotched with reddish purple, wings with a linear appendage at base ± as long as the claw, keel curved and rounded at apex, pouched near base. Pods c 2.5×0.5cm, softly pubescent outside, glabrous within.

Bhutan: C—Thimphu district (near Shodu and Barshong); **Chumbi**. Open hillsides and river banks, 3500−3800m. May.

The record of *C. gerardiana* Royle from Bhutan (71) refers to this species.

2. C. sukiensis Schneider; *C. nepalensis* Kitamura

Similar to *C. jubata* but stems less densely spinous, apparently only the rachises of the first leaves on shoots persisting and hardening, leaflets 3−7 pairs, not inrolled, softly appressed pubescent; calyx c 1.3cm divided in upper third into triangular teeth; petals c 2cm yellow, wings with long linear appendages; pods 3−4.5cm pubescent outside and inside.

Bhutan: C—Ha, Thimphu, Punakha and Bumthang districts, **N**—Upper Bumthang Chu district. Stony or sandy river banks, 2750−3650m. June.

3. C. spinifera Komarov
Similar to *C. jubata* and more especially *C. sukiensis* but leaflets 2−4 pairs, glabrous or finely pubescent; stipules subulate 5−6mm, strongly spinous; calyx c 1.2cm shortly toothed; petals c 2cm, standard obovate, wings with small rounded auricles at base; pods glabrous within.
Sikkim: locality unknown.

66. SPONGIOCARPELLA Yakovlev & Ulzijkhutag

Perennial herbs, woody at base, acaulescent or short stemmed. Leaves odd-pinnate, leaflets entire, stipels absent; stipules leafy, adnate to rachis. Flowers usually solitary on axillary peduncles. Calyx campanulate, pouched at base, divided above into 5 subequal teeth. Petals long-clawed, standard with a small tooth on either side near middle, wings not pouched. Stamens diadelphous. Pods oblong, continuous within, valves twisting at maturity, spongy within, seeds reniform.

1. S. purpurea (Li) Yakovlev; *Chesneya purpurea* Li, *C. nubigena* auct. p.p. non (D. Don) Ali, *C. polystichoides* (Handel-Mazzetti) Ali subsp. *bhutanica* Ohashi, *Caragana crassicaulis* sensu F.B.I. p.p. non Baker, *Astragalus larkyaensis* auct. p.p. non Kitamura
Rootstock thickly woody, branching at ground level. Stems not more than 5cm, densely covered with non-spiny persistent leaf rachises. Leaves 4−8(−15)cm, leaflets 13−31, ovate-oblong, 3−7×1−3mm, acute, base rounded, densely appresssed pubescent above, ± glabrous beneath; stipules 1−1.5cm, entire or divided into 2−6 linear segments on each side. Flowers ± sessile. Calyx c 1.5cm divided in upper third into triangular teeth, white pubescent. Petals 2−3cm, yellow sometimes tinged purple, or entirely purple, standard obovate, c1.5cm broad, notched or bifid at apex. Pods ellipsoid, 1.5−2×0.5cm, pubescent.
Bhutan: C—Ha, Thimphu and Bumthang districts, N—Upper Mo Chu, Upper Mangde Chu, Upper Bumthang Chu and Upper Kulong Chu districts; **Sikkim.** On open rocky slopes, 3650−4570m. May−July.

67. ASTRAGALUS L.

Perennial herbs. Leaves odd-pinnate, leaflets entire, stipels absent, stipules free or connate. Flowers in racemes, sometimes 1−3 on short peduncles. Calyx campanulate, teeth subequal. Petals clawed, standard obovate or oblong, wings slightly pouched near middle and with a rounded or oblong auricle at base, keel somewhat shorter than wings rounded or acute distally. Stamens diadelphous. Pods oblong, ± inflated, wholly or partially 2-celled by a longitudinal membrane from ventral suture.

67. ASTRAGALUS

1. Stipules very large, 3.5−7cm long .... **1. A. stipulatus**
+ Stipules smaller, 2.25cm or less long 2

2. Flowers yellow, petals sometimes tipped reddish-purple or brown 3
+ Flowers reddish, mauve, blue or white but not yellow 7

3. Acaulescent herb, stems densely tufted up to 3cm .... **2. A. acaulis**
+ Herbs with well-developed stems more than 3cm 4

4. Leaflets 25 or more 5
+ Leaflets 9−13 6

5. Stipules membranous, ovate, connate 5−12mm broad .... **3. A. concretus**
+ Stipules herbaceous, lanceolate, free 2−3mm broad .... **Species 4 & 5**

6. Small plants 7−10cm; leaflets not more than 10×6mm . **6. A. lessertioides**
+ Taller plants 30−40cm; leaflets commonly 25×10mm ....**7. A. tongolensis**

7. Stipules 10−15mm (sometimes as little as 6mm in *A. kongrensis*)
   Species 8 & 9
+ Stipules 2−5mm 8

8. Petals 10−16mm; flowers solitary or few, up to 7 in racemes
   Species 10 − 12
+ Petals 7−12mm; racemes 6−30-flowered (sometimes as few as 3 in
   A. tenuicaulis but then petals small) 9

9. Short tufted herbs, stems 5−15cm .... **Species 13 & 14**
+ Sprawling herbs with stems 15−30cm .... **Species 15 & 16**

1. A. stipulatus Sims
Stems 0.6−2m. Leaves 20−40cm, leaflets 31−45, oblong, obtuse at both ends, 1−4.5×0.5−1.5cm, ± glabrous; stipules ovate 3.5−7×1.5−3.5cm, sheathing stem. Flowers numerous, orange-red in racemes c15cm (elongating to 45cm in fruit), peduncles 6−8cm. Calyx c1cm, divided to middle into narrow linear-lanceolate teeth, usually brownish pubescent. Petals c 1.5cm, standard obovate, blade c 10×6mm, wings linear curved c 8×2.5mm, auricle oblong, keel ovate c 7×3mm. Pods narrowly ovoid, 2.5−3.5×0.5−0.7cm, curved and tapering to the apex, divided longitudinally within, 6−8-seeded on each side, seeds reniform c 2mm compressed, blackish.

Bhutan: C—Thimphu, Punakha, Tongsa and Mongar districts; **Sikkim.** On shingle by streams, 1830−2750m. June−August.

2. A. acaulis Baker

Acaulescent, stems densely tufted up to 3cm, bearing persistent, membranous, broadly ovate, connate stipules 8−12×6−10mm. Leaves 2.5−15cm, leaflets 15−25 pairs, narrowly lanceolate 5−10×1−2mm, acuminate, base rounded or cuneate, white pubescent when young, leaf rachis subpersistent but not hardened. Flowers 1−3 on short peduncles. Calyx tube 7−8mm, teeth lanceolate 3−5mm, white pubescent. Petals yellow, 2.5−3cm, standard broadly elliptic, blade 1.3−2×1−1.5cm, wings narrowly oblong 15−17×3mm, keel elliptic 15×5−7mm. Pods oblong 2.5−5×1−1.5cm, slightly curved, glabrous, 10−15-seeded.

Bhutan: C—Bumthang district (Tripte La), **N**—Upper Mo Chu and Upper Bumthang Chu districts; **Chumbi; Sikkim.** On sand and gravel on open hillsides, 3950−4265m. May−June.

3. A. concretus Bentham; *A. vicioides* Baker non Ledebour, *A. xiphocarpus* auct. non Bunge

Stems up to 1m. Leaves 10−20cm, leaflets 17−31, elliptic-oblong, 1−3×0.5−1.2cm, obtuse or subacute, base rounded or cuneate, sparsely appressed especially beneath; stipules ovate 1−2×0.5−1.2cm, membranous, acuminate, connate, straw-coloured or brownish. Flowers numerous in axillary racemes on peduncles 2−6cm, bracts linear, 3−4mm, caducous. Calyx tube 6−7mm, blackish pubescent, teeth very short (c1mm). Petals yellow, 1.3−1.5cm, standard obovate 6−7mm broad, wings narrowly oblong, blade c 7−8×2mm, keel ovate 5−6×3mm. Pods ovoid 2.5−3×0.5−0.7cm, glabrous, narrowed at both ends, the basal stalk ± as long as calyx.

Bhutan: C—Bumthang and Mongar districts; **Chumbi; Sikkim.** Steep banks in Fir forest, 2900−3500m. July−August.

4. A. chlorostachys Lindley

Similar to *A. concretus* but stipules green, lanceolate, 7−10×2−3mm, spreading, not embracing stem, bracts linear-lanceolate, 6−7mm, persisting at flowering time; pods oblong 1−1.3×0.5cm narrowed at both ends, basal stalk twice as long as calyx.

Bhutan: C—Thimphu district (near Thimphu) and Bumthang district, **N**—Upper Mo Chu district. Roadsides and open bamboo forest, 2100−3050m. July.

FIG. 44. **Leguminosae.** a, *Lablab purpureus:* pod. b, *Stracheya tibetica:* pod. c−g, *Crotalaria cytisoides:* c, leaf and inflorescence; d, standard; e, wing; f, keel; g, pod. h−n, *Astragalus bhotanensis:* h, leaf and inflorescence; i, flower; j, standard; k, wing; l, keel; m, pod; n, transverse section of pod. o−s, *Oxytropis lapponica:* o, inflorescence; p, flower; q, standard; r, wing; s, keel. t−x, *Caragana jubata:* t, portion of flowering shoot; u, leaf, v, standard; w, wing; x, keel. Scale: c, h × ½; a, t × ⅔; g, o, v, w, x × 1; u × 1⅓; m × 1½; b, d, e, f, q, r, s × 2; i, j, k, l, p × 2½; n × 3.

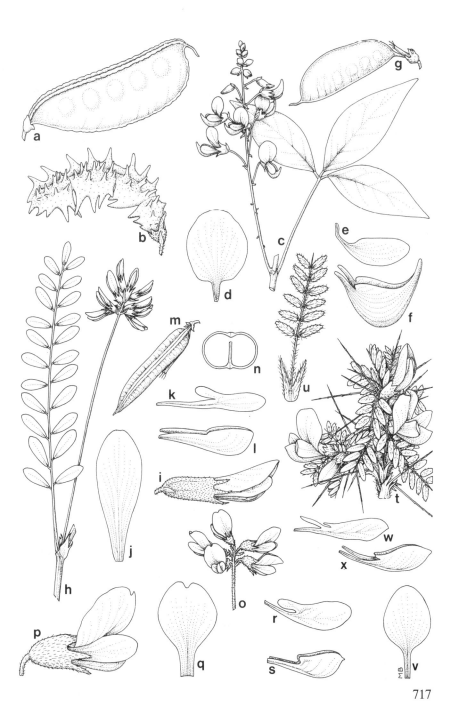

5. A. floridus Bunge

Similar to *A. concretus* but stems greyish with a mixture of black and white appressed hairs; stipules lanceolate 5−8×1−2mm free; bracts linear-lanceolate 3−4mm persistent; calyx borne obliquely, almost pouched above attachment with pedicel, tube 4−5mm, teeth 1−2mm, densely appressed black pubescent; standard elliptic, 4−5mm broad, minutely black streaked near midrib; pods ellipsoid 1−1.5×0.7−0.9cm narrowed at both ends, basal stalk ± as long as calyx, blackish pubescent, unilocular, 3−5-seeded.

Bhutan/Chumbi border: between Sharna and Tremo La; **Sikkim:** Thangu. 4265m. July.

6. A. lessertioides Bunge

Stems decumbent or erect, 7−10cm, glabrous or blackish pubescent. Leaves 3−7cm, leaflets 9−13, oblong-elliptic 6−12×4−7mm, obtuse, base rounded or cuneate, sparsely black pubescent beneath; stipules ovate 6−12×3−7mm, black pubescent. Racemes condensed 4−20-flowered, peduncles 7−20cm, blackish pubescent. Calyx c 6mm, divided to middle into narrowly triangular teeth, black pubescent. Petals 1−1.3cm, yellow, sometimes tipped brownish or purple, often drying blackish, standard obovate 6−8mm broad, wings oblong, blade 4−5×1.5mm, keel ovate 4×2mm. Pods ellipsoid c 1.7×0.5cm, basal stalk ± as long as calyx, black pubescent.

Bhutan: N—Upper Bumthang Chu district (Pangotang); **Sikkim:** Tungu and Alookthang. On grassy hillsides, 3960−4570m. July−August.

7. A. tongolensis Ulbrich

Stems 30−40cm ± erect. Leaves 7−12cm, leaflets 7−13, ovate, 1.7−3×0.7−1.5cm, acute, base rounded, glabrous above, pubescent beneath with white and sometimes black hairs; stipules ovate-elliptic, 1.5−2×0.7−1cm, erect, acute, herbaceous, pubescent. Racemes 10−15-flowered, crowded at end of peduncles 4−11cm. Calyx c 7mm including broadly triangular teeth c 3mm, black pubescent inside and outside. Petals yellow, 1.5−1.8cm, wing and keel long clawed. Ovary blackish pubescent, style glabrous.

Bhutan: N—Upper Pho Chu district (Gafoo La) and Upper Bumthang Chu district (Marlung). River banks amongst rocks and grass, 4265−4420m. July.

The Bhutanese material possibly represents a distinct subspecies of this W Chinese species. Leaves and flowers are somewhat smaller than in the latter but it agrees in leaflet and stipule shape and distribution of hair on the calyx.

8. A. kongrensis Baker

Rhizomatous subacaulescent herb, stems rarely up to 10cm. Leaves

8−18cm, leaflets 13−21(−31), broadly elliptic, 7−13×5−9mm, obtuse or emarginate, base rounded or cuneate, sparsely pubescent on both surfaces with soft white hairs; stipules ovate-lanceolate 10−12×3−5mm, membranous, pubescent. Flowers 5−12 in short racemes, peduncles 7−12cm blackish pubescent. Calyx tube 5−6mm, black pubescent, teeth narrowly triangular 3−4mm. Petals 1.5−1.7cm, purple, standard obovate 7−9mm broad, emarginate, wings oblong, blade 8−9×2−3mm, auricles linear, keel obovate c 8×4mm. Ovary densely pubescent with straight hairs, basal stalk ± as long as calyx tube.

Bhutan: N—Upper Kulong Chu district (Me La); **Sikkim:** Kongra Lama. River banks and open grassy hillsides, 3800−4265m. June−August.

9. A. bhotanensis Baker. Fig. 44 h−n

Prostrate or sprawling herb, stems 20−100cm appressed white pubescent. Leaves 7−15cm, leaflets (13−)21−29, elliptic-oblanceolate, 6−15×3−6mm, obtuse, base cuneate, sparsely appressed white puberulous beneath, stipules ovate-acuminate 10−15×3−4mm. Flowers 10−20 in dense subumbellate racemes, peduncles 7−17cm. Calyx blackish pubescent, tube 5−6mm, teeth linear-lanceolate 2−2.5mm. Petals purplish 10−13mm, standard obovate-spathulate c 5mm broad, wings oblong, blade c 4×1.5mm, sometimes pale or white, keel ovate c 4×3mm. Pods narrowly ovoid 1.5−2×0.4−0.5cm, inflated, subsessile, glabrous, longitudinally sub-bilocular, seeds 20−30.

Bhutan: C—Ha, Thimphu and Tongsa districts. Roadsides and cultivated ground, 2100−3100m. April−June.

10. A. donianus DC.; *A. pycnorhizus* Bentham

Stems prostrate or ascending 15−30cm, glabrous or sparsely pubescent at first. Leaves (1.5−)2.5−5cm, leaflets 7−13, obovate-oblong (2.5−)5−6×(1.5−)3−4mm, emarginate, base rounded, glabrous above, appressed pubescent beneath; stipules ovate, c 2mm. Flowers 1−2 on peduncles ± as long as leaves. Calyx 7−8mm, divided to middle into ovate-acuminate teeth, black pubescent. Petals deep brownish purple, 1.3−1.8cm, standard suborbicular, blade 1.1−1.3cm long and broad, wings oblong 9−10×2−4mm, keel lunate 10−12×4−5mm. Ovary pubescent, style bearded near the apex on the inner side. Pods oblong, 1.7−2.2×0.8−1cm, inflated, subsessile, acuminate at apex into a long fine point 8−10mm, finely pubescent, 12−16-seeded.

Bhutan: C—Thimphu, Bumthang and Sakden districts, **N**—Upper Kulong Chu district (Me La); **Sikkim.** Open grassy hillsides, 2133−4400m. July−August.

11. A. balfourianus Simpson

Similar to *A. donianus* but stems appressed pubescent with black and white hairs; leaves (2−)3−4cm, leaflets 7−13(−21) elliptic or oblanceolate, 5−10×2−5mm, acute, obtuse or emarginate, mucronate, cuneate or rounded at base; stipules lanceolate, 3−4mm; flowers 1−2(−7) on peduncles up to 8cm; calyx teeth linear, 5−6mm; style bearded at apex; pods ellipsoid c 1.5×0.7cm, narrowed at both ends, basal stalk ± as long as calyx tube.

Chumbi: near Kajugompa. Open hillsides, 3200m. July.

12. A. zemuensis W.W. Smith

Similar to *A. donianus* and *A. balfourianus* but stems up to 6cm; leaves 5−7cm, leaflets elliptic, 6−7mm, white hairy; stipules 3−4mm, white hairy; peduncles ± equalling leaves, 2−4-flowered; petals c 10mm; pods large, 3−4×1−1.2cm.

Sikkim: Zemu Valley. 3660m.

13. A. strictus Bentham

Tap-rooted perennial. Stems prostrate or ascending up to 15cm, appressed white pubescent. Leaves 2−4cm, leaflets 13−19, elliptic 4−6×2−2.5mm, obtuse or acute, base cuneate, glabrous or bearing a few hairs at margins above, appressed white pubescent beneath; stipules lanceolate, 4−5mm. Racemes short, dense, 10−30-flowered, on peduncles c 2.5mm. Calyx tube 3−3.5mm, teeth narrowly lanceolate 2−2.5mm, appressed black pubescent. Petals pale purple or mauve, 7−8mm, standard obovate c 5mm broad, emarginate, wings oblong, blade c 5×1.5mm, keel lunate c 4.5×2mm. Pods oblong c 6mm, somewhat recurved, basal stalk ± as long as calyx tube, 6−8-seeded.

Bhutan: C—Thimphu district, N—Upper Bumthang Chu and Upper Kulong Chu districts; **Sikkim**. On flat open grassland, 3500−4265m. May−June.

A larger form appears to exist in Chumbi with stems up to 30cm, leaflets 10−12×4−5mm, stipules ovate c 8mm, petals 10mm, pods 8−9×2.5−3mm, black pubescent, 1−3-seeded.

14. A. rigidulus Bunge

Similar to *A. strictus* but rhizomatous, stems scarcely up to 5cm, glabrous; leaves 3.5−7.5cm, leaflets 17−21, ovate-elliptic, 4−7×3mm, obtuse or subacute, appressed white pubescent; stipules deltoid, 3−5mm, membranous; racemes 5−10-flowered on peduncles up to 2cm; calyx 5−6mm divided to middle into lanceolate teeth; petals 10−12mm, standard spathulate; pods oblong-ellipsoid, 10−15×6mm, sessile, glabrous, 6−8-seeded.

Bhutan: C—Thimphu district (Taglung La); **Sikkim:** Tungu. On peaty soil, 3960−4400m. May.

15. A. sikkimensis Bunge

Stems prostrate, slender, up to 30cm, appressed pubescent. Leaves 3.5–6cm, leaflets 17–19 elliptic or narrowly obovate 4.5–9×2–4mm, obtuse, often emarginate, base cuneate, glabrous above, appressed white pubescent beneath; stipules narrowly ovate c 3mm. Racemes 8–14-flowered on peduncles 1–5cm. Calyx c 5mm, divided in upper ⅓ into acuminate teeth, black pubescent. Petals bluish-white or mauve, 10–12mm, standard elliptic, c 5mm broad, wings oblong, blade c 6×2mm, keel lunate c 6×2.5mm. Ovary densely pubescent on stalk c 2mm. Pods ovoid 7–10× 4–5mm, turgid, blackish pubescent, unilocular, 2–3-seeded.

Bhutan: C—Ha district (near Ha), N—Upper Kulong Chu district (Me La); **Sikkim.** On open hillsides, 2750–3800m. June–July.

16. A. tenuicaulis Bunge

Similar to *A. sikkimensis* but more slender; leaflets 11–15, obtuse; racemes 4–8-flowered, shorter than leaves; calyx sparsely black pubescent, teeth ⅓–¼ as long as tube; pods oblong c 7×3mm, sessile, shortly pale hirsute or hispid, 4–6-seeded.

Sikkim: Lachoong and Karponang. On river banks, 2100–2700m.

68. OXYTROPIS DC.

Perennial herbs, acaulescent or shortly stemmed. Leaves odd-pinnate, leaflets entire, stipels absent, stipules free or adnate to rachis. Flowers in axillary spikes or racemes. Calyx campanulate, teeth subequal. Petals long-clawed, standard folded forward over wings and keel, wings pouched in lower half and auricled at base, keel usually shorter than wings and ending in a short sharp beak. Stamens diadelphous. Pods inflated, often longitudinally 2-celled by membranous ingrowth of dorsal suture, seeds solitary or 5–12, reniform.

1. Flowers few up to 12 in ± globose inflorescences .... **1. O. lapponica**
+ Flowers numerous in oblong inflorescences .... **Species 2 & 3**

1. O. lapponica (Wahlenberg) Gay. Med: *Sedgno*. Fig. 44 o–s

Stems 7–20cm. Leaves 5–10cm, leaflets 11–25, narrowly ovate-lanceolate, 5–10×2–2.5mm, acuminate, base rounded, appressed pubescent especially beneath; stipules ovate 6–8mm, acuminate, connate, leaf opposed, sheathing but free from stem at base. Racemes 5–12-flowered, ± compact, peduncles 4–12cm, bracts narrowly lanceolate 5–7mm. Calyx tube c 4mm, teeth linear-lanceolate, 2–4mm, appressed black pubescent. Petals yellow or purple, 1–1.2cm, standard suborbicular, blade 8×8mm, wings narrowly obovate 7×3mm, keel elliptic c 5×3mm

76. LEGUMINOSAE

ending in short (c 0.5mm) downwardly curved beak. Pods ellipsoid, 8−12×3mm, unilocular, blackish pubescent at first, 5−6-seeded.

Bhutan: N—Upper Mo Chu district (near Lingshi); **Sikkim:** Lhonak. On sandy loam, 3800−4570m. July−August.

The typical var. **lapponica** has purple flowers but the yellow-flowered plants have been segregated as var. **xanthantha** Baker (Med: *Sedkar*) both appear to be equally common.

An acaulescent specimen from Sikkim (Lhonak), brownish pubescent throughout, with leaflets 5−7×3−4mm, and stipules adnate to base of petiole; calyx tube c 2mm, teeth ± 4mm; petals purple 6−7mm, has been doubtfully identified with *O. humifusa* Karelin & Kirilow, a W Himalayan species with dense white indumentum and larger flowers. A similar specimen from Lingshi may have been wrongly identified as *O. microphylla* (Pallas) DC., another W Himalayan species.

A specimen from Upper Bumthang Chu district and another from Sikkim have persistent stipules and leaf rachises and red flowers; these, although similar to *O. lapponica,* may represent a distinct taxon.

2. O. sulphurea Ledebour

Acaulescent or with stems up to 10cm, spreading brownish pilose throughout. Leaves 8−20cm, leaflets 21−31, ovate, 7−15×3−7mm acute, base rounded; stipules ovate 6−12×4−7mm. Peduncles 7−25cm, flowers numerous in racemes up to 6cm, bracts narrowly lanceolate 10−15×1−2mm. Calyx 8mm divided to middle into narrowly lanceolate teeth, blackish pubescent. Petals yellow, 1−1.3cm, standard ovate, blade c 7mm broad, wings oblong c 6×2mm, keel ovate c 4×3mm. Ovary glabrous, ovules c12.

Chumbi: Chugya. 4570m. July.

3. O. sericopetala Fischer

Similar to *O. sulphurea* but densely whitish appressed pubescent throughout; leaflets ovate-elliptic 6−25×2−7mm; stipules lanceolate 10−15mm, acuminate; racemes many-flowered, 3−8cm, on peduncles 12−15cm, bracts linear acuminate 2−4mm; calyx tube c 4mm, teeth linear-lanceolate 4−5mm, white pubescent; petals 7−12mm, purple, silky villous outside, standard ovate c 6mm broad, wings oblong, blade 6×2mm, keel ovate 4.5×2mm; pods ellipsoid c 6×4mm, densely silky, 1-seeded.

Sikkim: locality unknown.

69. GUELDENSTAEDTIA Fischer

Low growing perennial herbs with thickened rootstocks. Leaves odd-pinnate, leaflets emarginate or 2-lobed, stipules adnate to petiole.

Flowers solitary or 2−3. Calyx 5-toothed, upper 2 broader. Standard obovate, wings with a hooked appendage at base, keel short, upcurved. Pods linear, ± inflated, 4−6-seeded, thinly septate within.

1. G. himalaica Baker; *G. santapauii* Thothathri

Stems 2−10cm, pubescent. Leaves 2−15cm, leaflets 9−17, obovate 3−10×2−7mm, emarginate, base rounded or cuneate, brownish pubescent with long straight hairs; stipules ovate 6−7mm, membranous. Peduncles 1.5−9cm. Calyx 4−5mm, reddish. Petals usually dark purple occasionally whitish, standard 8−10mm, rounded or bilobed, wings 7−8mm, keel 3.5−4mm. Pods c1.5cm, sparsely pubescent.

Bhutan: C—Ha, Thimphu, Bumthang and Mongar districts, **N**—Upper Mo Chu and Upper Bumthang Chu districts; **Chumbi; Sikkim.** Open hillsides, 2800−4250m. May−July.

70. HEDYSARUM L.

Rhizomatous perennial herbs. Leaves odd-pinnate, leaflets entire, stipels absent, stipules connate, scarious. Flowers in axillary racemes. Calyx campanulate, divided to middle into 5 subequal teeth. Petals clawed, wings with linear appendages at base, keel upwardly curved, ± acute distally. Stamens diadelphous. Pods compressed, constricted along both margins and dividing into several 1-seeded segments, indehiscent.

1. H. sikkimense Baker

Stems erect or ascending, 8−30cm, sparsely pubescent. Leaves 5−10cm, leaflets 15−25, ovate-elliptic, 5−13×2.5−6.5mm, obtuse, ± emarginate, base rounded, glabrous above, sparsely white pubescent beneath, hairs sometimes confined to margin and midrib; stipules oblong, 5−15mm, bifid at apex, brown, sheathing. Racemes 12−15-flowered on peduncles c 8cm. Calyx 6−6.5mm, teeth lanceolate, brownish pubescent. Petals purple, 1.3−1.7cm, standard obovate 5−6mm broad, wings narrowly oblong, blade 9−11×2mm with linear appendages at base ± as long as claw, keel spathulate 10−12×4−5mm. Pods constricted into 1−3 elliptic segments 7−12×6−8mm, membranous, thinly pubescent, reticulately veined.

Bhutan: N—Upper Mo Chu district (Lingshi) and Upper Kulong Chu district (Me La); **Sikkim.** Open grassland, 3800−4000m. June−July.

71. STRACHEYA Bentham

Acaulescent perennial herbs. Leaves odd-pinnate, stipels absent, stipules scarious, sheathing. Flowers in crowded, shortly peduncled racemes. Calyx campanulate, divided to middle into 5 subequal teeth. Petals clawed, wings and keel with short rounded auricles at base. Stamens diadelphous. Pods oblong, curved, compressed, with a row of broad teeth on each margin and along the middle on both sides, breaking transversely into several 1-seeded, indehiscent segments.

1. S. tibetica Bentham. Fig. 44b

Leaves 2–5(–7)cm, leaflets 7–17, elliptic, 5–10×2–4mm, obtuse or acute, base cuneate, glabrous above, appressed white pubescent along margins and midribs beneath; stipules ovate, c5mm. Racemes 3–6-flowered on peduncles up to 10mm. Calyx 6–8mm, teeth lanceolate, greyish pubescent. Petals purple, 1.5–1.7cm, standard obovate c10mm broad, wings oblong, blade 8–10×3mm, keel narrowly obovate 12×5mm, rounded distally. Pods 2–3×1cm, breaking transversely into 2–5 segments, finely appressed pubescent.

Sikkim: Tsomgo Lake; **Chumbi:** near Phari. On sandy soil and open grassy meadows, 3350–4400m. May–July.

72. LOTUS L.

Perennial herbs. Stems prostrate or ascending. Leaves 4–5-foliate, 3 leaflets crowded at apex of short leaf rachis and 1–2 stipule-like at its base, true stipules minute or absent. Flowers umbellate on axillary peduncles, bracts trifoliate. Calyx tube campanulate, teeth 5, subequal. Petals free from staminal tube, standard suborbicular, wings obovate, auricled at base, adherent to keel, keel incurved and beaked, both petals united along lower margin and partly along upper margin. Stamens diadelphous, alternate filaments broadened at apex. Pods linear, terete, ± straight, thinly septate between seeds.

1. L. corniculatus L. Eng: *Birdsfoot Trefoil*

Stems 15–30cm, sparsely pubescent. Terminal leaflets oblanceolate or obovate, 1–2×0.5–1cm obtuse, mucronate, base cuneate, margins entire, glabrous or with a few hairs at base, lower leaflets ovate, 7–15×5–10mm, acute, base rounded; leaf rachis 5–10mm, sparsely pubescent. Flowers 4–8 in a compact umbel, peduncle 5–10cm. Calyx tube c 4mm, teeth triangular, ± as long as tube, finely acuminate. Petals yellow or tipped with red, standard 12×9mm, wings c10×5mm, rounded distally, keel c10mm, upcurved near middle, c 5mm broad near base. Pods 20–25×2.5mm, many-seeded.

Bhutan: C—Bumthang district (Byakar). Roadside bank near cultivation, 2700m. June.

Originally from Europe, now a widespread agricultural weed.

73. VICIA L.

Annuals climbing by means of tendrils. Leaves even-pinnate, leaflets entire, rachis ending in a simple or branched tendril, stipels absent; stipules triangular, fan-shaped or semi-sagittate with 1−5 teeth, persistent. Flowers solitary or 2−20 in axillary clusters or racemes. Calyx campanulate, 5-toothed. Petals clawed, wings adherent to keel. Stamens diadelphous or monadelphous. Style pubescent at apex. Pods compressed, oblong or linear, non-septate, 2-valved; seeds suboglobose or compressed.

1. Leaflets 2−6 pairs; flowers solitary or 2, subsessile or on peduncles 1.5−4cm .... **Species 1 & 2**
+ Leaflets 6−8 pairs; flowers 2−20 in racemes, peduncles 2.5−6cm
Species 3 & 4

1. V. sativa L.; *V. angustifolia* L. Fig. 42 g&h.

Stems up to 50cm, glabrous or sparsely pubescent at first. Leaf rachis ending in a branched tendril, leaflets 3−6 pairs, linear, elliptic or oblanceolate, 1−3.5×0.2−0.9cm, acute, obtuse or emarginate, mucronate, base cuneate, sparsely pubescent beneath; stipules 5−9mm, usually with a round central gland. Flowers solitary, subsessile. Calyx tube 5−6mm, teeth lanceolate 2−5mm. Petals purplish, 1.3−1.8cm, standard broadly ovate 8−9mm broad, wings oblong-obovate, blade 6−7×3mm, with slender basal auricles, keel curved and rounded distally 4×2mm. Pods 3−5×0.4−0.5cm, leathery, minutely pubescent at first; seeds 5−10, dark brown, subglobose c 2.5mm diameter.

Bhutan: S—Phuntsholing district, **C**—Ha to Tongsa districts; **Chumbi.** By streams and on grassy banks, 250−2500m. February−June.

The Bhutanese plants, though variable, belong to var. **angustifolia** L. No material has been seen from E Himalaya of var. *sativa* which has leaflets 1.5cm broad, petals 2−2.5cm and pods 6−8mm broad.

2. V. tetrasperma Moench

Similar to *V. sativa* but smaller; tendrils usually simple; leaflets 2−4 pairs, linear or elliptic, 10−20×2−3.5mm, obtuse or acute; flowers 1−2 on peduncles 1.5−4cm; calyx 2−2.5mm; petals 4−5mm, purple; pods 10−15×4mm, glabrous; seeds 3−5, compressed, 2−2.5mm diameter.

Bhutan: C—Thimphu district (near Thimphu and Paro). Weed of cultivated ground, 2200−2500m. April.

3. V. hirsuta (L.) Gray

Stems 30−60cm, sparsely pubescent with fine soft hair. Leaf rachis ending in a branched tendril, leaflets 6−8 pairs, linear, 4−15×0.5−1.5mm, truncate or emarginate, mucronate, glabrous; stipules oblong or lanceolate

4−5×0.7−1mm, eglandular, with 1−2 subulate branches ± as long attached laterally in lower half. Flowers 2−6 in racemes, peduncles 2.5−3cm. Calyx 2.5−3mm. Petals white or cream-coloured, 4−5mm. Pods oblong 7−10×3.5−4mm, acute at base and apex, pubescent; seeds 2−5 ± compressed, 2−2.5mm diameter, black.

Bhutan: S—Phuntsholing district, **C**—Thimphu, Tongsa and Tashigang districts. Weed of cultivation, 250−2350m. February−November.

4. V. tibetica Fischer
Similar to *V. hirsuta* but leaflets larger, elliptic, 7−20×3−6mm, rounded at apex and base, mucronate, pubescent and prominently veined beneath, stipules fan-shaped 10−12mm, usually with 5 radiating sharp teeth; racemes 10−20-flowered; calyx 5−6mm; petals purplish, 1−1.3cm; pods (immature) elliptic, c 12×3mm, apparently 2-seeded.

Bhutan: N—Upper Mo Chu district (Laya). Weed of cultivation, 3850m. September.

74. LATHYRUS L.

Annual herbs climbing by means of tendrils. Leaves even-pinnate, leaflets few, entire, leaf rachis ending in a branched tendril, stipels absent, stipules semisagittate. Flowers solitary or few in axillary, long-peduncled racemes. Calyx campanulate, 5-toothed. Petals short-clawed, standard erect or reflexed, wings free or slightly adherent to abruptly upcurved keel. Stamens diadelphous. Style pubescent on inner side. Pods compressed, non-septate, seeds 2−10, globose.

1. L. odoratus L. Eng: *Sweet Pea*
Stems 1.5−2m, pubescent at first. Leaflets 1 pair, elliptic or oblanceolate, 3−7×1.5−3cm, acute, base cuneate, sparsely pubescent beneath, petioles 2.5−5cm, narrowly winged; stipules ovate-lanceolate, 1.5−2.5cm, medifixed. Flowers fragrant, peduncles 10−25cm. Calyx tube 7−8mm, teeth triangular 3−5mm. Petals white, red or purple, 2−3.5cm, standard suborbicular, 3.5−4cm broad, wings oblong or obovate, blade 1.7−2.5×1−1.3cm, keel almost semicircular 2×1cm. Pods oblong, 5−6×0.8−1cm, pubescent.

Bhutan: Phuntsholing. Cultivated, 250m. April−July.
Cultivated in gardens for its decorative flowers.

75. PISUM L.

Annual herbs climbing by tendrils. Leaves even-pinnate, leaflets few, usually toothed, rachis ending in a branched tendril, stipels absent, stipules large, semicordate. Flowers 1−3 on short axillary peduncles. Calyx campanulate 5-toothed. Petals shortly clawed, standard adherent to wings by longitudinal folds near base, keel strongly upcurved. Stamens diadelphous. Style ± at right angles to ovary, bearded above. Pods oblong, inflated; seeds 4−8 globose.

1. P. sativum L. Eng: *Field Pea, Garden Pea*

Leaves 7−15cm, leaflets 2−4 pairs, ovate-elliptic, 2−4.5×1−2.5cm, acute or obtuse, base cuneate, margin shallowly dentate near apex, glabrous, stipules obliquely elliptic 3−5×1.5−2.5cm, rounded and auriculate at base, dentate in lower half, often purplish streaked. Peduncles 2−6cm. Calyx tube 4−5mm, teeth lanceolate 5−7mm. Petals blue-purple, 1−1.5cm, standard erect, suborbicular 1.5−1.7cm broad, pale coloured, wings obliquely ovate or oblong, blade 10×7mm, keel lunate 7×5mm. Pods 4−5×0.7−1.2cm, seeds sometimes darkly streaked.

Bhutan: C—Tashigang district (Kharsong to Tashigang, 117). Cultivated. March−August.

The above description relates to var. **arvense** (L.) Poiret which is widely cultivated for its edible seeds in E Himalaya.

76. CICER L.

Annual. Leaves odd-pinnate, leaflets small, margins serrate, stipels absent, stipules foliaceous, toothed. Flowers solitary, axillary. Calyx oblique, 5-toothed. Petals clawed, wings free, narrowly appendaged at base, keel broad, sharply upcurved. Stamens diadelphous. Pods inflated, seeds 1−3.

1. C. arietinum L. Hindi: *Chana;* Eng: *Chickpea.* Fig. 42 l−n

Stems 20−45cm, usually branched from base. Leaves 3−7cm, leaflets 11−13, ovate-elliptic 7−15×3−10mm acute, base cuneate, margin sharply serrate in upper half, softly pubescent on both surfaces; stipules ovate 4−5mm. Peduncles 1.5−2cm, erect at first then deflexed. Calyx tube 2−3mm, teeth lanceolate 5−6mm. Petals purplish 8−10mm, standard suborbicular 7−8mm broad, wings spathulate, blade 5×3mm, keel squarish c 3×3mm. Pods ellipsoid, 1.5−2×0.7−1cm, pubescent; seeds angularly ovoid or globose, 7−10×4−8mm, white, yellowish or brown, beaked at hilum end.

Bhutan: C—Thimphu district (Taba); **Sikkim.** 2400m. April−−July. Cultivated for its edible seeds.

76. LEGUMINOSAE

77. PAROCHETUS D. Don

Prostrate perennial herb, rooting at lower nodes. Leaves digitately 3-foliate, stipules free or shortly adnate to base of petiole. Flowers solitary or 2 on axillary peduncles. Calyx campanulate, unequally 5-toothed, upper 2 teeth connate to near apex. Standard obovate, wings oblong, keel shorter than wings and slightly hooked distally. Stamens diadelphous. Pods linear-oblong, ± inflated, undivided within, seeds 8−20.

1. P. communis D. Don. Fig. 41 e−j
Leaflets broadly obovate, 0.5−2.5×0.5−2.5cm, ± retuse, base cuneate, margin subentire or crenate, glabrous above, appressed pubescent beneath, sometimes with a dark-coloured band near middle; petioles 2.5−12(−20)cm; stipules lanceolate, c 5mm. Peduncles somewhat longer than petioles, with a ring of connate bracts c 5mm near middle. Calyx 6−8mm, toothed to middle. Petals blue, standard obovate, 15−22mm, margin ± reflexed, wings and keel 10−15mm. Pods 15−20×3−4mm, seeds rounded c 1.5mm, blackish.

Bhutan: S—Chukka and Deothang districts, **C**—Thimphu to Sakden districts; **Sikkim.** Moist places and streamsides, 1500−3960m. March−September.

78. MELILOTUS Miller

Annual or biennial herbs. Leaves pinnately trifoliate, stipules adnate to petioles. Racemes axillary, many-flowered, pedunculate. Flowers minute. Calyx tube campanulate, teeth 5, subequal. Petals free from staminal tube, standard narrowly obovate, wings oblong, auricled at base, standard and wings adherent to obtuse keel. Stamens diadelphous. Pods minute, ± compressed, 1 or few-seeded, scarcely dehiscent.

1. M. indica (L.) Allioni; *M. parviflora* Desfontaines
Stems erect, (5−)20−30cm, sparsely pubescent. Leaflets oblanceolate or obovate, 7−20×3−20mm, obtuse or retuse, base cuneate, margin denticulate in upper half, sparsely pubescent beneath at first; petioles 1−3cm; stipules lanceolate 2−10mm. Racemes elongate, (1.5−)4−8cm, bracts subulate, c1mm. Calyx 1−2mm, teeth triangular. Petals yellow, 2.5−3mm. Pods ellipsoid, 3−4×2−2.5mm, finely reticulate, usually 2-seeded.

Bhutan: S—Samchi district (Samchi) and Phuntsholing district (Phuntsholing). Weed of cultivated ground, 250−500m. February−March.

79. TRIGONELLA L.

Annual or perennial herbs. Leaves pinnately 3-foliate, leaflets toothed or denticulate, stipules adnate to petioles. Flowers solitary or few in umbels or more numerous in small racemose heads. Calyx tube campanulate, teeth subequal, about as long as tube. Petals free from staminal tube, standard obovate or oblong, wings narrowly oblong, often auricled at base above claw, keel shorter or longer than wings, rounded or acute at apex. Stamens ± diadelphous. Pods elongated, elliptic or linear, ± compressed, continuous within, dehiscent.

1. Flowers solitary or up to 3(−6) in an umbel .... **Species 1 & 2**
+ Flowers 5−15 in short racemose heads .... **Species 3 & 4**

1. T. incisa Bentham; *T. polycerata* sensu F.B.I. non L.
Annual. Stems prostrate or ascending 10−45cm. Leaflets obovate 3−15×3−7mm, obtuse, base cuneate, margin coarsely and sharply serrate, sparsely appressed pubescent; petioles up to 2cm; stipules lanceolate, 4−7mm, toothed and angular at base. Flowers 1−3 in umbels, peduncles up to 1.5cm. Calyx tube c 3mm, teeth lanceolate c 2mm. Petals yellow, standard obovate 5−7mm, wings oblong 4−5mm, auricled above basal claw, keel oblong ± as long as wings, rounded at apex. Pods linear, 40−50×2mm, ± compressed, finely but prominently reticulate, 10−20-seeded.
Bhutan: S—Phuntsholing. Weed of cultivated ground, 250m. February.

2. T. foenum-graecum L.
Similar to *T. incisa* but stems erect, 15−45cm; leaflets oblanceolate or obovate, 1−3×0.5−1.5cm, obtuse, margin finely dentate or serrate, glabrous, petals 0.5−3cm; stipules lanceolate 4−5mm; flowers 1−2, ± sessile, axillary; calyx tube 4−5mm, teeth almost as long, pale pubescent; petals white or cream-coloured, standard narrowly obovate 12−15mm, wings oblong, minutely auricled at top of basal claw, keel half as long as standard, rounded at apex; pods compressed, linear, curved, 7−10×0.5cm, tapered at apex into a long (c 3cm) seedless beak, smooth, finely striate; seeds 10−20.
Sikkim: Darjeeling.
Cultivated for its edible aromatic seeds which are a constituent of curries.

3. T. emodi Bentham
Perennial. Stems erect 20−40cm, appressed pubescent, leaflets obovate or obcordate 10−15×5−10mm, obtuse, mucronate, base rounded or cuneate, margins denticulate, glabrous or finely pubescent beneath at first; petioles 0.5−1.5cm; stipules lanceolate 5−8mm, coarsely dentate. Racemes 5−10-flowered, peduncles 3−5cm. Calyx tube c 3mm, teeth narrowly

76. LEGUMINOSAE

lanceolate c 2mm, pubescent. Petals yellow, standard obovate or suborbicular, 10−12mm, wings oblong 8−10mm, shortly auricled at base above claw, keel c 8mm upwardly curved and acute at apex. Pods elliptic, 12−25×4mm compressed, finely reticulate, 2−5-seeded.

Bhutan: N—Upper Kuru Chu district (Dengchung). Banks at edge of cultivation, 2133m. July.

4. T. corniculata (L.) L.

Similar to *T. emodi* but annual, stems up to 55cm; leaflets oblong to obovate, 10−30×8−10mm, obtuse or retuse, base cuneate, margin denticulate in upper part; petioles 1−5cm; stipules lanceolate, 5−8mm, coarsely toothed; racemes 8−15-flowered, peduncles 3−6cm; calyx c 3mm, glabrous or with a few hairs between teeth; standard obovate, 6−8mm, emarginate, wings oblong c 4mm, keel almost as long as standard; pods linear 10−15×2−3mm, curved, compressed, 3−5-seeded.

Sikkim: locality unknown (73).

80. TRIFOLIUM L.

Annual or perennial herbs. Leaves pinnately or digitately 3-foliate, stipules adnate to petiole. Flowers in ± dense subglobose heads, axillary. Calyx unequally 5-toothed. Petals adnate to staminal tube, ± persistent in fruit, standard and wings narrow, keel straight shorter than wings. Stamens diadelphous. Pods minute, 1 or few-seeded, scarcely dehiscent.

1. Leaves pinnately 3-foliate; flower heads c 7mm diameter, yellow
 1. T. dubium
+ Leaves digitately 3-foliate; flower heads 2−3cm diameter, white or purplish .... **Species 2 & 3**

1. T. dubium Sibthorp; *T. minus* Smith

Sprawling annual herb 5−30(−45)cm. Leaves pinnately 3-foliate, leaflets obovate 5−10×3−7mm, rounded or emarginate with a few minute teeth near apex, base cuneate, glabrous above, sparsely pubescent beneath; petioles 3−4mm; stipules ovate ± as long as petiole. Flowers 5−15(−20) in heads c 7mm diameter, peduncles 1−2.5cm. Calyx c 2mm, upper 2 teeth shorter than others. Petals yellow 3−3.5mm, standard narrowly ovate, folded forward and enclosing wings and keel. Pods ellipsoid c 2mm, seeds 1−2.

Sikkim: Darjeeling.

Introduced European weed.

2. T. repens L.
Creeping perennial, stems frequently rooting at nodes. Leaves digitately 3-foliate, leaflets obovate or obcordate 1−3×0.7−2cm, apex rounded or emarginate, base cuneate, margin finely denticulate, glabrous, usually with whitish V-shaped marking on upper surface; petioles 4−16cm; stipules membranous, sheathing, upper parts lanceolate, 1.5−2cm, subulate at apex. Flower heads 2−2.5cm diameter, peduncles 6−30cm, pedicels 2−4mm. Calyx tube 2−3mm, teeth lanceolate 3−4mm, upper pair longer than lower ones. Petals white, 7−10mm. Pods oblong, 4−5mm, seeds 3−4.

Bhutan: C—Tongsa district (Tongsa and Shamgong) and Bumthang district (Byakar); **Sikkim:** Darjeeling. Grassy slopes and roadside banks, 2000−2700m. April−June.

Introduced from Europe through agriculture.

3. T. pratense L.
Similar to *T. repens* but stems appressed pubescent, 20−75cm, prostrate or ascending; leaflets elliptic or obovate, 2−3×0.7−1.5cm, obtuse or acute, base cuneate, margin minutely denticulate, glabrous above, sparsely appressed pubescent beneath; petioles 5−20cm; stipules ovate c 1.5×1cm, upper part triangular and tapering to a fine point; flower heads c 3cm diameter, ± sessile among uppermost leaves; calyx tube c 3mm, teeth triangular or lanceolate, 3−5mm, lowest one longest, bearing long sparse spreading hairs; petals 1.2−1.5cm, narrow, purplish pink; pods c 3mm, 1-seeded.

Bhutan: C—Bumthang district (Byakar); **Sikkim:** Darjeeling. Roadside banks, 2700m. June.

Introduced through agriculture.

81. CROTALARIA L.

Annual or perennial herbs or shrubs. Leaves simple or digitately 3-foliate, leaflets entire, stipules small or absent. Flowers in terminal, leaf-opposed or interpetiolar racemes. Calyx tube campanulate, teeth subequal or the upper 2 smaller. Standard suborbicular, with paired thickenings above basal claw, keel shorter, sharply incurved and beaked, wings wrinkled or scaly in middle. Stamens monadelphous, tube split dorsally, filaments alternately long with basifixed anthers and short with versatile anthers. Pods oblong or elliptic, ± inflated, style usually persisting and forming a short hooked point at apex; seeds few to many.

1. Leaves trifoliate 2
+ Leaves simple 3

2. Racemes usually leaf-opposed, rarely terminal .... **Species 1 & 2**
+ Racemes always terminal .... **Species 3 & 4**

3. Racemes terminal 4
+ Racemes interpetiolar or leaf-opposed 6

4. Tall herbs or shrubs 0.75−2m tall .... **Species 5−7**
+ Herbs or small to medium-sized shrubs 20−50(−100)cm 5

5. Calyx covered with long (1.5−5mm) hairs .... **Species 8−10**
+ Calyx covered with short (less than 1mm) hairs .... **Species 11−13**

6. Stems winged .... **14. C. alata**
+ Stems unwinged .... **Species 15−17**

1. C. cytisoides DC.; *C. psoralioides* D. Don non Lamarck, *Priotropis cytisoides* (DC.) Wight & Arnott. Sha: *Benanglebee, Pekpekpu.* Fig. 44 c−g

Shrub 1−5m, stems densely appressed sericeous. Leaflets elliptic 3−6(−8)×1−2(−3)cm, acute or acuminate, base cuneate or attenuate, glabrous above puberulous beneath; petioles 2−4cm; stipules minute subulate. Flowers 10−30 in racemes, usually leaf-opposed, sometimes terminal. Calyx broadly campanulate 6−7mm divided to middle into triangular teeth, upper pair shorter. Petals yellow, standard suborbicular c 10mm diameter, claw 3mm, wings oblong-elliptic 13×4mm, lower part of keel c 12×6mm, beak ± as long. Pods oblong-elliptic (2−)3−3.5×1−1.2cm scarcely inflated, finely pubescent; seeds 3−6, reniform c 4×3mm, brown.

Bhutan: S—Samchi to Deothang districts, **C**—Tongsa district; **Sikkim**. In Warm broad-leaved forests, 200−2250m. April−November.

2. C. bracteata Roxb.

Similar to *C. cytisoides* but leaflets generally larger, 4−9×2−4cm; racemes more dense, c 50-flowered; pods ovoid-ellipsoid, c 2.5×0.8cm, much inflated, densely brown hirsute.

Bhutan: locality unknown.

3. C. pallida Aiton; *C. mucronata* Desveaux, *C. striata* DC.

Similar to *C. cytisoides* but leaflets obovate, 3−7×1.5−4cm, apex rounded, emarginate and shortly mucronate, base cuneate; racemes terminal, elongate, c 50-flowered; petals yellow, finely lined with reddish purple; pods oblong 4−5×0.7cm, inflated, ± cylindrical; seeds numerous, reniform, c 3mm.

Bhutan: S—Samchi, Phuntsholing and Gaylegphug districts; **Sikkim**. 270−400m. November−May.

Fibre is prepared from stems (146).

4. C. trifoliastrum Willdenow

Erect perennial herb 30−100cm, pubescent. Leaves 3-foliate, leaflets obovate, 1−3×0.5−1.5cm, obtuse or retuse, mucronate, base cuneate, glabrous above, appressed pubescent beneath; petioles 0.7−3cm; stipules subulate, 3−5mm, spreading. Racemes terminal, 20−30-flowered. Calyx c 6mm divided to middle into triangular teeth, pubescent. Petals c 1cm, yellow. Pods broadly ovoid, c 6×5mm, appressed pubescent; seeds 2.

Bhutan: S—Deothang district (Wamrung, 117). 2300m.

5. C. tetragona Andrews

Shrub 1−2m. Stems squarish, angular, appressed pubescent. Leaves simple, lanceolate, 10−25×1−3cm, gradually acuminate, base rounded, margin entire, sessile or on petioles up to 5mm, appressed pubescent; stipules linear-lanceolate, 3−5mm, recurved. Flowers 5−10 in loose terminal or axillary racemes. Calyx finely brown tomentose, 2−2.5cm, divided almost to base into lanceolate teeth. Petals yellow, often spotted or streaked brownish purple outside, standard broadly ovate 2.5−3×2cm, wings oblong, 1.7−2×0.6cm, keel angular c 2cm. Pods oblong, 3.5−6×1.25−1.7cm, obtuse, inflated, densely brown tomentose; seeds 10−15, brown, c 4mm.

Bhutan: C—Punakha (Wangdu Phodrang), Tongsa (Dakpai) and Mongar districts (Lhuntse); **Sikkim**. On hot open hillsides and river banks, 1200−1500m. April−November.

6. C. juncea L. Eng: *Sunn* (34), *Indian Hemp* (34)

Similar to *C. tetragona* but annual; stems up to 2m, ± terete, grooved, appressed pubescent; leaves elliptic 7−12×0.5−3cm, obtuse or acute, base cuneate; stipules minute, erect; calyx 1.5cm, divided almost to base into lanceolate teeth, brown tomentose; standard c 2×1.5cm, keel c 1.5×0.8cm; pods oblong c 3×1cm, brown velvety with stiff hairs.

Sikkim: cultivated in the Terai (34).

Stems yield a useful fibre for making ropes, canvas and fishing nets (126).

7. C. spectabilis Roth; *C. sericea* Retzius non Burman f., *C. leschenaultii* DC.

Annual. Stems 75−130cm, ± angular, glabrous. Leaves oblanceolate to obovate, 5−15×3−8cm, apex rounded, mucronate, base cuneate, glabrous above, appressed pubescent beneath; petioles 4−8mm; stipules ovate c 10×5mm, acuminate. Racemes elongate 15−40cm, bracts ovate 10−15×5mm, acuminate. Calyx 10−12mm divided to middle into ovate-triangular teeth, glabrous. Petals yellow, standard suborbicular c 2cm, across, wings oblong c 17×7mm, keel c 12×7mm, pale pubescent along lower edge. Pods oblong or club-shaped, 4−5×1.5−2cm, glabrous.

Sikkim: Tista. 300m. August−September.

8. C. capitata Baker; *C. bhutanica* Thothathri

Low sprawling shrub, stems 15–40cm, appressed brown pubescent. Leaves oblanceolate or obovate, 1.2–3×0.5–1cm, acute, base cuneate, sessile, sparsely pilose above, appressed brown pubescent beneath; stipules minute, deciduous. Flowers 3–20 in short dense terminal racemes. Calyx c 12mm densely long brown sericeous, deeply divided into lanceolate teeth, upper pair broader. Petals blue-violet, standard suborbicular c 1cm across, paler than wings and keel, wings oblong c 10×3mm, keel c 12×4mm, upwardly curved near middle, pale puberulous along upper margin. Pods obovoid, c 10×5mm, inflated, glabrous.

Bhutan: C—Thimphu to Tashigang districts, N—Upper Mo Chu and Upper Kuru Chu districts; **Sikkim.** On open grassy banks and gravel, 1825–3000m. July–September.

9. C. calycina Schrank

Annual. Stems erect 30–60(–100)cm, appressed brownish hirsute. Leaves lanceolate-elliptic, 4–12×0.7–2cm, acuminate, base cuneate, ± sessile, sparsely pilose above, appressed brown pubescent beneath; stipules c 4mm, subulate. Racemes lax, terminal, 3–10-flowered, bracts and bracteoles lanceolate 1–1.5cm, pubescent. Calyx 2.5–3cm divided ± to base into lanceolate teeth 3–5mm broad, densely covered with long brown silky hairs. Petals yellow, standard elliptic c 15×10mm, wings obovate-oblong c 15×5mm, keel c 18×5mm upwardly curved in lower third. Pods oblong, 2–2.5×0.7cm, glabrous, black when mature.

Bhutan: S—Deothang district (Kheri Gompa), C—Thimphu, Punakha and Tashigang districts. On dry grassy banks, 750–1370m. July–September.

10. C. sessiliflora L.

Similar to *C. calycina* but racemes denser and spike-like, bracts linear-lanceolate c 10mm; calyx c 1.5cm, accrescent, divided almost to base into lanceolate teeth, densely pale or brown hirsute; petals blue, standard obovate c 10×7mm, wings oblong c 12×4mm, keel c 12×5mm, upcurved below middle; pods oblong, c 1×0.5cm, glabrous, black when mature.

Bhutan: S—Gaylegphug district (Tama), C—Thimphu and Punakha districts, N—Upper Mo Chu district (Gasa); **Sikkim.** On sandy turf, 1370–2370m. August–September.

11. C. occulta Bentham

Annual, stems c 60cm densely appressed pubescent at least when young. Leaves elliptic-oblanceolate, 3–6×1–2cm acute, base cuneate, sessile, glabrous above, sparsely pilose beneath; stipules apparently absent. Racemes lax, 5–20-flowered, bracts and bracteoles linear c 2mm. Calyx 1.5–2cm divided ± to base into 5 teeth, upper 2 oblong 5–6mm broad, lower ones lanceolate c 2.5mm broad, densely appressed brownish pubescent. Petals blue-violet, ± as long as calyx teeth. Pods glabrous.

Bhutan: S—Deothang district (Wamrung, 117). 2330m. September.

12. C. albida Roth

Herb or subshrub 30−60cm, stems appressed pale pubescent. Leaves oblanceolate, 2−5×0.3−1.5cm, obtuse, mucronate, base attenuate, sessile, sparsely pilose above, appressed pale pubescent beneath; stipules minute subulate, deciduous. Racemes 5−15-flowered, bracts subulate c 2mm. Calyx 7−10mm, finely appressed pubescent, deeply divided into 5 teeth, upper pair broader, 2.5−3mm broad, lower ones lanceolate, c 1.5mm broad. Petals yellow, standard 7−8×5mm, sometimes purple streaked, wings oblong c 7×2.5mm, keel c 7×4mm upcurved at middle. Pods oblong, 1.5−2×0.5cm, glabrous.

Bhutan: S—Chukka district, **C**—Punakha, Tongsa and Mongar districts; **Sikkim.** On grassy banks, 1200−1750m. June−October.

13. C. linifolia L.f.; ?*C. montana* Roth

Similar to *C. albida* but pubescence generally brownish; leaves linear-oblanceolate sometimes obovate; calyx 8−9mm, dark brown pubescent, two upper teeth connate nearly to apex; petals ± as long as calyx; pods ovoid-ellipsoid 5−6×4mm, glabrous.

Sikkim: locality not known.

14. C. alata D. Don; *C. bialata* Schrank

Undershrub, stems 20−50(−100)cm, winged on both sides from decurrent stipules, pale pubescent. Leaves elliptic or obovate, 3−7× 1−4cm, obtuse or acute, mucronate, base cuneate, appressed pale pubescent on both surfaces; stipules triangular c 7mm. Racemes c 5-flowered borne on upper internodes. Calyx c 12mm, upper teeth broader than lower ones, brownish sericeous. Petals yellow, standard broadly elliptic 12×10mm, wings oblong c 10×4mm, keel c 10×4mm, upcurved about middle. Pods oblong 3−4×1cm, glabrous, black when mature.

Bhutan: S—Deothang district, **C**—Thimphu, Punakha and Tongsa districts; **Sikkim.** On dry banks, 1450−2450m. April−September.

15. C. ferruginea Bentham

Herbs or subshrubs, stems often trailing, up to 100cm, appressed or spreading brown pubescent. Leaves elliptic to obovate, 3−7×1−3cm, obtuse or acute, mucronate, base rounded or cuneate, appressed brownish sericeous on both surfaces; stipules lanceolate 4−7mm. Racemes 2−7-flowered, leaf opposed, bracts narrowly elliptic 4−5mm. Calyx 1−1.5cm divided almost to base into lanceolate teeth, appressed brown sericeous. Petals yellow, ± as long as calyx, standard broadly elliptic, keel upcurved in lower third. Pods oblong, 2.5−3×0.8−1cm, glabrous, black when mature.

Bhutan: S—Gaylegphug district (S of Tama), C—Tongsa district (Shamgong); **Sikkim.** On grassy banks, 1350−2150m. September−November.

16. C. humifusa Bentham

Similar to *C. ferruginea* but stems only up to 30cm, ± prostrate; leaves elliptic or obovate 2−3×1.5−2cm, obtuse or acute, mucronate, sparsely brown pubescent; stipules subulate c 4mm; racemes 2−5-flowered; calyx 5−6mm, covered in spreading (c1mm) brownish hairs; petals ± as long as calyx; pods ovoid-ellipsoid c 5×3.5mm, glabrous.

Bhutan: S—Gaylegphug district (S of Tama), C—Tongsa district (Shamgong); **Sikkim.** On shady grassy banks, 1350−1650m. September−November.

17. C. acicularis Bentham

Similar to *C. ferruginea* and *C. humifusa* but stems up to 50cm, prostrate, spreading brownish pilose; leaves ± broadly elliptic 1−2.5×0.7−1.5cm, obtuse, base rounded or cordate, spreading brown pilose especially on upper surface, pale ± glaucous beneath, stipules linear-lanceolate, c 3mm, often recurved; racemes 5−15-flowered; calyx 5−6mm, covered with long (2−3mm) brown hairs; petals ± as long as calyx; pods oblong c 7×3mm, glabrous.

Sikkim: Darjeeling terai.

82. EUCHRESTA Bennett

Low growing shrubs with tuberous rootstocks. Leaves odd-pinnate, leaflets coriaceous, entire, stipels absent, stipules small. Flowers in terminal or leaf-opposed racemes, bracts small, deciduous. Calyx campanulate, teeth short and broad. Standard narrow, not clawed, wing and keel petals broadly clawed, subequal. Stamens diadelphous. Pods oblong, stalked, somewhat fleshy, indehiscent, 1-seeded.

1. E. horsfieldii (Leschenault) Bennett; *E. horsfieldii* var. *bhutanica* Ohashi

Shrub 70−150cm, branchlets brown pubescent at first. Leaves 5(−7)-foliate, rachis grooved and narrowly winged, leaflets elliptic 15−18×5−9cm, acute or acuminate, base rounded or cuneate; stipules subulate c 2mm. Racemes erect 4−12cm, pedicels 3−6mm, bracts 2−3mm. Calyx 8−10mm, teeth broadly triangular, acute. Petals white, standard narrowly obovate 18×5mm, emarginate, base attenuate, wings and keel petals with claws c 7mm, blades 9×4−5mm, wing petals conspicuously rugulose. Pods oblong, 12−18×8−10mm, glabrous, purplish.

Bhutan: C—Punakha district (Gon Chungnang). 1800m. May.

83. PIPTANTHUS Sweet

Deciduous shrubs. Leaves digitately 3-foliate, petiolate, leaflets entire, sessile, stipules connate. Flowers in short terminal racemes. Calyx campanulate, teeth 5, upper 2 ± connate. Petals long-clawed, subequal, standard suborbicular, wings oblong, keel slightly incurved, rounded at apex. Stamens free. Pods linear-oblong, shortly-stalked, compressed, continuous within.

1. P. nepalensis (Hook.) Sweet; *P. laburnifolius* (D. Don) Stapf, *P. bombycinus* Marquand. Nep: *Gahate-phul* (34)

Shrub, 1−2.5m, young shoots densely pubescent. Leaflets ovate-elliptic, 5−8×1.5−2.5cm, acuminate, base cuneate, glabrous or pubescent above, ± densely appressed pubescent beneath; petioles 1−3.5cm; stipules boat-shaped, 7−10(−18)mm, bifid at apex, deciduous leaving annular scars on stems. Calyx tube 8−9mm, teeth lanceolate 8−10×3mm, lower ones becoming reflexed. Petals yellow, 2−3cm, standard erect c 2cm broad, brown streaked at base, wings and keel 7−8mm broad. Pods 8−12× 1−1.7mm, thinly coriaceous, shortly pubescent. Seeds 4−9 ± reniform, c 7×4mm, blackish.

Bhutan: C—Ha to Mongar districts, N—Upper Mo Chu district; **Sikkim.** On open hillsides and amongst scrub, 2300−3650m. April−May.

Two specimens from E Bhutan differ in having leaflets white pubescent above, densely pubescent or tomentose and prominently veined beneath. In these respects they approach the western Chinese *P. tomentosus* Franchet but lack the shorter keel of that species.

84. THERMOPSIS Brown

Perennial herbs with woody rootstocks. Annual stems erect, branching from base. Leaves digitately 3-foliate; stipules free, similar to leaflets, persistent. Flowers in pairs or whorls of 3 forming terminal racemes, bracts leaf-like. Calyx tube campanulate, teeth free, subequal, as long as tube. Petals long-clawed, standard erect, broadly ovate or suborbicular, shorter than the slightly curved keel. Stamens free. Pods oblong, inflated, coriaceous, continuous within.

1. T. barbata Royle. Dz: *Losi Metok*

Stems 15−30cm, densely pale hirsute in flower, elongating to 50cm and becoming less hirsute in fruit. Leaflets elliptic-oblanceolate, 2−3(−4)×0.5−1(−1.5)cm, acute, base cuneate, lateral leaflets decurrent on petiole 3−5mm, ± densely covered with long pale coloured hairs. Calyx tube c1cm, teeth lanceolate, 10×3mm, acuminate. Petals dark purple,

76. LEGUMINOSAE

standard 2−2.3cm long and broad, emarginate, wings and keel petals oblong 2.5−2.8×0.8−1cm, rounded at apex, auricled at base. Pods 3−5×1.25−2cm, sparsely pale hirsute, seeds 3−6, ovoid, c 5×4mm, brown.

Bhutan: C—Ha, Thimphu and Punakha districts, **N**—Upper Mo Chu and Upper Kulong Chu districts; **Sikkim; Chumbi**. Open grassy hillsides and gravel, 3400−4250m. May−July.

85. SPARTIUM L.

Shrub with slender ± leafless branches. Leaves simple, caducous. Flowers in terminal racemes, showy. Calyx membranous, very oblique, spathe-like, teeth minute. Petals yellow, standard large, recurved, keel longer than wings, upwardly curved and shortly beaked, adnate to staminal tube. Stamens monadelphous, alternately short and long. Pods oblong-linear, compressed, ± septate within, 10−15-seeded.

1. S. junceum L.

Stems erect 1−3m, glaucous. Leaves oblanceolate 2−3×0.5cm, obtuse, mucronate, base attenuate to petiole c 0.5cm, sparsely appressed pubescent; stipules minute, adnate to petiole. Racemes 10−20-flowered. Calyx ovate c 10mm, split almost to base along upper side. Standard obovate or suborbicular 2×1.5−2cm, wings obovate c1.5×0.8−0.9cm, keel c 2.25× 0.5cm, finely pubescent along lower margin. Pods leathery 6−8×0.7cm.

Bhutan: C—cultivated as an ornamental eg. at Paro, 2400m.

Family 77. PODOSTEMACEAE

by D.G. Long

Dwarf, annual, gregarious, moss-like herbs with green basal thallus (?root). Stems short, scattered on thallus, with alternate, distichous, scale-like stem leaves. Flowers minute, soliary, terminal on leafy shoots, bisexual, zygomorphic, enclosed by and bursting through sheath-like spathe. Sepals 2, connate at base. Petals absent. Stamens 2, filaments united. Ovary superior, 2-celled; styles 2; ovules numerous on central placenta. Fruit a ribbed capsule with numerous minute seeds.

1. HYDROBRYUM Endlicher

Description as for Podostemaceae.

1. H. griffithii (Griff.) Tulasne. Fig. 45 a−e

Thallus forming leathery, ± circular, lobed patches on rocks 10−

20(−30)cm diameter, green, becoming brown, submerged or emergent, bearing scattered tufts of 2−5 filamentous leaves up to 1cm, and scattered short suberect to prostrate flowering stems 2−3mm. Stem-leaves 4−7, ovate, concave and boat-shaped, 0.8−1.7×0.4mm. Spathe cylindric, becoming toothed at mouth when flower emerges. Sepals linear. Stamens with filaments united almost to apex and attached to base of sepals. Ovary ellipsoid, c 2mm, on short pedicel; stigmas bilobed, purplish. Capsules narrowly compressed-ellipsoid, c 2.5mm, pale brown, with 12 thick ribs.

Bhutan: S—Gaylegphug district (Gale Chu), **C**—Punakha district (Between Bhotokha and Rinchu) and Tashigang district (Ghunkarah); **Sikkim:** Kurseong. On rocks in fast-flowing streams in Warm broad-leaved forests, 1220−1450m. November.

Family 78. OXALIDACEAE

by D.G. Long

Annual or perennial herbs, with rhizomes, bulbs or bulbils, or small trees. Leaves alternate or rosetted, compound, trifoliate or 1-pinnate, stipulate or exstipulate. Flowers in basal or axillary peduncled umbels or panicles, bisexual, actinomorphic. Sepals 5, free or weakly coherent at base. Petals 5, free or weakly coherent at base. Stamens 10 in 2 whorls, filaments free or united at base, outer whorl sometimes without anthers. Ovary superior, 5-celled; styles 5, free with small capitate stigmas, ovules 1−10 per cell, axile. Fruit a capsule or berry; seeds usually with aril.

Many Oxalidaceae display 'trimorphic heterostyly' where the flowers of a single species may be of three types — with long styles, and medium and short stamens, medium styles with long and short stamens, and short styles with long and medium stamens.

1. Trees; leaves odd-pinnate; fruit a berry .... **1. Averrhoa**
+ Herbs; leaves trifoliate or even-pinnate; fruit a capsule 2

2. Leaves trifoliate .... **2. Oxalis**
+ Leaves even-pinnate .... **3. Biophytum**

1. AVERRHOA L.

Evergreen trees. Leaves odd-pinnate with subopposite leaflets, exstipulate, leaflets entire, pinnately veined. Flowers in axillary panicles. Sepals and petals weakly coherent at base. Filaments free, those of outer whorl thickened at base, tapering above and without anthers. Ovary with 3−7 ovules per cell. Fruit a 5-lobed berry.

1. A. carambola L. Hindi: *Kamaranga* (48), *Kamrak* (34); Eng: *Carambola Apple Tree*. Fig. 45 j−1

Tree to 14m. Leaves with 3−6 pairs of leaflets; petioles 2−4cm. Leaflets ovate, 2−5×1−2.5cm, acuminate, base cuneate (oblique on lateral leaflets), pubescent on veins, pale beneath. Panicles 2−5cm, pubescent. Sepals red, ovate, c 3mm. Petals obovate, c 7mm. Ovary ellipsoid c 2.5mm. Berry up to 12×6cm, ovoid or ellipsoid, ribbed. Seeds with fleshy aril.

Sikkim: Darjeeling district (Jalpaiguri). July−November.

Native possibly of Java, cultivated for its edible fruit which occurs in sweet and sour forms (34). Leaves and fruits used medicinally (126).

2. OXALIS L.

Perennial herbs with creeping stems, or stemless with rhizomes or bulbs. Leaves often folding at night, 3-foliate, stipulate or exstipulate; leaflets subsessile. Flowers in umbels on axillary or basal peduncles. Petals coherent above the claw. Longer filaments sometimes with a dorsal tooth. Ovary with 1−10 ovules per cell. Fruit a capsule; seeds dispersed by a fleshy elastic aril.

1. Plants with creeping leafy stems; bulbs absent; flowers yellow
    1. O. corniculata
+ Plants stemless with bulbs or underground rhizomes, leaves all basal; flowers white, yellow or pinkish 2

2. Peduncles 1-flowered, petals white or yellow sometimes pink at base 3
+ Peduncles 2−many-flowered, petals pinkish throughout 4

3. Leaves 1−2, sparsely pilose; leaflets obcordate with narrow sinus; petals 9−12mm .... **2. O. leucolepis**
+ Leaves 3−7, pilose-hairy; leaflets fish-tail shaped (obtriangular) with broad shallow sinus; petals 14−17mm .... **3. O. griffithii**

FIG. 45. **Podostemaceae** and **Oxalidaceae**. **Podostemaceae**. a−e, *Hydrobryum griffithii*: a, flowering thallus; b, tuft of leaves; c, flowering shoot; d, sepals and stamens; e, capsule (redrawn after J. Jap. Bot. 39(6) 1964). **Oxalidaceae**. f−i, *Biophytum reinwardtii*: f, habit; g, tip of leaf; h, flower dissected (2 petals removed); i, capsule. j−l, *Averrhoa carambola*: j, leaf and inflorescence; k, flower dissected (2 petals removed); 1, fruit. m & n, *Oxalis leucolepis*: m, habit; n, flower dissected (2 petals removed). Scale: 1 × ½; f × ⅔; j × ¾; a, g, m × l; n × 2; k × 4; h, i × 5; b, c, d × 8; e × 10.

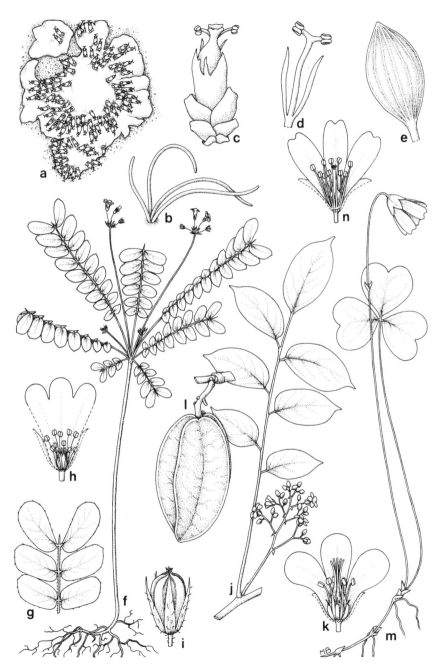

4. Leaflets obcordate with narrow sinus, appressed-pilose and gland-dotted beneath .... **4. O. corymbosa**
+ Leaflets fish-tail shaped, with broad sinus, glabrous and not gland-dotted beneath .... **5. O. latifolia**

1. O. corniculata L. Hindi: *Amrul* (146)

Stems prostrate up to 30cm, rooting at nodes, with erect leafy branches, pilose; bulbs absent. Leaves alternate or ± clustered. Leaflets broadly obcordate 5−15×7−18mm, divided in upper half, lobes rounded, base cuneate, appressed pilose; petioles 1−4cm; stipules adnate to petiole, c 2mm. Peduncles axillary, 3−8cm, 1−5-flowered. Pedicels up to 15mm. Sepals linear-elliptic 3.5−5×1−1.5mm. Petals yellow spathulate, emarginate, mostly 5−7mm. Capsules 10−20×2−2.5mm, puberulous, on deflexed pedicels.

Bhutan: S—Samchi, Phuntsholing, Chukka, Gaylegphug and Deothang districts, **C**—Ha, Thimphu, Punakha and Tongsa districts; **Sikkim**. On disturbed soil on roadsides and in cultivation, 250−2450m. February−August.

A cosmopolitan and variable weed of unknown origin, many varieties have been described; these intergrade with each other and are of minor taxonomic significance.

Foliage eaten as a vegetable and as a cure for scurvy (146).

2. O. leucolepis Diels; *O. acetosella* sensu F.B.I. p.p. non L. Fig. 45 m & n

Stemless herb with slender underground creeping rhizome bearing distant thickened scales; bulbs absent. Leaves 1−2, arising basally; leaflets often reddish beneath, obcordate, 0.7−2.2×0.9−3cm, divided to ⅓, lobes rounded, with narrow acute sinus, base broadly cuneate, sparsely pilose at least at base; petioles slender, 5−16cm. Flowers solitary on slender basal peduncle 7−17cm, usually exceeding leaves. Sepals ovate-lanceolate, 5−6mm. Petals white, finely streaked pink or purplish, spathulate, 9−12mm, notched at apex. Capsules (immature) subglobose, 3mm diameter, 5-lobed.

Bhutan: C—Thimphu, Tongsa, Bumthang and Mongar districts, **N**—Upper Mangde Chu district; **Sikkim:** Lachung, Changu, Rathong Chu etc.; **Chumbi.** Mossy ground in Hemlock and Fir forests, 2400−3900m. June−August.

Records of *Oxalis acetosella* L. from Sikkim and Bhutan (e.g. 69,80) probably all refer to this species; the former occurs in the NW Himalaya and differs in its thicker rhizomes with densely crowded scales at ground level, larger less deeply bilobed leaflets and larger flowers.

3. O. griffithii Edgeworth & Hook. f.; *O. acetosella* L. subsp. *griffithii* (Edgeworth & Hook. f.) Hara

Similar to *O. leucolepis* but pilose-hairy throughout; rhizomes stout, densely scaly at ground-level; leaves 3–7; leaflets fish-tail shaped (obtriangular), bilobed with broad shallow sinus; flowers larger, petals white or yellow, sometimes pink towards base, 14–17mm, truncate or emarginate; capsules ovoid, 6×5mm.

Bhutan: S—Tongsa district (Jirgang Chu and Kinga Rapden), Mongar district (Sana), Tashigang district (E side of Donga La) and Sakden district (Taktoo); **Sikkim:** Tonglo, Senchal, Lachung etc.; **Chumbi.** Mossy banks in Cool broad-leaved forests, 2000–3300m. March–May.

4. O. corymbosa DC.; *O. martiana* Zuccarini

Stemless herb with basal bulb 1.5–2.5cm diameter consisting of numerous ovoid bulbils, outer bulbil scales 3-nerved. Leaves numerous, basal; leaflets obcordate, 1–3×1.5–3.5cm, bilobed to ⅕, lobes rounded, sinus narrow, acute, base cuneate, appressed-pilose and gland-dotted beneath; petioles 7–20cm, hairy. Peduncle up to 30cm bearing 2–12 flowers in compound umbels. Pedicels 1–3cm. Sepals elliptic, c 5mm, with two narrow thickened glands near apex. Petals pink, spathulate, 12–15mm. Capsule not seen.

Bhutan: S—Phuntsholing district (Phuntsholing); **Sikkim:** Rishap. Weed of cultivation in subtropical zone, 250–900m. February–June.

Native of tropical S America.

5. O. latifolia Humboldt, Bonpland & Kunth

Similar to *O. corymbosa* but outer bulbil scales many-nerved; leaflets fish-tail shaped (obtriangular), up to 4×6cm, with broad shallow sinus, glabrous and not gland-dotted; umbels not compound; sepals with 2 large hastate apical glands; petals reddish-pink.

Bhutan: C—Thimphu district (Thimphu); **Sikkim:** Kalimpong and Darjeeling. Weed of disturbed ground and rice fields in temperate zone, 2150–2450m. June–July.

Native of tropical America.

3. BIOPHYTUM DC.

Annual herbs, with erect simple stem. Leaves even-pinnate, in a whorl at apex of stem, stipulate. Flowers in peduncled umbels borne amongst leaves. Petals coherent above the short claw. Shorter filaments thickened at base. Ovules 3–6 per cell. Seeds dispersed by a thin white elastic aril.

1. B. reinwardtii (Zuccarini) Klotzsch. Fig. 45 f–i

Stem 8–15cm, pubescent. Leaves 10–20 in a whorl; rachis ending in short subulate point; leaflets 6–12 pairs, terminal pairs largest, obovate,

7—13×4—7mm, apiculate, base oblique, glabrous; leaflets decreasing in size towards base, lower ones oblong, symmetric; petioles 1—2cm, pubescent; stipules subulate. Peduncles 3—6cm, with septate, gland-tipped hairs; umbels 4—6-flowered. Pedicels 3—5mm. Bracts c1mm. Sepals 2—3mm, 4—8-veined. Petals elliptic, 6—8mm, yellow. Capsules 2—3×2mm, with septate gland-tipped hairs.

Sikkim: Darjeeling terai. Subtropical terai forests, 150—300m. December—May.

Family 79. GERANIACEAE

by E.J.F. Campbell & D.G. Long

Herbs. Leaves opposite or alternate, lobed or palmately dissected, palmately veined, stipulate. Flowers solitary, or in few-flowered cymes, or umbellate, bisexual, actinomorphic (zygomorphic in *Pelargonium*). Sepals 5, free (united at base in *Pelargonium*). Petals 5, free, sometimes alternating with 5 nectaries. Stamens 10, usually all fertile, rarely 5 without anthers, filaments free or united at base. Ovary superior, 5-celled, ovules 1—2 per cell, styles 5 fused into a basal elongating beak. Capsule 5-lobed, mericarpic, 5-celled, each cell 1-seeded, mericarps dehiscing with part of beak breaking and coiling upwards elastically from the central column.

1. Flowers actinomorphic; sepals all equal, not spurred; all stamens fertile, filaments free; native plants .... **1. Geranium**
+ Flowers zygomorphic; uppermost sepal larger and bearing a nectariferous spur adnate to pedicel; some stamens without anthers, filaments connate at base; introduced cultivated plants ... **2. Pelargonium**

1. GERANIUM L.

Med: *Bonga Karpo* applies generally to *Geranium* species; Med: *Migme Sangey* applies to roots of several species.

Perennial herbs with ± woody rhizomes. Flowers solitary or in few-flowered cymes, rarely umbellate, actinomorphic. Sepals all equal, not spurred. Petals alternating with nectaries. Stamens all fertile, filaments not united at base.

1. Flowers 3—10 in umbels, borne on short pedicels up to 1.5cm
 .... **1. G. polyanthes**
+ Flowers solitary or in loose 2—3-flowered cymes, borne on long pedicels 1.5—7cm 2

2. Peduncles always 1-flowered .... **2. G. nakaoanum**
+ Peduncles mostly 2−3-flowered 3

3. Styles and filaments black or very dark purple in both living and dry conditions 4
+ Styles and filaments pink or purplish when living, becoming brownish when dry, never black 5

4. Plant often rooting at nodes; sepals 6−9mm; petals c15mm
3. G. procurrens
+ Plant not rooting at nodes; sepals 9−14mm; petals up to 30mm
4. G. lambertii

5. Petals 5−6(−8)mm, scarcely longer than sepals .... **5. G. nepalense**
+ Petals 12−15mm, much longer than sepals 6

6. Basal leaves 3−8cm across, lobes shallowly dissected with short ovate segments; sepals 10−13mm; petals white or pale pink ..... **6. G. refractum**
+ Basal leaves 0.5−3cm across, lobes deeply dissected into narrow linear segments; sepals 6−10mm; petals deep pink or purple .... **7. G. donianum**

1. G. polyanthes Edgeworth & Hook. f.

Erect or decumbent little-branched herb, 17−35cm. Basal leaves suborbicular, 3.5−5(−7.5)cm across, divided into 5−7 oblong or obovate lobes, each lobe with a few rounded or apiculate teeth, sparsely pubescent; petioles 10−15cm, spreading glandular-pilose; upper leaves on shorter petioles; stipules ovate, 5−8mm, connate. Peduncles 3−10cm, each bearing a loose to dense 3−10-flowered umbel subtended by sessile leaf-like bracts. Pedicels slender 3−15mm. Sepals 6−7(−10)mm, with mucro 1mm. Corolla pinkish-purple 1.5−2cm across. Filaments pink or white. Fruit including beak 2−2.5cm.

Bhutan: C—Thimphu, Punakha, Tongsa, Bumthang and Mongar districts, N—Upper Mo Chu district; **Sikkim; Chumbi.** Damp ground in Fir and Juniper forests, 2900−3600m. June−September.

2. G. nakaoanum Hara

Erect or decumbent slender little-branched herb, 6−20cm. Leaves mostly basal, suborbicular, 1.5−3.0cm across, divided into 7−9 obovate lobes, the apex of the lobes deeply 3-toothed, sparsely pubescent; petioles 2−17cm, subglabrous or with short bristle-like hairs; upper leaves usually on shorter petioles; stipules oblong-lanceolate, 3−4mm, free. Peduncles

4–15cm, each bearing one flower. Pedicels slender 1.5–6.5cm. Sepals 5–7mm, with mucro 0.5mm. Corolla bright pink-mauve, 2.5–3.5cm across.

Bhutan: C—Bumthang district (Badar La), **N**—Upper Bumthang Chu district (Pangothang); **Sikkim:** Yampung, Megu, Bikbari. Grassy mountain slopes, 3350–4200m. May–July.

Sometimes confused with *G. donianum* which differs in its larger leaves 2.5–4.5cm across, 2-flowered peduncles and corollas 2.8–3.6cm across.

3. G. procurrens Yeo; *G. grevilleanum* sensu F.B.I. p.p. non Wall., *G. lambertii* Sweet var. *backhousianum* (Regel) Hara. Fig. 46 a–c

Fairly stout but decumbent herb, stems 30–70cm rooting at nodes. Basal leaves suborbicular, 4–8cm across, cordate, divided into 5 rhombic lobes, lobes subacutely to acutely toothed in upper half, finely pubescent; upper petioles 4–9cm, sparsely glandular-pubescent; stipules oblong-lanceolate, 4–9mm, free at base and apex but connate at intermediate nodes. Peduncles 3–12cm, 2-flowered, not nodding. Pedicels slender, 1–5cm mostly glandular pubescent. Sepals 6–8mm, with mucro 1–3mm. Corolla red-purple with a blackish middle and radiating black veins, 3–4cm across. Filaments, anthers and style black. Fruit including beak 2.2–2.8cm.

Bhutan: C—Thimphu, Punakha, Tongsa, Bumthang and Mongar districts, **N**—Upper Mo Chu and Upper Kulong Chu districts; **Sikkim:** Lachung and Lachen. Hemlock and Fir forests, 2440–3600m. July–September.

Records of the W Himalayan *G. wallichianum* D. Don from Bhutan (117) probably refer to this species.

4. G. lambertii Sweet; *G. grevilleanum* Wall., *G. chumbiense* Knuth. Med: *Ligadur, Gadur*

Similar to *G. procurrens* but not rooting at nodes, leaf lobes ovately toothed, upper petioles thickly pubescent; stipules lanceolate, acute, 8–20mm, free; pedicels with mostly eglandular hairs, nodding; sepals 9–14mm; corolla pale blue violet to bright lilac, rarely with darkish veins, 4–6cm across.

Bhutan: C—Thimphu district (Paro, Thimphu, Pajoding and Barshong), Tongsa district (Rinchen Chu) and Bumthang district (Dhur Chu); **Sikkim; Chumbi.** Clearings in Blue Pine and Fir forests and Juniper scrub, 2350–4200m. July–September.

FIG. 46. **Geraniaceae, Tropaeolaceae, Linaceae** and **Zygophyllaceae. Geraniaceae.** a–c, *Geranium procurrens:* a, portion of flowering shoot; b, dissected flower (petals and two sepals removed); c, dehiscing fruit. **Tropaeolaceae.** d, *Tropaeolum majus:* upper portion of stem with leaves, flowers and young fruit. **Linaceae.** e & f, *Reinwardtia indica:* e, portion of flowering shoot; f, ovary. g–i, *Anisadenia pubescens:* g, habit; h, dissected flower (petals removed); i, young fruiting calyx. **Zygophyllaceae.** j–m, *Tribulus terrestris:* j, portion of flowering shoot; k, flowers; l, developing fruit; m, mature fruit seen from above. Scale: a, d, g, j × ½; e × ⅔; c × 1; k, l, m × 2; b, f × 2½; h, i × 3.

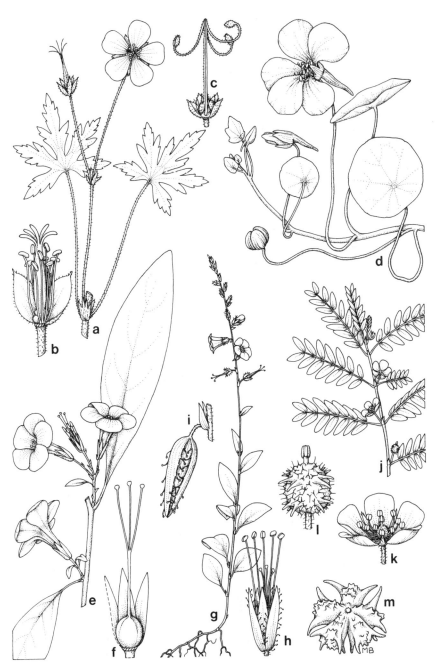

5. G. nepalense Sweet

Sprawling to decumbent, little to much branched herb 9–40cm. Basal leaves cordate, 1.5–4.5cm across, deeply divided into 3–5 ovate to rhombic lobes, which are unequally toothed, pubescent; petioles 5–8cm, very pubescent; upper leaves on shorter petioles; stipules linear lanceolate, 2–10mm, free. Peduncles 2–6cm, 2-flowered. Pedicels eglandular, 8–18mm. Sepals 4–5mm, with mucro 0.5mm. Corolla bright rose with darker veins, 13–15mm across. Fruit including beak 13–18mm.

Bhutan: S—Chukka, Gaylegphug and Deothang districts, **C**—Ha, Thimphu, Tongsa, Bumthang, Mongar and Tashigang districts, **N**—Upper Kulong Chu district; **Sikkim; Chumbi.** Roadsides in Warm and Cool broad-leaved and Blue Pine forests, 1400–3000m. February–July.

6. G. refractum Edgeworth & Hook. f.

Erect, stout herb 15–70cm. Basal leaves cordate 3–8cm across, divided into 5–7 rhombic lobes, with shallowly dissected teeth, shortly pubescent; petioles 8–30cm, very sparsely pubescent; upper leaves on shorter petioles; stipules broadly oblong-ovate, 6–16mm, free. Peduncles 2.5–11cm, 2-flowered, sometimes crowding quite closely into a loose umbel. Pedicels 1–3.5cm, clothed with glandular and eglandular hairs. Sepals 10–13mm, with mucro c 4mm. Corolla white or pale pink, 2.5–3cm across, reflexed. Filaments pink, anthers black. Fruit including beak 3–3.5cm.

Bhutan: C—Thimphu and Tongsa districts, **N**—Upper Mo Chu, Upper Pho Chu, Upper Bumthang Chu and Upper Kulong Chu districts; **Sikkim; Chumbi.** Mountain meadows, 3650–4600m. June–August.

7. G. donianum Sweet; *G. collinum* sensu F.B.I. p.p. non Willdenow, *G. stenorrhizum* Stapf

Similar to *G. refractum* but usually much branched, 10–30cm tall. Basal leaves suborbicular, 2.5–4.5cm across, lobes deeply incised into long linear segments with acutish tips; petioles 4–18cm; stipules linear-lanceolate, acute, 4–8mm. Peduncles 3–8.5cm. Pedicels 1–6.5cm. Sepals 8–10mm with mucro 0.5mm. Corolla deep pink-purple, 2.8–3.6cm across, not reflexed. Fruit including beak 2cm.

Bhutan: C—Ha, Thimphu, Tongsa, Bumthang and Tashigang districts, **N**—Upper Mo Chu, Upper Bumthang Chu and Upper Kulong Chu districts; **Sikkim; Chumbi.** Fir and Juniper forests, 2400–4100m. June–October.

Similar to *G. nakaoanum* which differs in its smaller size and 1-flowered peduncles.

2. PELARGONIUM Aiton

Similar to *Geranium* but flowers in dense umbels, zygomorphic; sepals united at base, uppermost sepal larger and bearing an elongated nectariferous spur adnate to pedicel; 2 upper petals usually larger than others; nectaries absent; stamens 10, usually only 5—7 fertile.

1. P. zonale (L.) Aiton

Herb 50—70cm, woody at base. Basal leaves palmately lobed, 3—8cm across, shallowly 9-lobed, lobes slighly toothed; upper leaf petioles 2.5—3cm; stipules broadly ovate to suborbicular, mucronate, c 8mm free. Peduncles 6—15cm, each bearing a dense 22—30-flowered umbel. Pedicels stout 2.5—3.5cm. Sepals 4—11mm, acute to mucronate. Corolla scarlet, 3—4cm across. Fruit including beak 2.5—3cm.

Bhutan: C—Mongar district (Mongar). Cultivated in garden, 1800m. June.

Native of S Africa cultivated as an ornamental.

Family 80. TROPAEOLACEAE

by D.G. Long

Prostrate or climbing herbs with juicy sap. Leaves alternate, simple, palmately veined, peltate, exstipulate. Flowers solitary, axillary, bisexual, zygomorphic. Sepals 5, free, uppermost modified into a nectar-spur. Petals 5, clawed, upper 2 dissimilar to lower 3. Stamens 8 in 2 whorls, unequal, free, curved. Ovary superior, 3-celled; style simple, with 3-lobed stigma; ovules 1 per cell, axile. Fruit dividing into three 1-seeded indehiscent segments (mericarps).

1. TROPAEOLUM L.

Description as for Tropaeolaceae.

1. T. majus L. Eng: *Nasturtium, Indian Cress.* Fig. 46 d

Annual creeping and climbing herb. Leaves orbicular 3—10cm diameter, 7—9-veined, entire or shallowly sinuately-lobed, puberulous beneath; petioles 6—30cm, sometimes twining. Pedicels 6—18cm. Sepals ovate-elliptic, 1.2—2.0×0.4—0.7cm, acute; spur slender, 2—3cm. Petals yellow, orange or red, upper pair obovate 2.5—3.5×1.7—2cm, shortly clawed, lower 3 suborbicular, 1.5—2cm, each with a distinct laciniate claw. Fruit 1—1.5cm diameter, striate.

Bhutan: C—Mongar district (Mongar). Cultivated in garden, 1980m. June—July.

Native of S America cultivated for the ornamental flowers. Leaves, flowers and fruit may be eaten as salad.

Family 81. ZYGOPHYLLACEAE

by A.J.C. Grierson & D.G. Long

Herbs, often woody at base. Leaves opposite, evenly 1-pinnate, stipulate. Flowers solitary, axillary, bisexual, actinomorphic. Sepals 5, free. Petals 5, free. Stamens 10, borne on hypogynous disc. Ovary superior, 5-celled, style simple, bearing 5 stigmatic lobes; each cell with numerous, axile, superposed ovules. Fruit a 5-angled spinous capsule.

1. TRIBULUS L.

Description as for Zygophyllaceae.

1. T. terrestris L. Fig. 46 j−m

Prostrate annual or biennial herb, woody at base, with spreading branches 15−80cm; softly hirsute throughout. Leaves in unequal pairs, larger 2−5×1−2cm, smaller 1−3cm or sometimes absent; leaflets 5−7 pairs, oblong, 4−14×2−5mm, obtuse or minutely mucronate, base obliquely rounded, entire; petioles 3−8mm; stipules triangular, 3−5mm, persistent. Pedicels 6−15mm. Sepals lanceolate 5−6mm, persistent. Petals yellow, obovate, 6−8×4mm. Ovary ellipsoid, 2mm, hairy. Capsules subglobose, angular, pubescent, splitting into 5 wedge-shaped several-seeded segments (mericarps), each with 2 long and several short spines on outer face.

Bhutan: C—Punakha district (Punakha); **Sikkim.** Grassy ground in dry open valleys, 1400m. May−September.

Family 82. BALANITACEAE

by A.J.C. Grierson

Spiny trees or shrubs. Leaves alternate, bifoliate, exstipulate. Flowers bisexual in dense axillary cymes. Sepals and petals each 5 free, imbricate. Stamens 10, inserted around a lobed disc. Ovary 1-celled, globose, ovule solitary, style short simple. Fruit a fleshy drupe.

1. BALANITES Delile

Description as for Balanitaceae.

1. **B. aegyptiaca** (L.) Delile; *B. roxburghii* Planchon

Shrub or tree up to 10m, spines ascending 1−4.5cm, simple or branched.

Leaflets oblanceolate or obovate 1.5−4.5×0.5−2.5cm obtuse, base attenuate, entire, appressed pubescent at first; petiole 1−5mm; leaflets sessile or on petiolules up to 5mm. Cymes subumbellate 1.5−3.5cm borne on or at base of spines, pedicels c 7mm, flowers yellowish-green. Sepals oblong c 3×1.5mm pubescent outside, woolly within. Petals oblong c 4× 1.5mm glabrous outside, woolly within. Stamens c 3mm, filaments c 1.5mm. Ovary c 2mm across, styles c 1mm. Drupes ovoid 4−6×2−4cm, yellow when ripe, flesh foetid; stone ellipsoid, somewhat angular 4−6×1−1.5cm.

Sikkim terai: locality not known (80).

Family 83. LINACEAE

by A.J.C. Grierson & D.G. Long

Annual or perennial herbs or low shrubs. Leaves alternate, simple, pinnately veined or 3-veined from base, stipulate or exstipulate. Flowers in racemes or spike-like panicles, or axillary clusters or cymes, rarely solitary, actinomorphic, bisexual. Sepals 5, free. Petals 5, free. Stamens 5, alternating with minute staminodes, connate at base into a glandular ring; anthers versatile. Ovary superior, 3−5-celled (or 6−10-celled with secondary septa); styles 3−5, free or shortly connate at base, ovules 1−2 per cell, axile. Fruit a 1−10-seeded capsule (indehiscent in *Anisadenia*).

1. Subshrubs to 1m; flowers in axillary cymes or clusters, rarely solitary; petals yellow 1.5−5cm .... **1. Reinwardtia**
+ Herbs (sometimes woody at base) to 30cm; flowers in racemes or narrow panicles; petals white, pink or blue, 0.3−1.5cm 2

2. Rhizomatous herbs; leaves elliptic, 7−35mm broad, stipulate; calyx glandular-hairy; styles 3 .... **2. Anisadenia**
+ Non-rhizomatous herbs; leaves linear, 1−4mm broad, exstipulate; calyx eglandular; styles 5 .... **3. Linum**

1. REINWARDTIA Dumortier

Shrubs. Leaves entire or serrulate; stipules minute, caducous. Flowers in axillary and terminal fascicles or cymes, sometimes solitary. Corolla funnel-shaped, showy. Styles 3−5, filiform, free or connate at base. Capsules globose, 6−8-seeded.

83. LINACEAE

1. Leaves obovate-oblanceolate, acute or obtuse, entire or minutely serrulate; flowers solitary or few, axillary or terminal; petals 1.5–3cm .... **1. R. indica**
+ Leaves elliptic, acuminate, serrate; flowers in terminal cymes; petals 3–5.5cm .... **2. R. cicanoba**

1. R. indica Dumortier; *R. trigyna* Planchon. Fig. 46 e & f

Shrub to 1m. Leaves obovate-oblanceolate, 3–9×1–3cm, acute or obtuse, base attenuate, margins entire or minutely serrulate, glabrous; petioles 2–20mm; stipules c1mm. Flowers solitary or few in axillary and terminal clusters or cymes. Bracts 1.5–3mm, glandular-toothed. Pedicels 2–7mm, pubescent. Sepals lanceolate, 7–10mm. Petals yellow, obovate, 1.5–3cm. Styles 3, 1.4–2.2cm, connate at base. Capsules subglobose, 6–7mm.

Bhutan: S—Deothang district (Deothang, Demri Chu and Chungkar), **C**—Tashigang district (Jiri Chu); **Sikkim**. Warm broad-leaved forest margins, 900–1800m. October–December.

2. R. cicanoba (D. Don) Hara; *R. tetragyna* (Bentham) Planchon

Similar to *R. indica*, differing in its larger, elliptic leaves 5–13×1.5–4cm, acuminate, margins serrate; flowers more numerous in terminal cymes; petals larger 3–5.5cm; styles 3–4(–5).

Sikkim: Punkabari, Kodabari and Sureil. Warm broad-leaved forests, 1050–1500m. October–December.

2. ANISADENIA Meisner

Perennial rhizomatous herbs with simple or little-branched stems. Leaves alternate but sometimes ± clustered at top of stem, entire; stipules membranous, axillary. Flowers sessile or on short pedicels in a terminal spike-like raceme or panicle. Sepals coriaceous, striate, outer 3 glandular-bristly. Corolla funnel-shaped. Ovary 3-celled, styles 3, ovules 2 per cell. Fruit indehiscent, 1-seeded, deflexed, enclosed by persistent glandular-bristly calyx.

1. Leaves 4–8×1.5–3cm, ± clustered at top of subglabrous stem; petals 8–10mm .... **1. A. saxatilis**
+ Leaves 1.5–4×0.7–2cm, scarcely clustered on pubescent stem; petals 13–15mm .... **2. A. pubescens**

1. A. saxatilis Meisner; *A. khasyana* Griff.

Leafy shoots simple, erect, 15–25cm, stems subglabrous. Leaves ± clustered at top of stems, elliptic, 4–8×1.5–3cm, acute, base cuneate, appressed pubescent beneath; petioles 0.5–3cm; stipules ovate, striate, 4–6mm. Racemes (or panicle) 7–14cm, rachis pubescent. Flowers sessile or on simple or branched pedicels 1–4mm; bracts lanceolate, striate, caducous. Sepals lanceolate 4−5mm; glandular bristles c1mm. Petals white or pink, 8−10mm, obovate, emarginate. Styles c6mm. Fruit oblong c 2mm.

Bhutan: C—Thimphu district (Dotena), Punakha district (Thinleygang), Tongsa district (Tongsa Dzong) and Mongar district (Denchung), **N**—Upper Mo Chu district (Tamji and Gasa); **Sikkim:** Rongbe, etc. Cool broad-leaved and Evergreen oak forests, 1800−2440m. July−September.

2. A. pubescens Griff. Fig. 46 g−i

Similar to *A. saxatilis* but stems decumbent, sometimes branched, pubescent, leaves scarcely clustered, smaller, 1.5−4×0.7−2cm, pubescent on both surfaces; petals larger 13−15mm.

Bhutan: S—Deothang district (Wamrong and Chungkar), **C**—Mongar district (Kuru Chu N of Lhuntse) and Tashigang district (Rongtong and Tashiyangtsi); **Arunachal Pradesh:** Nyam Jang Chu. Open slopes and Evergreen oak forests, 1980−2850m. July−August.

3. LINUM L.

Annual or perennial glabrous herbs. Leaves narrow with single mid-vein or 3-veined from base, exstipulate but older leaves with swollen basal glands. Flowers in cymes. Sepals ovate. Corolla broadly funnel-shaped. Ovary 10-celled, styles 5, free, ovules 1 per cell. Capsule globose, 10-seeded.

1. Annual herb 25−40cm, little-branched at base; leaves 2−4mm broad, 3-veined from base; petals 10−13mm .... **1. L. usitatissimum**
+ Perennial herb 5−30cm, much branched at base; leaves 1−1.5mm broad, 1-veined; petals c3mm .... **2. L. nutans**

1. L. usitatissimum L. Eng: *Flax, Linseed*

Erect annual herb 25−40cm, simple or branched at base and with a few branches near apex. Leaves linear-lanceolate, 15−25×2−4mm, 3-veined from base to apex, margins smooth. Flowers in loose leafy cymes. Pedicels slender 1−2cm (up to 4cm in fruit). Sepals ovate 5−6mm, acute, margins membranous, sometimes ciliate. Petals blue, obovate, 10−13mm. Ovary globose, styles slender, c 3mm. Capsules 7−8mm diameter.

Sikkim: Darjeeling. Cultivated, 2100m. February to May.

Cultivated for the production of linseed oil and manufacture of linen.

2. L. nutans Maximowicz

Similar to *L. usitatissimum* but a smaller herb 5−30cm, much branched at base; leaves shorter and narrower, 6−12×1−1.5mm, 1-veined; pedicels

83. LINACEAE

4–8mm (–2cm in fruit); sepals ovate, c 2.5mm obtuse; petals shortly exserted, c 3mm; styles c 1.2mm; capsules c 5mm diameter.

Sikkim: locality unknown; **Chumbi:** N of Phari. By mountain streams, 4000–4600m. June.

Family 84. EUPHORBIACEAE

by D.G. Long

Trees, shrubs (sometimes succulent, rarely climbing) or herbs, often with milky sap, indumentum of simple or stellate hairs, rarely peltate scales. Leaves alternate, sometimes opposite or whorled, simple, sometimes palmately lobed; venation pinnate or palmate at base; usually stipulate. Monoecious or dioecious; flowers small, solitary or in clusters, spikes, racemes, panicles or cymes, or (in *Euphorbia* and *Pedilanthus*) in cup-like cyathia, unisexual, actinomorphic. Calyx of 3–6(–12) free or partly united segments, petals 3–6(–10), or often absent. Stamens 2–many, free or variously united into a column or bundles. Ovary superior, 2–4(–15)-celled, styles 2–4, free or united, simple or bifid; ovules 1–2 per cell, axile. Fruit usually a capsule splitting into 2–3(–8) segments, each 1–2-seeded; occasionally fleshy or leathery and indehiscent; seeds often arillate.

1. Succulent, often leafless trees or shrubs with angular stems and stipular spines .... **1. Euphorbia**
+ Trees, shrub or herbs, if succulent then stems terete and without stipular spines 2

2. Leaves opposite or whorled 3
+ Leaves alternate 8

3. Leaves pinnately veined to base 4
+ Leaves palmately 3–9-veined at base 7

4. Young shoots and leaves stellate-scaly .... **15. Croton**
+ Plants glabrous or with simple hairs 5

5. Flowers in cup-like gland-fringed cyathia containing one ovary and several stamens; perianth absent .... **1. Euphorbia**
+ Flowers not in cyathia, sexes in separate plants or inflorescences; sepals present 6

6. Herb with opposite leaves .... **20. Mercurialis**
+ Shrub with leaves alternate, opposite or whorled .... **28. Lasiococca**

7. Indumentum of fascicled hairs, without minute gland-dots ... **23. Trewia**
+ Indumentum of stellate hairs mixed with minute red or yellow gland-dots .... **24. Mallotus**

8. Leaves pinnately veined to base (lowermost lateral veins weaker than those above) 9
+ Leaves palmately 3–9-veined at base (often pinnately veined above but basal lateral veins stronger and distinct from others) 40

9. Flowers in cyathia, either cup-like and gland-fringed, or strongly zygomorphic and 6-lobed, perianth absent 10
+ Flowers not in cyathia, sepals and sometimes petals present 11

10. Shrubs or herbs; leaves not succulent and coriaceous; cyathia cup-like, not strongly zygomorphic .... **1. Euphorbia**
+ Shrubs; leaves succulent and coriaceous; cyathia strongly zygomorphic
2. Pedilanthus

11. Young growth and leaf undersides with stellate or peltate scales or with a mixture of fascicled hairs and minute glossy gland-dots 12
+ Young growth and leaf undersides with simple hairs or glabrous, not gland-dotted 14

12. Leaves linear 0.8–1.7cm broad, peltate-scaly beneath ....**27. Homonoia**
+ Leaves oblong-lanceolate, ovate or elliptic, 3–10cm broad, stellate scaly (at least when young) or with a mixture of fascicled hairs and gland-dots 13

13. Leaves and young shoots stellate-scaly .... **15. Croton**
+ Leaves and young shoots with fascicled hairs and minute gland-dots
26. Macaranga *(M. gamblei)*

14. Cultivated shrub wih leaves variegated green, yellow and red
16. Codiaeum
+ Native shrubs or trees, leaves not variegated 15

15. Inflorescences leaf-opposed, males in cymes, females solitary or few in fascicles .... **31. Suregada**
+ Inflorescences axillary or terminal, flowers solitary or in fascicles, racemes, spikes or panicles (in *Aporosa* and *Baccaurea* spikes often on old wood) 16

84. EUPHORBIACEAE

16. Flowers of one or both sexes solitary or in axillary fascicles or clusters, never forming racemes or spikes (or flowering shoots sometimes partly leafless in *Bridelia*) 17
+ Flowers of one or both sexes clustered or not, forming leafless racemes or spikes, sometimes branching to form panicles 31

17. Leaves distinctly serrate; apex of petiole 2-glandular .... **25. Cleidion**
+ Leaves entire (rarely minutely serrulate in *Drypetes*); petiole eglandular
18

18. Leaves becoming opposite or whorled towards branch-ends; pedicels and flowers with gland-tipped hairs .... **28. Lasiococca**
+ Leaves alternate throughout; hairs never gland-tipped 19

19. Petals present; flower clusters with conspicuous scarious bracts; fruit a 1–2-seeded drupe; lateral veins often closely parallel .... **3. Bridelia**
+ Petals absent; flower clusters without conspicuous scarious bracts; fruit capsular or fleshy, 3–16-seeded (except in *Drypetes*); lateral veins not closely parallel 20

20. Sepals 4; leaves coriaceous, asymmetric at base; fruit 2-seeded, leathery
9. Drypetes
+ Sepals 5–6 (4 in male flower of *Phyllanthus sikkimensis*); leaves often membranous, base symmetric or asymmetric; fruit capsular or fleshy, 3–16-seeded 21

21. Male flowers present 22
+ Female flowers or fruit present 27

22. Stamens all free; plants monoecious or dioecious 23
+ Stamens with some or all filaments connate into a column; monoecious
24

23. Monoecious; petioles 1–2mm .... **4. Phyllanthus** (*P. glaucus*)
+ Dioecious; petioles 4–10mm .... **5. Flueggea**

24. Sepals free to base; stamens 3–8 25
+ Sepals united into a turbinate, minutely lobed tube, or into a flattened disc with inflexed lobes; stamens 3 26

25. Herbs, shrubs or trees; leaves 0.5–4×0.1–2cm; anthers free or connate without enlarged connectives .... **4. Phyllanthus**
+ Shrubs or trees; leaves mostly 4.5–18×1.5–9.5cm; anthers united, capped by prominent connectives .... **6. Glochidion**

756

26. Calyx tube of male flowers turbinate, minutely lobed .... **7. Breynia**
+ Calyx tube of male flowers disc-like, with inflexed lobes .... **8. Sauropus**

27. Dioecious; female sepals 5, free; fruit fleshy, 5–8-seeded ... **5. Flueggea**
+ Monoecious; female sepals (4–)6, free or united; fruit fleshy or capsular 3–16-seeded 28

28. Styles connate into a stout column, sometimes globose or broad and flat; fruit depressed, woody, 3–15-lobed and seeded ...... **6. Glochidion**
+ Styles free or united into a slender branched column; fruit globose, fleshy or capsular, not prominently lobed, 3–16-seeded 29

29. Female calyx not disc-or cup-like, scarcely enlarged in fruit
   4. Phyllanthus
+ Female calyx disc- or cup-like, much enlarged in fruit 30

30. Styles united into column on ovoid ovary .... **7. Breynia**
+ Styles free, borne on outer edge of obconic ovary .... **8. Sauropus**

31. Leaves with narrow cordate base; pedicels and flowers with gland-tipped hairs   **28. Lasiococca**
+ Leaves with cuneate or rounded base; pedicels and flowers without gland-tipped hairs 32

32. Flowers few in racemes; leaf bases asymmetric .. **9. Drypetes** *(D. indica)*
+ Flowers numerous in spikes, racemes or panicles; leaf bases not or weakly asymmetric 33

33. Petioles mostly 0.3–2cm 34
+ Petioles mostly 2–14cm 35

34. Petioles 1.3–2cm, 2-glandular at apex; male spikes dense, catkin-like, stamens 2; female spikes very short .... **10. Aporosa**
+ Petioles 0.3–0.9cm, eglandular at apex; male spikes slender, not catkin-like, stamens 2–5; female spikes elongate .... **11. Antidesma**

35. Leaves entire or shallowly sinuate 36
+ Leaves serrate or serrulate 37

36. Petioles eglandular at apex; spikes borne on trunk and main branches; calyx 4–5-lobed; stamens 5 .... **12. Baccaurea**
+ Petioles on many leaves 2-glandular at apex; spikes borne amongst leaves; calyx 2–3-lobed; stamens 2–3 .... **34. Sapium**

37. Petioles 7–14cm, eglandular; monoecious .... **19. Claoxylon**
+ Petioles 2–9cm, on most leaves 2-glandular at apex; dioecious 38

38. Subshrub; flowers in narrow panicles
    32. Baliospermum *(B. corymbiferum)*
+ Trees or large shrubs; flowers in pendulous or erect spikes 39

39. Evergreen tree or shrub; flowers in pendulous spikes; stamens numerous .... **25. Cleidion**
+ Deciduous tree; flowers in erect spikes; stamens 2–3
    34. Sapium *(S. insigne)*

40. Twining shrub; leaves oblong ....**33. Pterococcus**
+ Herbs, shrubs or trees, never twining; leaves ovate, lanceolate, elliptic or suborbicular 41

41. Some or all leaves palmately lobed 42
+ Leaves never palmately lobed 46

42. Indumentum stellate mixed with sessile gland dots; leaves shallowly lobed with discoid glands on veins near base; native trees ...**24. Mallotus**
+ Plants without stellate hairs or sessile gland dots; leaves shallowly to deeply lobed, with basal glands absent or only on apex of petiole; cultivated and naturalised trees and shrubs 43

43. Leaves deeply lobed to within 6–10mm of petiole apex, lobes entire
    17. Manihot
+ Leaves shallowly to deeply lobed to within 15mm or more of petiole apex, lobes serrate or entire 44

44. Dioecious; petioles 2-glandular at apex; calyx 2-lobed; fruit fleshy, indehiscent .... **14. Vernicia**
+ Monoecious; petioles eglandular or 1-glandular at apex; calyx 3–6-lobed; fruit capsular 45

45. Leaf lobes entire; petioles eglandular; flowers in cymes, petals present, stamens 6–10; capsules smooth .... **13. Jatropha**
+ Leaf lobes serrate; petioles 1-glandular; flowers in panicles, petals absent, stamens many; capsules prickly .... **29. Ricinus**

46. Indumentum of stellate hairs (or fascicled hairs in *Macaranga),* mixed or not with minute red or yellow gland dots 47
+ Indumentum never stellate or fascicled, minute gland dots absent 50

47.	Indumentum stellate or fascicled (at least in part) mixed with minute red or yellow gland dots 48
+	Indumentum stellate, minute gland dots absent 49
48.	Indumentum of stellate hairs; stamens 25−50 .... **24. Mallotus**
+	Indumentum of fascicled and simple hairs; stamens 2−20 **26. Macaranga**
49.	Monoecious; leaves serrate or serrulate; male flowers with calyx 5-lobed and 5−6 petals; fruit a capsule .... **15. Croton**
+	Dioecious; leaves entire; male flowers with calyx 4-lobed, apetalous; fruit indehiscent .... **30. Endospermum**
50.	Leaf margins entire, base cuneate or deeply cordate 51
+	Leaf margins serrate, crenate or dentate, base rounded, truncate or shallowly cordate 53
51.	Leaves lanceolate with cuneate 3-veined base; petioles c2mm; flowers in short axillary racemes .... **8. Sauropus** *(S. repandus)*
+	Leaves broadly ovate or suborbicular, base cordate, 5−7-veined; petioles 3−11cm; flowers in terminal cymes or panicles 52
52.	Monoecious shrub; petioles eglandular; flowers in cymes; fruit capsular **13. Jatropha** *(J. curcas)*
+	Dioecious tree; petioles 2-glandular at apex; flowers in panicles; fruit fleshy, indehiscent .... **14. Vernicia**
53.	Leaf base with 2 lanceolate or subulate stipels .... **22. Alchornea**
+	Leaf base without stipels (sometimes with rounded glands) 54
54.	Herbs or cultivated shrubs; petioles eglandular; flowers in dense (sometimes short) spikes; sepals 4, petals absent .... **21. Acalypha**
+	Native shrubs or trees; petioles often glandular at apex; flowers in racemes or panicles; sepals 3 or 5−6, petals present or absent 55
55.	Flowers in elongate lax racemes or panicles borne on old wood; sepals 3, petals present, stamens 30−35 .... **18. Ostodes**
+	Flowers in short racemes or lax or dense panicles in axils of leaves; sepals 5−6, petals absent, stamens 12−20 .... **32. Baliospermum**

1. EUPHORBIA L.

Annual or perennial herbs, or succulent cactus-like shrubs or small trees with milky juice. Leaves alternate or opposite, rarely in whorls of 3; stipules absent or present. Monoecious; flowers in solitary, clustered or umbellate cyathia, each cyathium composed of a cup-like involucre, its margin bearing 4−5 glands and 4−5 small petal-like lobes, each cup containing a single central female flower surrounded by several minute male flowers. Female flowers with a short pedicel and 3-celled ovary bearing 3 free or united, simple or bilobed styles; perianth absent. Male flowers with a single stamen, perianth absent. Fruit a 3-valved trigonous capsule.

1. Succulent shrubs or trees, branches angular, winged, bearing paired stipular spines 2
+ Herbs or shrubs; branches terete, unarmed 4

2. Scrambling shrub, branches 0.7−1cm broad, with dense spines 1−2.8cm .... **3. E. milii**
+ Erect shrubs or small trees, branches 2−4.5cm broad, with short distant spines 0.2−0.7cm 3

3. Branches (4−)5-angled, wings not shallowly sinuate; cymes subsessile, condensed 5−12mm; styles minutely bilobed .... **1. E. royleana**
+ Branches 3(−4)-angled, wings sinuately lobed; cymes stalked, lax, 15−30mm; styles deeply bilobed .... **2. E. antiquorum**

4. Cultivated shrubs with showy bracts 5
+ Native herbs, sometimes with showy bract-like upper leaves 6

5. Leaves alternate, 8−20cm; cyathia leaf-opposed ...... **4. E. pulcherrima**
+ Leaves in whorls of 3, 4−8cm; cyathia terminal .... **5. E. leucocephala**

6. Perennial herbs; leaves alternate below, whorled at stem apex; leaf base symmetric 7
+ Annual herbs; leaves opposite throughout, with oblique base 12

7. Dwarf herb 2−11(−20)cm, much-branched from base; leaves oblong-obovate 8−15×3−8mm; styles 1−1.4mm, free and recurved to base .... **11. E. stracheyi**
+ Robust herbs 13−100cm, simple or branching only in upper half; leaves elliptic, ovate-elliptic, linear-lanceolate or oblong, 10−130×5−26mm; styles 1.5−3.6mm, united in lower ⅓−½, erect or recurved only in upper part 8

8. Upper leaves and bracts strongly red- or orange-pigmented, especially when dry; young stems and umbel rays glabrous except for tufts of hair below base of petioles .... **6. E. griffithii**

\+ Upper leaves and bracts green or yellow-green, sometimes weakly reddish-tinged in autumn or when dry; young stems and umbel rays pubescent, or if glabrous then completely so 9

9. Stems 60–100cm, unbranched or with a few short branches near apex; ovary warted; styles deeply bifid .... **9. E. longifolia**
\+ Stems 13–80cm, with several to many long branches in upper half; ovary smooth; styles simple but with bilobed stigmas 10

10. Stems 40–80cm; stem leaves linear-lanceolate, 6–9cm long; young shoots and umbel rays glabrous .... **7. E. sikkimensis**
\+ Stems 13–35(–60)cm; stem leaves oblong, oblong-lanceolate or ovate-elliptic, 2.0–4.2cm long; young shoots and umbel rays crispate pubescent 11

11. Stem leaves ovate to ovate-elliptic, 1.2–2.4cm broad; umbel rays 4–5; bracts 0.9–1.4cm broad .... **8. E. luteo-viridis**
\+ Stem leaves oblong or oblong-lanceolate, 0.5–0.9(–1.3)cm broad; umbel rays 6–10; bracts 0.5–0.9cm broad .... **10. E. himalayensis**

12. Stem leaves 15–45mm long 13
\+ Stem leaves 3–8mm long 14

13. Stems yellowish spreading-pilose; leaves acute; cyathia in globose heads 4–8mm, on leafless peduncles 3–7mm .... **12. E. hirta**
\+ Stems thinly whitish appressed pubescent; leaves obtuse; cyathia in short cymes 5–15mm, subtended by 2 or more small leaves, on peduncles 8–30mm .... **13. E. hypericifolia**

14. Capsules weakly angled, evenly hairy throughout, erect and scarcely exserted from cyathial cup .... **14. E. thymifolia**
\+ Capsules distinctly angled, with hairs restricted to angles, deflexed and long-exserted from cyathial cup .... **15. E. prostrata**

1. E. royleana Boissier. Tongsa: *Lu Shing;* Sha: *Nimthomozoo;* Nep: *Sheonri*

Succulent, erect cactus-like shrub 2–3m, or tree to 8m. Older stems terete, woody; branches whorled, erect, pith septate, (4–)5-angled, angles with entire or shallowly sinuate wings 1–1.5(–2)cm broad, when young wings bearing deciduous paired short stipular spines 2–3mm. Leaves borne at branch tips, deciduous (often absent at flowering time), membranous, obovate 8–10.5×2–2.5cm, obtuse and apiculate, attenuate at base, glabrous, entire, subsessile. Cymes subsessile, condensed, 5–12mm, borne above axils of fallen leaves, each composed of 1–4 cyathia with broad, rounded marginal appendages and glands. Styles minutely bilobed. Capsules trigonous, c 7mm.

Bhutan: C—Punakha district (Wangdu Phodrang), Tongsa district (Mangde Chu valley), Mongar district (Shongar Chu) and Tashigang district (Dangme Chu). Dry valley slopes and cliffs, sometimes with Chir Pine, 915–2130m. April–May.

2. E. antiquorum L.

Similar to *E. royleana* but branches with 3(−4) sinuately lobed wings; spines stouter 5–7mm; cymes lax, 15–30mm, stalked; styles deeply bilobed; capsules c10mm.

Sikkim: Darjeeling, cultivated (34). February.

Native of S India (16); cultivated as a hedge-plant (34).

3. E. milii Des Moulins; *E. splendens* Hooker. Eng: *Crown of Thorns*

Diffuse or scrambling shrub to 1m, branches 0.7–1cm broad, 5–6-winged, with dense stipular spines 1–2.8cm. Leaves few, membranous, obovate, 2–3.5×1–1.5cm, mucronate, glabrous, base attenuate. Cymes lax, on peduncle 3–5cm, each cyathium subtended by 2 broadly ovate showy red bracts 10–12mm broad.

Bhutan: S—Phuntsholing district (Phuntsholing). 250m.

Cultivated as an ornamental.

4. E. pulcherrima Klotzsch; *Poinsettia pulcherrima* (Klotzsch) Graham. Eng: *Poinsettia*

Shrub 2–3m; branches stout. Leaves alternate, lower green, ovate, 13–20×7–13cm, acute, base cuneate, margins entire or coarsely sinuate-dentate, pubescent beneath; petioles red 2.5–6cm; upper leaves bright red, elliptic, 5–12×2–4cm, acuminate, entire. Cyathia green, solitary on short leaf-opposed stalks 2–4mm, c 7mm broad, with a yellow, ring-like gland on one side; anther stalks red; ovary green with 3 deeply bifid styles connate at base. Capsule c1cm diameter, on deflexed red stalk c1cm.

Bhutan: S—Phuntsholing district (Phuntsholing); **Sikkim:** Darjeeling. Cultivated in gardens and on roadsides, 230–1200m. April–May.

Native of tropical America cultivated for its showy foliage.

FIG. 47. **Euphorbiaceae.** a & b, *Euphorbia griffithii:* a, apex of flowering shoot; b, cyathium. c & d, *Pedilanthus tithymaloides:* c, portion of flowering shoot; d, vertical section of cyathium. e–h, *Bridelia retusa:* e, portion of flowering shoot; f, male flower; g, female flower; h, portion of fruiting stem. i–k, *Phyllanthus reticulatus:* i, portion of flowering shoot; j, male flower; k, female flowers. l–o, *Phyllanthus emblica:* l, portion of flowering shoot; m, male flower; n, female flower with male bud borne at base; o, transverse section of fruit. p–s, *Glochidion velutinum:* p, portion of flowering shoot; q, male flower; r, female flower; s, fruit. Scale: e, i, o × ½; a × ⅔; l × ¾; c, p × 1; h, s × 1½; d × 2; b, f, g × 4; j, k, q × 5; r × 6; m, n × 8.

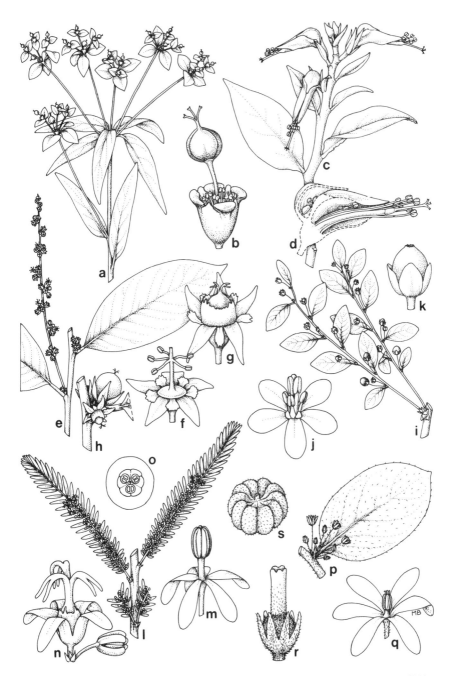

84. EUPHORBIACEAE

5. E. leucocephala Lotsy
Slender, sometimes climbing shrub differing from *E. pulcherrima* in its lower leaves in whorls of 3, ovate-lanceolate, $4-8 \times 2-3.5$cm, obtuse and minutely apiculate, base rounded, entire, sparsely pubescent beneath; petioles $1-2.5$cm; upper leaves pink or white, oblanceolate, $1-1.5$cm, apiculate; cyathia terminal, c 2mm across, bearing 2 glands; capsules c 5mm.

Bhutan: S—Phuntsholing district (Phuntsholing). Cultivated in nursery, 250m. February.

Native of tropical America cultivated as an ornamental.

6. E. griffithii Hook. f.; *Tithymalus griffithii* (Hook.f.) Hara, *E. sikkimensis* Boissier subsp. *bhutanica* Fischer. Med: *Durjit*. Fig. 47 a&b

Perennial herb with creeping rhizomes and erect annual stems $40-80$cm, simple or with a few branches from upper leaf axils, glabrous except for a tuft of hair at each node; stems, upper leaves and inflorescence reddish-tinged, especially when dry. Stem leaves alternate, linear or lanceolate, $4-13 \times 0.8-2.2$cm, acute, base cuneate, glabrous, subsessile or on short petioles $2-4$mm, exstipulate; uppermost leaves $6-8(-20)$ in a whorl, each subtending an umbel-ray. Rays (peduncles) $2-4.5$cm, glabrous, each bearing a whorl of $3-4$ ovate, acute, orange or red bracts $1-2 \times 0.8-1.2$cm, enlosing a cyathium and often $3-4$ shorter but similar branches of the inflorescence. Cyathia campanulate $3-5$mm across, bearing $4-5$ orange or yellow semicircular glands at margin, alternating with 5 short ciliate lobes; ovary smooth, on curved pedicel; styles c 3mm, united in lower third, recurved above, each simple but with minutely bilobed stigma. Capsule globose-trigonous, c 5mm, smooth.

Bhutan: C—Ha, Thimphu, Tongsa, Bumthang and Mongar districts, **N**—Upper Mo Chu district (Gasa); **Chumbi**. In clearings and amongst scrub in Blue Pine, Oak and conifer/Rhododendron forests, $2300-3500$m. May—August.

The typical var. **griffithii** has glabrous leaves; the var. **bhutanica** (Fischer) Long (*E. sikkimensis* subsp. *bhutanica* Fischer), from near Paro, differs in having leaves softly pubescent beneath.

7. E. sikkimensis Boissier; *Tithymalus sikkimensis* (Boissier) Hurusawa & Tanaka

Similar to *E. griffithii* but stems glabrous throughout, without tufts of hair below base of petioles; bracts without red pigmentation.

Sikkim: Lachen and Lachung. $2740-3350$m. July.

A poorly-known species, from which *E. griffithii* may only be subspecifically distinct. The record from Thimphu, Bhutan (117) requires confirmation.

8. E. luteo-viridis Long; *E. himalayensis* sensu F.B.I. non (Klotzsch) Boissier

Similar to *E. griffithii* and *E. sikkimensis* but rootstock thick and woody, not creeping; leafy stems tufted, shorter, 17−30(−50)cm; leaves and inflorescences yellow-green, sometimes reddish-tinged in autumn; young shoots pubescent with crispate hairs; stem leaves ovate-elliptic, broad, 2.3−4.2×1.2−2.4cm, obtuse, sessile; rays of umbel 4−5, crispate-pubescent, 1−3cm; bracts suborbicular or broadly ovate, 0.9−1.4cm broad; styles 2.2−2.7mm; capsules c 6mm diameter.

Sikkim: Gnatong, Phallut, Sandakphu, Lachen, etc. Grassy mountain slopes, 3350−3960m. May−June.

9. E. longifolia D. Don; *Tithymalus longifolius* (D. Don) Hurusawa & Tanaka

Similar to *E. griffithii* and its allies but without red pigment, stems simple, 60−100cm, young stems and rays glabrous or sparsely pilose; stem leaves linear-lanceolate 6−11×1−1.8cm; bracts yellow, ovate, 1−2cm; ovary distinctly warted; styles 3−3.6mm, united in lower half, branches erect and deeply bilobed; capsule warted.

Bhutan: C—Tongsa district (Yuto La) and Tashigang district (Tashi Yangtsi and Yonpu La). In forest clearings, 1680−3350m. May−June.

10. E. himalayensis (Klotzsch) Boissier; *Tithymalus himalayensis* Klotzsch non *E. himalayensis* sensu F.B.I.

Perennial herb, usually dark green throughout, with stout woody rootstock (not creeping) and erect annual stems 13−35(−60)cm, with slender ascending branches in upper part, young shoots finely crispate-pubescent. Stem leaves oblong or oblong-lanceolate, 2−4× 0.5−0.9(−1.3)cm, obtuse or subacute, base rounded, glabrous, sessile; exstipulate; uppermost leaves 6−10 in a whorl, each subtending an umbel ray. Rays 2−4cm, densely crispate-pubescent, each bearing a whorl of 3 broadly ovate acute bracts 5−9mm broad, enclosing a cyathium. Cyathia campanulate, 4−5mm across, bearing 5 brown semicircular glands alternating with 5 short ciliate lobes; ovary smooth, on short erect pedicel; styles 1.6−2mm, connate in lower ⅓−½, branches recurved, simple with small capitate stigma. Capsule globose-trigonous, 5−6mm, smooth.

Bhutan: C—Ha district (Ha, Damthang and Chelai La) and Thimphu district (Pajoding); **Chumbi**. In alpine meadows, 2740−4000m. June−July.

11. E. stracheyi Boissier; *Tithymalus stracheyi* (Boissier) Hurusawa & Tanaka. Med: *Durjit*

Similar to *E. himalayensis* but a smaller often reddish-tinged plant with stems 2−11(−20)cm, branched only near base, rarely almost stemless, branches spreading or decumbent; leaves oblong-obovate 8−15×3−8mm, obtuse; umbel with 3−6 rays 0.5−1.5cm; bracts 6−8mm broad; glands dark red; ovary smooth or warted, with short styles 1−1.4mm, free to base, simple, recurved.

Bhutan: C—Sakden district (Orka La), N—Upper Mo Chu district (Laya), Upper Bumthang Chu district (Lubsing La and Sharrytseem) and Upper Kulong Chu district (Me La); **Sikkim:** Jongri, etc. Alpine meadows and in Juniper/Rhododendron scrub, 3960−4880m. May−June.

In exposed high-altitude localities a distinctive species; in more sheltered habitats it becomes larger and similar in habit to *E. himalayensis* which differs in its more erect habit, lack of pigment and styles united in the lower $\frac{1}{3}-\frac{1}{2}$.

12. E. hirta L.; *E. pilulifera* L., *Chamaesyce hirta* (L.) Millspaugh

Prostrate or decumbent annual herb, stems 9−30cm sometimes becoming woody at base, densely yellowish spreading-pilose, and minutely whitish appressed-pubescent. Leaves opposite, asymmetrically ovate-lanceolate, 2−4.5×1−1.5cm, acute, base oblique, rounded on lower margin, cuneate on upper, margins serrate, sparsely appressed hairy especially beneath, with 3 strong lateral veins arising above base; petioles 1−3mm; stipules minute, subulate. Cyathia clustered in globose heads 4−8mm diameter on leafless axillary peduncles 3−7mm or subsessile. Cyathial cup c 0.8mm across, bearing 4 minute red glands which lack a flattened limb, styles free, deeply bilobed. Capsule globose-trigonous, 1.2mm diameter, mintely pubescent.

Bhutan: S—Phuntsholing and Gaylegphug districts, C—Punakha, Tongsa and Mongar districts; **Sikkim.** Weed of roadsides, lawns and cultivated ground, 200−1370m. May−August.

13. E. hypericifolia L. agg.; including *E. parviflora* L., *E. indica* Lamarck, *Chamaesyce hypericifolia* (L.) Millspaugh

Similar to *E. hirta* but stems thinly appressed pubescent; leaves ± oblong, 1.5−2×0.7−0.9cm, obtuse, margins remotely serrulate, sparsely pubescent beneath; stipules minute, triangular; cyathia in short axillary cymes 0.5−1.5cm, subtended by 2 or more small leaves, on peduncle 0.8−3cm, glands with broad pink or white limb.

Bhutan: S—Deothang district (Kheri) (117), C—Punakha district (Punakha) and Tongsa district (Tama) (117); **Sikkim:** Tista valley. Weed of disturbed and cultivated ground, 300−1830m. July−August.

The Punakha plants probably belong to the species sometimes segregated as *E. parviflora* L., with pubescent capsules and smooth seeds. However the delimitation and number of segregate species in the Indian region is unclear and the aggregate name is retained; many authors restrict the name *E. hypericifolia* L. s. str. to a glabrous tropical American plant.

14. E. thymifolia L.; *Chamaesyce thymifolia* (L.) Millspaugh

Annual herb with prostrate reddish branching stems 5−30cm, young shoots whitish crispate-pubescent. Leaves opposite, often reddish-tinged, oblong, 4−8×1.5−4mm, obtuse, base oblique, rounded on one side, subcordate on other, margins crenulate-serrulate, glabrous or with scattered hairs beneath, distinctly 3-veined from base; petioles c 0.5mm; stipules lanceolate, 0.5−1mm. Cyathia reddish in small axillary clusters, subtended by several minute leaves, cyathial cup minute; styles free, bifid. Capsule trigonous-globose, c 1mm, evenly appressed-pubescent throughout, erect, very shortly stalked and scarcely exserted from cyathial cup.

Bhutan: S—Phuntsholing district (Torsa River) and Deothang district (Samdrup Jongkhar), **C**—Tashigang district (Gamri Chu); **Sikkim:** Jalpaiguri duars, etc. Weed of roadsides and gardens, 200−1080m. May−June.

15. E. prostrata Aiton; *Chamaesyce prostrata* (Aiton) Small

Very similar to *E. thymifolia* but leaves often smaller, 3−5×1.5−3mm, entire or obscurely serrulate near apex; capsule 1.4mm, more sharply angled, with hairs restricted to angles, long exserted from cyathial cup on deflexed pedicel.

Bhutan: C—Punakha district (Wangdu Phodrang). Weed of roadsides, 1450m. April.

2. PEDILANTHUS Poiteau

Succulent shrub with zigzag stems and milky juice. Leaves alternate, coriaceous, pinnately veined; stipules minute, knob-like. Flowers in cyathia arranged in terminal bracteate cymes; cyathia strongly zygomorphic, 6 lobed, dorsal lobe forming a rounded cap-like spur concealing 4 semicircular glands, remaining 5 lobes united into a tube enclosing flowers. Female flower solitary, without perianth, consisting of a slender pedicel and 3-celled ovary with styles united almost to apex, branches short and slender. Male flowers without perianth and with a single stamen. Fruit as in *Euphorbia*.

1. P. tithymaloides (L.) Poiteau. Eng: *Slipper-flower.* Fig. 47 c & d

Erect shrub to 1.3m, puberulous throughout. Leaves ovate, 4−9×1.5−4cm, subacute or bluntly acuminate, base rounded or cuneate, entire, lateral veins obscure; petioles 2−5mm. Bracts lanceolate 5−10mm. Cyathia bright red, tube 10−13mm, with anthers and styles exserted.

Bhutan: S—cultivated at Samchi and Phuntsholing. 200−500m. March.

Native of tropical America, cultivated as a hedge plant; plants with leaves variegated green and pink are grown as ornamentals.

3. BRIDELIA Willdenow

Trees or shrubs, sometimes climbing. Leaves alternate, distichous, entire with parallel lateral veins; stipulate. Monoecious or dioecious; flowers

84. EUPHORBIACEAE

surrounded by scarious bracts in axillary clusters, these sometimes forming spikes, leafless or bearing smaller leaves. Male flowers: calyx deeply 5-lobed; petals 5, minute, obovate; disc broad, cup-like; stamens 5, filaments united below into a column; pistillode present. Female flowers: similar to males but disc enclosing ovary; ovary 2-celled, each with 2 ovules; styles 2, free; stigmas deeply bifid. Fruit a 1−2-seeded drupe.

1. Plants without fruit 2
+ Plants with mature fruit 6

2. Leaves on vegetative shoots narrowly elliptic, small, mostly 6−10×1.5−3cm .... **4. B. tomentosa**
+ Leaves on vegetative shoots broadly elliptic to obovate, large, mostly 8−17×3−11cm 3

3. Leaves membranous, acuminate; lateral veins 7−11 pairs; shoots with many linear lenticels .... **5. B. pubescens**
+ Leaves thinly coriaceous, obtuse, subacute or acute; lateral veins mostly 10−20 pairs; shoots with ± rounded lenticels (except *B. stipularis* which has scattered elliptic lenticels) 4

4. Shoots glabrous but usually with dense, prominent rounded lenticels; secondary veins scarcely prominent beneath .... **1. B. sikkimensis**
+ Shoots pubescent or tomentose and with scattered round or elliptic lenticels; secondary veins distinctly prominent beneath 5

5. Shoots tomentose, with elliptic lenticels; lateral veins 10−13 pairs; climbing shrub .... **2. B. stipularis**
+ Shoots pubescent or glabrous, with rounded lenticels; lateral veins 15−20 pairs; tree .... **3. B. retusa**

6. Fruits all 1-seeded 7
+ Fruits mostly 2-seeded 8

7. Fruits ellipsoid 5−6mm long .... **1. B. sikkimensis**
+ Fruits oblong-ellipsoid 10−12mm long .... **5. B. pubescens**

8. Fruits broadly ellipsoid 12−13mm long .... **2. B. stipularis**
+ Fruits globose 4−8mm diameter 9

9. Fruits 6−8mm diameter .... **3. B. retusa**
+ Fruits 4−5mm diameter .... **4. B. tomentosa**

1. B. sikkimensis Gehrmann; *B. verrucosa* Haines, *B. montana* sensu F.B.I. p.p. non Willdenow. Nep: *Gayo* (34), *Lahara Gaijo* (117)

Semi-evergreen shrub 2−4m, sometimes climbing; young shoots glabrous but usually densely papillose with rounded lenticels. Leaves thinly coriaceous, obovate, rarely ovate, 8−18×5−11cm, obtuse or subacute, base rounded, pale and glabrous or minutely puberulous beneath, lateral veins 10−16 pairs, prominent beneath, secondary veins not prominent; petioles 8−13mm; stipules lanceolate, 4mm, early caducous. Monoecious; flower clusters small, 5−6mm diameter, bracts sericeous along mid-line. Flowers sessile; calyx glabrous, lobes 1.2mm in males, 1.8mm in females. Drupes ellipsoid, 1-seeded, 5−6×4mm.

Bhutan: S—Samchi, Phuntsholing, Sarbhang and Deothang districts, **C**—Punakha and Tongsa districts; **Sikkim:** foothills. Amongst shrubs at margins of Subtropical and Warm broad-leaved forests, 200−1610m. September−January.

Leaves used as fodder (48).

2. B. stipularis (L.) Blume. Nep: *Lahara Gayo* (34), *Kasreto*

Similar to *B. sikkimensis* but a large climbing shrub, shoots brownish-tomentose with scattered elliptic lenticels; leaves large on vegetative shoots, 11−17×6−12cm, subacute, with 11−14 pairs of lateral veins, brownish pubescent and with prominent secondary veins beneath; stipules triangular-acuminate, 6−12mm, sometimes persistent; inflorescences often spike-like, leafless or with small 3−8cm leaves; flowers large, males sessile, females on short thick pedicels 1−2mm, calyx lobes acuminate 3−4mm, sparsely pubescent; drupes broadly ellipsoid, 2-seeded, 12−13×10mm.

Bhutan: S—Samchi district (Torsa valley and Tamangdhanra forest), Sankosh district (Pinkhua) and Sarbhang district (Singi Khola); **Assam Duars:** Durunga; **Sikkim:** terai. Subtropical and Sal forests, 310−650m. September−October.

3. B. retusa (L.) Sprengel. Nep: *Gayo* (34), *Kuhir*. Fig. 47 e−h

Similar to *B. sikkimensis* and *B. stipularis* but a tree 6−12m; shoots pubescent with few scattered rounded lenticels; leaves usually elliptic, 8−14×3.5−8cm, acute, lateral veins more numerous, 15−20 pairs, finely pubescent beneath and with prominent secondary veins; flower clusters 7−9mm, male and female flowers on pedicels c1mm, calyx lobes 1.6−2mm, sparsely pubescent; drupes 2-seeded globose, 6−8mm.

Bhutan: S—Deothang district (Kheri), **C**—Tongsa district (Dakpai and Tama) and Tashigang district (Dangme Chu); **W. Bengal Duars:** Buxa; **Sikkim:** terai. Subtropical and Chir Pine forests, 400−1650m. August−September.

Foliage used as fodder; timber durable (34); bark used in tanning; fruit edible (16).

4. B. tomentosa Blume. Nep: *Muse Gayo* (34)
Shrub or small tree to 10m; young shoots finely reddish tomentose, with few rounded lenticels. Leaves on vegetative shoots membranous, elliptic-lanceolate, 6−10×1.5−3cm, acute or shortly acuminate, base cuneate, with 7−12 pairs of slender lateral veins, pubescent beneath; petioles 2−5mm; stipules lanceolate, 3mm, caducous. Flower clusters 3−4mm in axils of smaller leaves 3−6cm. Flowers minute on very short pedicels, calyx lobes 0.8−1mm, glabrous. Fruits globose, 2-seeded, 4−5mm diameter.

Bhutan: S—Samchi district (Samchi); **Sikkim:** terai and foothills. Subtropical forests, 150−600m. October−November.

5. B. pubescens Kurz
Small tree 10−15m; young shoots glabrous or sparsely pubescent, with dense linear lenticels. Leaves membranous, elliptic-obovate, 10−15(−22)×3−6.5(−11)cm, acuminate, base cuneate, with 7−11 pairs of slender lateral veins, pubescent beneath; petioles 4−10mm; stipules lanceolate, 4.5mm. Flower clusters c 1cm. Pedicels 1−3mm; calyx lobes 1.5mm, tomentose. Fruits oblong-ellipsoid, 1-seeded, 10−12×7mm.

Bhutan: S—Chukka district (Marichong); **Sikkim:** foothills and Tista valley etc. Subtropical forests, 300−1200m. May−July.

4. PHYLLANTHUS L.

Trees, shrubs or herbs. Leaves alternate, entire, pinnately veined, often distichous, stipulate. Monoecious; flowers in axillary fascicles or clusters, sometimes solitary. Male flowers: sepals 4−6, petals absent, disc-glands often present, stamens 3−5, all or some filaments connate (all free in *P. acidus* and *P. glaucus*), anther cells dehiscing by vertical or transverse slits. Female flowers: sepals (4−)6, disc-glands often present, ovary 3(−4)-celled, styles free or connate, bifid. Fruit a capsule or berry-like, 3−16-seeded.

1. Herbs 10−30cm, sometimes woody at base 2
+ Shrubs or trees 0.5−10m 4

2. Leaves linear-lanceolate; fascicles containing both male and female flowers; fruiting pedicels 5−7mm; capsules minutely warted; seeds papillose .... **1. P. virgatus**
+ Leaves oblong to obovate-elliptic; fascicles unisexual; fruit sessile or on short pedicels 1−2mm, capsules coarsely warted or smooth; seeds ribbed 3

3. Shoots puberulous; leaves oblong, with hispidulous margins and prominent veins; fruit coarsely warted, sessile; seeds with 12−15 transverse ribs .... **2. P. urinaria**
+ Shoots glabrous; leaves obovate-elliptic, glabrous and with scarcely prominent veins; fruit smooth, on pedicels 1−1.5mm; seeds with 5−7 longitudinal ribs .... **3. P. debilis**

4. Leaves linear-oblong, 1−2.5mm broad; fruit fleshy 8−15mm diameter .... **4. P. emblica**
+ Leaves ovate to elliptic or obovate, 5−35mm broad; fruit dry or fleshy, 2−7mm diameter (17−22mm in *P. acidus*) 5

5. Cultivated tree; branches stout, leafless; branchlets deciduous, slender, leafy; fruit 17−22mm diameter .... **5. P. acidus**
+ Wild or cultivated shrubs; branches and branchlets slender, leafy, branchlets not deciduous; fruit 2−7mm diameter 6

6. Stamens 5, anthers elongate; styles long, stout, entire, or stigmas ± sessile on ovary; fruit 3- or 8−16-seeded 7
+ Stamens 3, anthers short; styles slender, bifid; fruit 6-seeded 8

7. Sepals 6; stigmas minute, sessile on ovary; fruit 8−16-seeded, borne on unthickened pedicel .... **6. P. reticulatus**
+ Sepals 5; styles simple, recurved; fruit 3-seeded borne on thickened pedicel .... **7. P. glaucus**

8. Flowers on long pedicels 10−25mm; male flowers with 4 fimbriate sepals, females with 5−6 fimbriate or dentate sepals 9
+ Flowers on short to long pedicels 2−18mm; male and female flowers with (5−)6 entire sepals 10

9. Leaves ovate-elliptic, 2−4.5×1−3cm; capsules papillose-puberulous .... **8. P. sikkimensis**
+ Leaves obliquely oblong-elliptic, 1.5−2.5×0.7−1.3cm; capsules glabrous .... **9. P. pulcher**

10. Female flowers and fruit on pedicels 8−18mm 11
+ Female flowers and fruit on pedicels 1−6mm 12

11. Leaves obovate; male flowers on pedicels 2−3mm; styles free almost to base .... **10. P. clarkei**
+ Leaves elliptic; male flowers on pedicels 8−10mm; styles united in lower half .... **11. P. griffithii**

12. Leaves 5–9mm; flowers solitary, on pedicels 1mm .... **12. P. parvifolius**
+ Leaves 9–20mm; male flowers in threes, on pedicels 2–3mm, female flowers 1–2 per axil, on pedicels 3–6mm .... **13. P. leschenaultii**

1. P. virgatus Forster f.; *P. simplex* Retzius

Annual or perennial glabrous herb with spreading or ascending branches 8–45cm from ± woody base, branches compressed or winged above. Leaves numerous, lanceolate, 10–25×3–8mm, acute, base rounded, subsessile, veins not prominent; stipules triangular-acuminate c 2mm, auriculate. Flowers 2–4 in axillary fascicles (containing both sexes); males on pedicels 1–2mm; females on slender pedicels 4–6mm. Sepals 6, elliptic, 0.8–1mm, acute. Fruiting pedicels 5–7mm, thickened at apex; capsules globose, 2.5–3.5mm, minutely warted; seeds papillose.

Bhutan: C—Punakha district (Tinleygang), Mongar district (Lingmethang) and Tashigang district (Dangme Chu); **Sikkim:** terai; **W. Bengal Duars:** Buxa. Weed of roadsides and cultivated ground, 160–1830m. June–September.

2. P. urinaria L.

Similar to *P. virgatus* but erect, stems with short branches throughout, young shoots puberulous; leaves oblong 6–14×3–6mm, obtuse or subacute, margins minutely hispidulous, veins prominent beneath; flowers 1–2 of only 1 sex in each axil, males towards apex, sessile; ovary coarsely warted; capsules c 2.5mm, coarsely warted, sessile; seeds with 12–15 transverse ridges.

Bhutan: S—Phuntsholing district (Phuntsholing), **C**—Punakha district (near Punakha), Tongsa district (Shamgong) and Tashigang district (Khari, S of Tashigang); **Sikkim.** Weed of roadsides and cultivated ground, 250–1300m. February–September.

3. P. debilis Willdenow

Similar to *P. virgatus* and *P. urinaria* but stems branched in upper part, young shoots glabrous; leaves obovate-elliptic, 12–20×5–13mm, acute, glabrous, veins scarcely prominent beneath; flowers 1–3 of only 1 sex in each axil, females towards apex, female pedicels 1–1.5mm (–2mm in fruit); female sepals narrowly obovate, 1.4–1.8mm; ovary and capsule smooth; seeds with 5–7 longitudinal ridges.

Bhutan: S—Gaylegphug district (Thewar Khola); **Sikkim:** foothills. Disturbed ground, 300–1000m. May–October.

4. P. emblica L.; *Emblica officinalis* Gaertner. Dz: *Omla;* Med: *Churoo;* Sha: *Chhorgeng;* Nep: *Amala* (34). Fig. 47 l–o

Deciduous shrub 1–3m or tree to 10m, young shoots slender, pubescent, bearing leaves and flowers. Leaves appearing with flowers, numerous, distichous, linear-oblong, 5−15×1−2.5mm, subacute, glabrous, subsessile; stipules triangular 1−1.2mm. Male flowers on short pedicels in dense, crowded clusters at base of young leafy shoots, female flowers few, sessile, borne above males. Male flowers yellow, 1.2−1.5mm diameter, sepals 6, disc-glands absent; stamens 3, filaments connate into short column, anthers vertical. Female flowers c 2mm, sepals 6; styles stout, exserted, branches bifid. Fruit a globose, fleshy, capsule c 2.5cm diameter, later splitting into 3 woody 2-celled valves, 6-seeded.

Bhutan: S—Phuntsholing, Sarbhang and Gaylegphug districts, C—Punakha, Tongsa and Mongar districts; **Sikkim.** Subtropical, Warm broad-leaved and Chir Pine forests, often on sunny slopes and in scrub on river banks, 460−1800m. March−April.

The fruits which are an important source of vitamin C (16) are edible and used medicinally (126); bark used in tanning; wood durable in water (48). A valuable firewood crop for arid regions (130).

5. P. acidus (L.) Skeels; *P. distichus* (L.) Mueller. Eng: *Star Gooseberry*

Deciduous tree to 15m; branches stout, leafless, with slender deciduous leafy branchlets towards ends. Leaves distichous, ovate, 3−5×1−2cm, acute, base rounded or broadly cuneate, glabrous; petioles 2−3mm; stipules lanceolate, c1mm. Flowers in dense clusters in slender leafy or leafless racemes. Male flowers numerous; pedicles slender, 1−3mm, sepals 4, orbicular, c 1mm, stamens 4, free. Female flowers few, sepals 4, c 1.5mm, ovary with or without staminodes, styles 4, free, bifid, recurved. Fruit yellow, depressed-globose, 1.7−2.2cm diameter, 6−8-lobed.

Sikkim: Jalpaiguri district. Cultivated in terai. March−April.

Native of tropical S America cultivated for its edible fruit, and green leaves eaten as spinach (16).

6. P. reticulatus Poiret; *P. dalbergioides* J. J. Smith, *Kirganelia reticulata* (Poiret) Gamble, *K. microcarpa* (Bentham) Hurusawa & Tanaka. Dz: *Dosem.* Fig. 47 i−k

Large, often climbing shrub with slender lenticellate branches, sometimes thorny. Leaves ovate or broadly elliptic, 1.4−4(−7.5)× 1−2(−3.5)cm, broadly acute or obtuse, base rounded, glabrous or pubescent; petioles 1−2mm; stipules narrowly triangular-subulate, c 1.5mm; on older branches sometimes forming enlarged, woody thorns. Flowers on slender leafy (sometimes leafless) side-shoots, 1−5 per axil, pedicels 3−6mm, sepals 6, c 1.5mm. Male flowers with 5−6 stamens, inner 3 united, anthers elongated, erect. Female flowers with globose ovary bearing 6−8 minute stigmatic lobes sessile on apex. Fruit 5−6mm, somewhat fleshy, crimson, 8−16-seeded; pedicels not thickened at apex.

Bhutan: S—Samchi, Phuntsholing, Sarbhang and Gaylegphug districts; **Sikkim:** foothills and terai. Amongst scrub in subtropical and terai forests, often on river banks, 200−600m. February−April.

Plants with larger leaves and almost leafless racemes are sometimes segregated as *P. dalbergioides* J. J. Smith; they have been collected in Mal Forest, Darjeeling.

7. P. glaucus Mueller; *Flueggiopsis glauca* (Mueller) A. Das, *Hemicicca glauca* (Mueller) Hurusawa & Tanaka. Dz: *Dosem;* Tongsa: *Prang Seng*

Similar to *P. reticulatus* but an erect unarmed shrub 3−4m; stipules 2.5mm, membranous; flowers in axillary clusters of 4−12, usually with 1 female flower per cluster; male flowers with 5 free stamens; female flowers with 3 conspicuous simple recurved styles; fruit 6−7mm, fleshy, 3-seeded, bearing stylar remains and borne on pedicel thickened at apex.

Bhutan: C—Thimphu, Punakha, Tongsa and Tashigang districts, N—Upper Kuru Chu district (Dunkar); **Sikkim.** Evergreen oak and Cool broad-leaved forests, 1700−2700m. April−May.

8. P. sikkimensis Mueller; *P. hamiltonianus* Mueller, *Eriococcus hamiltonianus* (Mueller) Hurusawa &Tanaka, *Reidia hamiltoniana* (Mueller) Cowan & Cowan

Shrub to 1m, branches terete, puberulous. Leaves ovate-elliptic, 2−4.5×1−3cm, acute, base rounded, pale and subglabrous beneath, on short petiole 1−2mm; stipules subulate. Male flowers on pedicels c1.5cm in axillary fascicles on lower part of branchlets, sepals 4, long-fimbriate, disc 4-lobed, stamens united into short column. Female flowers on pedicels 1.5−2.5cm towards branch-ends; sepals 5−6, dentate; styles free almost to base, each bifid almost to base. Capsule 2−3mm, papillose-puberulous; on elongate, curved pedicels to 3cm, thickened at apex.

Bhutan: S—Phuntsholing district (Phuntsholing); **Sikkim:** duars and foothills. Subtropical forests, 150−900m. May−June.

9. P. pulcher Mueller

Similar to *P. sikkimensis* but leaves obliquely oblong-elliptic, 1.5−2.5×0.7−1.3cm; male and female sepals deeply fimbriate; capsule smooth, glabrous.

Sikkim: Siliguri, cultivated. July−November.

Native of Malaysia.

10. P. clarkei Hook. f.

Wiry shrub 0.5−1m with slender leafy shoots rough with papillae on ridges. Leaves obovate, 1−2.5×0.4−1.2cm, obtuse, base cuneate, glabrous, pale beneath; petioles 1−1.5mm; stipules peltate, lacerate. Flowers 2−8 in axillary fascicles, males on short pedicels 2−3mm, females usually 1 per cluster on slender pedicel 12−18mm. Sepals (5−)6, obovate, entire, disc-glands present, stamens 3 connate into short column, anther cells rounded, separate. Ovary with slender styles free almost to base, deeply bifid. Fruit subglobose, 3−4mm, dry, 6-seeded.

Bhutan: C—Thimphu district (Dotena and Paro) and Punakha district (Wache, Tang Chu); **Sikkim**. River banks and rocky slopes in Warm and Cool broad-leaved forests, 1520−2550m. October−November.

11. P. griffithii Mueller

Similar to *P. clarkei* but branchlets smooth; leaves elliptic, 15−25× 7−10mm, acute or obtuse; male and female flowers on long pedicels 8−12mm; styles stout, united to middle, recurved and bilobed above.

Bhutan: S—Gaylegphug district (Gaylegphug (117) and lower Mangde Chu Valley). Subtropical forests, 270−280m. March.

12. P. parvifolius D. Don

Similar to *P. clarkei* but a taller shrub 1.5−2.5m, with short leafy branches, leaves distichous, closely arranged, obovate-elliptic, 5−9× 3−4mm, obtuse, base rounded; flowers solitary, on short pedicels 1mm; filaments free.

Bhutan: S—Sarbhang district (Noonpani above Sarbhang); **Sikkim:** Darjeeling foothills. Subtropical and Warm broad-leaved forests, 800−1800m. March.

13. P. leschenaultii Mueller

Similar to *P. clarkei* but a shrub to 1m, leaves obovate-elliptic, 9−20×4−13mm, obtuse, base rounded; male flowers often in threes, on pedicels 2−3mm, filaments free; female flowers 1−2 per axil, on pedicels 3−6mm, styles short, free, spreading, with slender recurved branches.

Bhutan: C—Tongsa district (between Pertimi and Tintibi Bridge, Mangde Chu); **Sikkim:** Ryang and Tista Valley. Subtropical forests, 250−980m. April.

The Bhutan collection differs from the typical form in having smaller leaves.

5. FLUEGGEA Willdenow

Evergreen or deciduous shrubs. Leaves alternate, pinnately veined, entire, stipulate. Dioecious; flowers in axillary clusters. Male flowers: sepals 5, petals 0, disc-glands present, stamens 5, filaments free, pistillode present with 3 simple styles. Female flowers: sepals 5, petals 0, disc-glands present, ovary (1−)3-celled, styles 3, bifid, recurved. Fruit a fleshy 5−8-seeded berry.

1. F. virosa (Willdenow) Voigt; *F. microcarpa* Blume, *Securinega virosa* (Willdenow) Baillon. Sha: *Geykang Shing;* Nep: *Darim Pate, Phalame* (34)

Shrub 2−3m, glabrous, branchlets ribbed. Leaves ovate 3−

10×1.5−5cm, acute or bluntly acuminate, base cuneate; petioles 4−10mm; stipules triangular, 1.5mm. Male flowers 15−40 in dense clusters in axils mostly of fallen leaves, pedicels filiform 2−5mm, sepals 1mm, dentate, anthers and sterile styles exserted. Female flowers 3−6 in axillary clusters, pedicels 2−6mm, ovary globose, styles united at base into short column, free parts spreading and often appressed to ovary, each bifid with subulate ± parallel lobes. Fruit globose, firm and green when young, becoming white and fleshy, 5−6mm diameter when ripe.

Bhutan: S—Phuntsholing, Chukka, Gaylegphug and Deothang districts, C—Tongsa, Mongar and Tashigang districts; **Sikkim**. Amongst scrub and on river banks in Subtropical, Chir Pine and Warm broad-leaved forests, 250−1400m. April−July.

The above description applies to subsp. **himalaica** Long. The more tropical subsp. **virosa** is known from the Tista Valley and Darjeeling duars and differs in its smaller (2−4×1−2.5cm), obovate, obtuse leaves. Records from Bhutan of *Securinega suffruticosa* (Pallas) Rehder, a NE Asiatic species from Mongolia, N China and Japan, refer to subsp. *himalaica*.

F. virosa resembles some species of *Phyllanthus,* notably *P. reticulatus* and *P. glaucus,* but these have shorter petioles 1−2mm. Wood durable and used to make agricultural implements (16).

6. GLOCHIDION J.R. & J.G.A. Forster

Evergreen trees or shrubs. Leaves alternate, pinnately veined, entire, stipulate. Monoecious; flowers in axillary clusters. Male flowers: sepals 6, petals absent, disc-glands absent, stamens 3−8, filaments connate into a column, dehiscing by vertical slits. Female flowers: sepals 6(−12), petals 0, ovary 3−15-celled, styles connate, usually columnar, lobed at apex, sometimes globose or broad and flat. Fruit a woody capsule, 3−15-lobed; seeds 3−15 each with a red aril-like coat.

1. Branchlets and undersides of leaves (at least veins) pubescent or tomentose 2
+ Branchlets and leaves glabrous (sparsely pubescent on veins beneath when young in *G. bhutanicum*) 7

2. Leaves broadly ovate-oblong, 10−18×4.5−9.5cm, rounded or often cordate at base; pedicels borne on short peduncles often supra-axillary
1. G. hirsutum
+ Leaves ovate, lanceolate, oblong-elliptic or obovate, 4.5−13×2−5cm, cuneate at base (or rounded in *G. velutinum*); pedicels sessile, axillary 3

3. Leaves narrowly oblong-elliptic or oblanceolate; stamens 4–12; fruit large, 15–20mm, 10–15-lobed or almost smooth 4
+ Leaves ovate, obovate, broadly elliptic or lanceolate; stamens 3; fruit small, 6–13mm, 4–12-lobed 5

4. Leaves 4.5–8(–11)cm, often pale-green when dry; minor veins conspicuous; fruit strongly 10–15-lobed, on pedicels 3–5mm
2. G. multiloculare
+ Leaves 7–12cm, brown when dry; minor veins inconspicuous; fruit almost smooth, on short pedicels 1–2mm .... **3. G. oblatum**

5. Leaves lanceolate, whitish beneath when dry; fruit 6–8mm diameter
6. G. acuminatum
+ Leaves ovate, obovate or broadly elliptic, brown beneath when dry; fruit 8–13mm diameter 6

6. Petioles 2–3mm; ovary with columnar style; male pedicels 3–8mm
4. G. velutinum
+ Petioles 3–5mm; ovary with broad, flat style; male pedicels 8–10mm
5. G. nubigenum

7. Leaves ovate or broadly elliptic, mostly 2.5–7cm broad, apex shortly and ± abruptly acuminate (sometimes more gradually acuminate in *G. khasicum*); leaf base often slightly oblique 8
+ Leaves lanceolate and gradually acuminate, or narrowly elliptic to oblong-obovate, acute or obtuse, mostly 1.5–2.5cm broad (sometimes up to 3.5cm in *G. multiloculare* and *G. sphaerogynum*); leaf base symmetric 11

8. Leaves 8–15×3–7cm; male pedicels 9–16mm; female flowers with short narrow columnar style with 4–6 minute erect lobes 9
+ Leaves 4.5–11×2.5–4cm; male pedicels 4–7mm; female flowers with long slender minutely 3-lobed columnar style, or very short style as broad as ovary with 5 short stout spreading lobes 10

9. Lateral veins distinctly prominent beneath; petioles 3–5mm; stamens 3; fruits often borne in dense clusters of 4–12, 8–10mm diameter, 4-lobed .... **7. G. assamicum**
+ Lateral veins scarcely prominent beneath; petioles 5–10mm; stamens 4–6; fruit 1–2 per axil, 10–18mm diameter, 8–12-lobed
8. G. lanceolarium

10. Female flowers and fruit sessile; style a long narrow column, shortly 3-lobed at apex .... **9. G. khasicum**
+ Female flowers and fruit on pedicels 4—6mm; style as broad as ovary with 5 stout spreading lobes .... **10. G. bhutanicum**

11. Leaves narrowly elliptic or oblong-obovate, acute or obtuse, whitish beneath 12
+ Leaves lanceolate, gradually acuminate, green or pale green but never white beneath 13

12. Male pedicels 5—8mm; stamens 4—12; female pedicels 3—5mm; female sepals broad, spreading, c2mm; fruit 18—24mm diameter, 10—15-lobed
2. G. multiloculare
+ Male pedicels 3—4mm; stamens 3; female pedicels 9—15mm; female sepals minute, appressed, 0.5mm; fruit 8mm diameter, 3—4-lobed
11. G. thomsonii

13. Lateral veins conspicuous, very prominent beneath; petioles 4—9mm; male pedicels 5—7mm; style subglobose, not lobed
12. G. sphaerogynum
+ Lateral veins obscure, slightly prominent beneath; petioles 3—4mm; male pedicels 7—11mm; style columnar, 4—6-lobed ...... **13. G. daltonii**

1. G. hirsutum (Roxb.) Voigt

Shrub or small tree to 5m, branchlets brownish tomentose. Leaves broadly ovate-oblong, 10—18×4.5—9.5cm, acute or subacute, base shallowly cordate or rounded, often drying brown beneath, pubescent beneath; petioles 4—8mm. Flowers pedicellate, borne on short stout peduncles 2—4mm, often supra-axillary. Male pedicels 10—12mm, slender; stamens 4—5. Female pedicels 2—5mm, stout, style a short hairy column, 4—5-lobed. Capsules subglobose, 10—12mm, very shallowly 6—8-lobed.

Sikkim: foothills and terai. Subtropical forests, 150—450m. May—July.

2. G. multiloculare (Willdenow) Mueller

Shrub or small tree, branchlets glabrous or occasionally puberulous, sometimes prickly with thickened stipules. Leaves narrowly oblong-elliptic or oblanceolate, 4.5—8(—11)×1.5—2.5(—3.5)cm, subacute or bluntly apiculate, base cuneate, pale and glabrous or rarely pubescent beneath, often yellow-green above when dry, veins and reticulations conspicuous; petioles 1—4mm. Male pedicels 5—8mm, slender, sepals 6, stamens 4—12. Female pedicels 3—5mm, stout, sepals 8—12, style a hollow grooved cone on top of ovary. Fruit strongly depressed-globose, 1.5—2cm diameter, strongly 10—15-lobed, apex deeply depressed with disc-like stylar remains.

Sikkim: foothills and terai. Subtropical forests, 150—300m. April—June.

3. G. oblatum Hook. f.

Similar to *G. multiloculare* but shoots finely tomentose; leaves elliptic, 7−12×2−4.5cm, acute or subacute, brownish when dry, pubescent on veins beneath, reticulations not conspicuous; female pedicels 1−2mm; style a short column, minutely 6−7-toothed; fruit subglobose 1.5−2cm, very indistinctly 5−6-lobed, apex not depressed.

Bhutan: S—Phuntsholing district (Phuntsholing, 117); **Sikkim:** foothills and terai. Subtropical forests, 100−250m. May.

4. G. velutinum Wight; *G. heyneanum* (Wight) Beddome. Sha: *Kotokmo Shing*. Fig. 47 p−s

Large shrub or tree to 12m, shoots densely pale tomentose. Leaves ovate, broadly elliptic or obovate, 4.5−8(−13)×2−5(−6.5)cm, acute or bluntly apiculate, base cuneate or rounded, dark brown when dry, densely pubescent beneath; petioles 2−3mm. Male flowers on pedicels 4−8mm; stamens 3. Female pedicels 2−5(−8)mm; styles forming a slender pubescent column obscurely toothed at apex. Capsules depressed-globose, 9−13mm, conspicuously 8−12-lobed, pubescent.

Bhutan: C—Punakha district (Punakha, Rinchu, Tinleygang and Tang Chu) and Tongsa district (Mangde Chu valley below Pertimi); **Sikkim:** terai and foothills. In dry scrub in Subtropical and Chir Pine forests, 150−1500m. February−May.

G. heyneanum (Wight) Beddome, often treated as a distinct species, appears to be a robust form with larger leaves and longer female pedicels.

5. G. nubigenum Hook. f.

Similar to *G. velutinum* but less pubescent throughout; leaves ovate-elliptic, 7−13×3.5−6cm, acuminate; base cuneate; petioles 3−5mm; style broad and flat, with 5−6 minute raised lobes; fruit strongly depressed, c 1cm diameter, deeply 8−10-lobed, with central broad, flat stylar remains.

Bhutan: S—Chukka district (Chima Khoti), **C**—Tongsa district (S of Tongsa) and Tashigang district (Damoitsi); **Sikkim:** Rungbi, Darjeeling, etc. Warm broad-leaved forests, 1500−2100m. April−June.

6. G. acuminatum Mueller. Nep: *Latikath* (34)

Similar to *G. velutinum* but less pubescent throughout; leaves lanceolate, 7−12×2−3.5cm, acuminate, often green above and pale beneath when dry; female flowers with short columnar style with 4−5 short erect lobes; fruit small, 6−8mm diameter, deeply 4−6-lobed with minute persistent columnar style.

Bhutan: S—Deothang district (Raidong), **C**—Punakha district (between Mishichen and Khosa); **Sikkim:** Kurseong, Lebong, etc. Warm broad-leaved forests, 1000−2000m. April−June.

7. G. assamicum (Mueller) Hook. f.; *G. assamicum* (Mueller) Hook. f. var. *brevipedicellatum* Hurusawa & Tanaka. Nep: *Haldi Kath* (117), *Lati Mauwa* (34)

Shrub or small tree 3–5m, shoots glabrous. Leaves membranous, elliptic 9–15×4–7cm, abruptly acuminate to fine apex, base obliquely cuneate, often rounded on one side, glabrous, brownish beneath when dry, lateral veins prominent beneath; petioles 3–5mm. Male pedicels 9–16mm, stamens 3. Female pedicels 2–5mm, style a short erect column minutely 4-toothed. Capsules 4–12 per axil, densely clustered, depressed-globose, 8–10mm diameter, 4-lobed, pubescent, with minute persistent style.

Bhutan: S—Chukka district (Raidak valley), Sarbhang district (Lao Pani and Phipsoo) and Gaylegphug district (117); **Sikkim:** terai and foothills. Subtropical forests, 150–600m. October–April.

8. G. lanceolarium (Roxb.) Voigt. Nep: *Bangikath* (34)

Similar to *G. assamicum* but leaves coriaceous, elliptic-oblong, often greyish when dry, base cuneate or attenuate, lateral veins obscure, not prominent beneath; petioles 5–10mm; male flowers with 4–6 stamens; female flowers sessile, style a minute 5–6-lobed column; capsules 1–2 per axil, large, 1.5–2cm diameter, 6–8-lobed.

Sikkim: Darjeeling district (34). Warm broad-leaved forests, to 1500m. December–March.

Often confused with *G. assamicum*; the Sikkim records require confirmation.

9. G. khasicum (Mueller) Hook. f.

Shrub 3–5m, shoots glabrous. Leaves coriaceous, elliptic, 6–11× 3–4cm, abruptly (sometimes ± gradually) acuminate, base cuneate, often greyish-green when dry, veins weakly prominent beneath; petioles 3–5mm. Male pedicels 6–9mm, stamens 3. Female flowers sessile, style a long-exserted column, with 3 erect blunt lobes. Capsule depressed globose, 7–9mm, 4–6-lobed, with persistent style 3–4mm.

Bhutan: S—Chukka district (Chukka), **C**—Tongsa district (S of Shamgong); **Sikkim:** Kurseong, Sittong and Sureil. Warm broad-leaved forests, 1200–1800m. April–June.

10. G. bhutanicum Long. Sha: *Kotokmo Shing*

Similar to *G. khasicum* but leaves membranous, 5–7×2.5–3.5cm, brown above and pale green beneath when dry; lateral veins 5–7 pairs, prominent beneath; female flowers on pedicels 4–6mm, style 1.4–1.6mm, as broad as ovary, deeply divided into 5 thick slightly spreading lobes; fruit unknown (ovary 5-celled).

Bhutan: S—Deothang district (between Samdrup Jongkhar and Riserboo). Cool broad-leaved forest, 2150m. June.

Endemic to Bhutan.

11. G. thomsonii Hook. f. Nep: *Latikath* (34)

Small tree, branchlets glabrous. Leaves elliptic, 5−7×2−2.5cm, acute or obtuse, base cuneate, glabrous, whitish beneath; petioles 3−4mm. Male flowers few, on pedicels 3−4mm, stamens 3. Female flowers with longer pedicels 9−15mm, sepals minute, triangular, 0.2mm, style as broad as ovary, divided into 3−4 rounded lobes. Capsules depressed globose, 8mm diameter, 3−4-lobed.

Sikkim: terai and foothills. Subtropical forests. April−July.

Distinct from other Bhutan & Sikkim species in having female flowers on longer pedicels than the males, and by the minute female sepals.

12. G. sphaerogynum (Mueller) Kurz. Nep: *Malchina, Rokte*

Small tree 6−8m, branchlets glabrous. Leaves thinly coriaceous, lanceolate, 9−15×2.5−4cm, long acuminate, often slightly curved, base cuneate, glabrous, green when dry; petioles 4−9mm. Male pedicels slender, 5−7mm, stamens 3. Female flowers densely clustered, pedicels 2−3mm, sepals rounded, style subglobose, unlobed. Capsules c1cm diameter, deeply 8−10-lobed, with subglobose style in depressed apex.

Bhutan: S—Samchi district (Deo Pani Khola), Sankosh district (Pinkhua), Gaylegphug district (Tatapani) and Deothang district (Raidong); **Sikkim:** Darjeeling district. Subtropical forests, 310−920m. March−June.

Foliage used as cattle fodder (146).

13. G. daltonii (Mueller) Kurz; *G. gamblei* Hook. f.

Similar to *G. sphaerogynum* but leaves 6−11×2−3cm, more shortly acuminate, brownish beneath when dry, lateral veins less prominent beneath; petioles 3−4mm; male pedicels longer, 7−11mm; female pedicels 1−2mm, ovary with columnar style 4−6-lobed at apex; fruit larger, 12−15mm diameter, 8−10-lobed but with persistent columnar style in depressed apex.

Sikkim: Darjeeling foothills, Riang, Mungpoo, Kurseong, Dulka Jhar, etc. Subtropical forests, 600−950m. August−March.

7. BREYNIA J.R. & J.G.A. Forster

Evergreen shrubs. Leaves alternate, pinnately veined, entire, stipulate. Monoecious; flowers solitary or in axillary clusters. Male flowers: sepals 6, united into turbinate tube with minute inflexed lobed, petals absent, stamens 3, filaments united into a column, anthers adnate to column. Female flowers: sepals 6, leathery, united into a shortly 6-lobed cup, accrescent in fruit, petals absent; ovary 3-celled, styles 3, united into a short column, branches spreading, 2-lobed. Fruit fleshy, indehiscent or 6-valved, borne on persistent perianth; seeds 4−6.

84. EUPHORBIACEAE

1. B. retusa (Dennstedt) Alston; *B. patens* (Roxb.) Bentham. Fig. 48 a–d
Shrub 0.5–3m, glabrous, branchlets narrowly winged. Leaves elliptic, 1.5–3×0.7–1.7cm, obtuse or subacute, base rounded, pale beneath; petioles 1–2mm; stipules triangular c1.5mm. Male flowers 1–3 per axil, pedicels 2–6mm, slender, deflexed; calyx tube narrowly funnel-shaped 2.5–3mm, with crenulate rim, lobes minute, rounded, borne within tube below rim; staminal column clavate. Female flowers solitary on short straight pedicel 1–4mm, calyx tube broadly funnel-shaped 2–3mm, with conspicuous reniform, subacute, spreading lobes; ovary 3-lobed with short columnar style and 3 bilobed branches. Fruit 8–10mm diameter with traingular apical pit, subtended by enlarged persistent calyx.

Bhutan: S—Gaylegphug district (between Gaylegphug and Tori Bari) and Deothang district (N of Deothang), **C**—Tongsa district (Shamgong) and Mongar district (Saleng); **Sikkim:** duars. Subtropical and Warm broad-leaved forests, 130–1800m. May–June.

Vegetatively very similar to *Sauropus quadrangularis* which differs in its disc-like (not turbinate) male calyx, and free styles borne on the outer margin of the ovary. Sap used medicinally in treatment of eye diseases (146).

8. SAUROPUS Blume

Evergreen shrubs. Leaves alternate, pinnately veined or 3-veined from base, entire, minutely stipulate. Monoecious; flowers solitary or in axillary clusters or short racemes. Male flowers: sepals 6, connate into a flattened disc with infolded lobes, petals absent, stamens 3, filaments united into a short column bearing sessile anthers. Female flowers: sepals 6, united only at base, often accrescent in fruit, ovary 3-celled, styles 3, bifid, incurved, borne on outer margin of obconic ovary. Fruit a fleshy or leathery capsule borne on persistent perianth, seeds 6.

1. Leaves 1.5–2.5cm .... **1. S. quadrangularis**
+ Leaves 3.5–16cm 2

2. Leaves 3-veined at base .... **5. S. repandus**
+ Leaves pinnately veined 3

3. Female flowers and fruit on long pedicels 2–6cm; leaves 10–20cm
   4. S. macranthus
+ Female flowers and fruit on short pedicels 0.8–1.3cm; leaves 3.5–9cm
   4

4. Leaves ovate to ovate-oblong, acute or obtuse; female calyx lobes obovate, obtuse; fruit sessile .... **2. S. androgynus**

8. SAUROPUS

+ Leaves ovate-lanceolate, acuminate; female calyx lobes linear-oblong; fruit shortly stalked .... **3. S. stipitatus**

1. S. quadrangularis (Willdenow) Mueller; *S. compressus* Mueller, *S. pubescens* Hook. f.

Shrub 1−2m, branchlets compressed and narrowly winged, glabrous or pubescent. Leaves elliptic, sometimes becoming ovate or obovate, 1.3−2.5×0.8−1.7cm, obtuse or subacute, base rounded, sometimes with pale median stripe, glabrous or pubescent beneath; petioles 1−2mm; stipules 1.5mm. Flowers 1−2(−4) per axil, on slender pedicels 2−5mm, males towards base of branches. Male calyx disc-like, c 3mm diameter, with emarginate lobes. Female calyx 4mm diameter, deeply divided into obovate subacute lobes. Fruit globose, 7−8mm, subtended by enlarged calyx 7−8mm.

Bhutan: locality unknown; **Sikkim:** Darjeeling foothills etc. Subtropical and Warm broad-leaved forests, 350−1800m.

Two varieties occur, the commoner, var. **compressus** (Mueller) Airy Shaw (*S. compressus* Mueller) glabrous throughout, and the more local var. **puberulus** Kurz (*S. pubescens* Hook. f.) differing in its pubescent shoots and leaf undersides, recorded from the Tista valley and Siliguri. Similar vegetatively to *Breynia retusa*; for differences see under that species.

2. S. androgynus (L.) Merrill; *S. albicans* Blume

Similar to *S. quadrangularis* but always glabrous, leaves ovate or ovate-oblong, 3.5−6×1.5−3cm; calyx c 4mm diameter almost unlobed; female calyx 5mm diameter; fruit 10−15mm diameter subtended by only slightly enlarged calyx 6−7mm diameter.

Bhutan: S—Gaylegphug district (Gaylegphug, 117); **Sikkim:** Darjeeling and foothills. Subtropical forests, 270−600m. June−August.

3. S. stipitatus Hook. f.

Similar to *S. quadrangularis* and *S. androgynus* but leaves ovate-lanceolate, 4−9×1.2−2.6cm, acuminate, base acute; female calyx 5mm diameter with linear-oblong lobes; fruit 10−12mm diameter, shortly (2−3mm) stalked, dehiscing stellately, borne on pedicel 11−13mm.

Sikkim: Darjeeling. Subtropical forests. May.

4. S. macranthus Hasskarl; *S. macrophyllus* Hook. f.

Glabrous shrub 2−3.5m. Leaves ovate-lanceolate or elliptic, 10−20×4−6.5cm, acuminate, base rounded; petioles 4−6mm; stipules subulate 5−7mm. Flowers 1−4 per axil, females uppermost. Male calyx 3−3.5mm diameter, deeply lobed, on filiform pedicel 4−7mm. Female flowers much larger, calyx 8−10mm diameter, deeply divided into obovate

lobes, on stout pedicels 2−6cm. Fruit depressed globose, c 1×2cm, subtended by calyx 12−14mm diameter.
Sikkim: Jalpaiguri district, Madarihat. Subtropical terai forests, 200m. January−April.

5. S. repandus Mueller; *S. trinervius* sensu F.B.I. p.p. non Mueller
Shrub 1−2m, glabrous. Leaves ovate-lanceolate, 6−11×2−3cm, acuminate, base cuneate, pale beneath, strongly 3-veined from above base; petioles c 2mm; stipules 3−4mm, subulate. Flowers in short bracteate racemes 3−6mm, on pedicels 5−10mm. Male flowers c 3mm diameter, calyx disc-like, unlobed. Female flowers c 8mm diameter, calyx divided to base into 3 long and 3 short ovate lobes. Fruit 2.5cm diameter, dehiscing stellately, borne on elongated and thickened pedicel c 2cm.
Bhutan: S—Samchi district (Chamorchi); **Sikkim:** Tista Valley, Mongpo and Namring. Subtropical forests, 220−1200m. May−August.
Records of *S. trinervius* Mueller (a Khasian species) from Sikkim (34,80) refer to this species; *S. trinervius* differs in its rounded leaf base and stellately-lobed male calyx.

9. DRYPETES Vahl

Evergreen trees. Leaves alternate, pinnately veined, base often oblique, entire or minutely serrulate, stipules minute. Dioecious; flowers in axillary fascicles or racemes, or solitary. Male flowers: sepals 4, concave, petals absent, stamens 3−12, free, dehiscing by vertical slits, inserted around disc. Female flowers: sepals 4, disc present or absent, ovary 2−3-celled, styles 2−3 short or absent, stigmas fleshy. Fruit an indehiscent, leathery, 2-seeded drupe.

1. Leaves thinly coriaceous, caudate-acuminate; stamens 4−8; female pedicels 2.5−3.5cm, stigmas borne on short styles .... **3. D. indica**
+ Leaves thickly coriaceous, shortly acuminate; stamens 3−4 or 8−12; female pedicels 0.1−0.6cm, stigmas sessile 2

2. Male flowers c10mm diameter, on pedicels c5mm, stamens 8−12; female pedicels 4−6mm .... **1. D. assamica**
+ Male flowers c4mm diameter, on pedicels 1−2mm, stamens 3−4; female pedicels 1−2.5mm .... **2. D. subsessilis**

1. D. assamica (Hook. f.) Pax & Hoffmann; *Cyclostemon assamicus* Hook.f.
Tree 10−15m. Leaves coriaceous, elliptic, 8−18×3−6cm, shortly acuminate, base oblique, rounded or broadly cuneate, entire, lateral veins spreading; petioles 5−10mm. Male flowers in axillary fascicles, c10mm

9. DRYPETES

diameter, on pedicels c 5mm; stamens 8—12. Female flowers solitary axillary, on pedicels 4—6mm, sepals c 5mm, ciliate, stigmas sessile. Drupes ellipsoid, c 2×1.5cm, shallowly furrowed, densely brown appressed pubescent.

Bhutan: S—Samchi district (Khagra Valley above Gokti) and Gaylegphug district (Tatapani); **Sikkim:** Rongbe, Sivoke etc. Subtropical forests, 500—600m. December—January.

2. D. subsessilis (Kurz) Pax & Hoffmann; *Cyclostemon subsessilis* Kurz
Similar to *D. assamica* but leaves oblong-elliptic 15—20×3.5—8cm, minutely serrulate near apex, lateral vein asending; male flowers c 4mm diameter, on very short pedicels 1—2mm, stamens 3—4; female flowers 1—2 per axil, on pedicels 1—2.5mm; fruit subsessile.

Sikkim: Rongbe and Balasun. Subtropical forests, 600m. April—May.

3. D. indica (Mueller) Pax & Hoffmann; *D. griffithii* (Hook. f.) Pax & Hoffmann, *D. lancifolia* (Hook. f.) Pax & Hoffmann, *Cyclostemon indicus* Mueller, *C. griffithii* Hook. f., *C. lancifolius* Hook. f. Nep: *Hare* (34)

Similar to *D. assamica* and *D. subsessilis* but leaves thinly coriaceous, ovate-lanceolate, 8—12×2.5—4cm, abruptly caudate-acuminate, petioles 2—4mm; flowers solitary or few in short racemes, males c 2.5mm diameter, stamens 4—8; females on long pedicels 2.5—3.5cm, stigmas borne on short styles; fruit ellipsoid, c 2.5cm.

Bhutan: C—Tongsa district (Dakpai near Shamgong); **Sikkim:** Lopchu, Kalimpong. Warm broad-leaved forests, 160—1650m. April—May.

10. APOROSA Blume

Evergreen trees. Leaves alternate, pinnately veined, entire or shallowly toothed; petiole 2-glandular at apex; stipules deciduous. Dioecious; male flowers sessile in clustered axillary bracteate catkin-like spikes, female flowers sessile in very short bracteate spikes. Male flowers clustered in axil of bract, sepals 4, free, stamens 2, free; pistillode minute. Female flowers: sepals 4, free, ovary ellipsoid, 2-celled, styles 2, short, 2-fid, stigmas papillose. Fruit a 1—2-seeded drupe.

1. A. octandra (D. Don) Vickery; *A. dioica* (Roxb.) Mueller, *A. roxburghii* Baillon. Nep: *Asare, Hare Kusum, Barikaunli* (34), *Chipli Khari* (34). Fig. 48 e—g

Tree 10—12m. Leaves thinly coriaceous, ovate to elliptic, 10—17×4—6.5cm, shortly acuminate or apiculate, base rounded or cuneate, with 2 glands at junction with petiole, entire or remotely denticulate, glabrous; petioles 1.3—2cm; stipules ovate, 6—7mm, deciduous. Male spikes 1–4 per axil, 1–3×0.3cm, yellow; bracts broad, rounded, ciliate, 2mm long; sepals thin, oblong, 1.4mm. Female spikes 5–10mm, sepals thick, triangular, 2mm; ovary 2.5–3mm, pubescent. Fruit ellipsoid, 10–12mm.

Bhutan: S—Sarbhang district (Singi Khola and near Phipsoo); **Sikkim:** foothills and terai, common. Subtropical forests, often on river banks, 300–600m. March–April.

11. ANTIDESMA L.

Evergreen or deciduous trees or shrubs. Leaves alternate, pinnately veined, entire, stipulate. Dioecious; flowers in slender, often branched racemes or spikes. Male flowers: calyx cup-shaped, 3–5-lobed, petals absent, disc present; stamens 2–5, free, anthers with thickened connective bearing 2 erect, rounded anther cells; pistillode usually present. Female flowers: calyx as in males, ovary ovoid, 1-celled, bearing 2 bifid stigmas. Fruit a compressed ellipsoid drupe with persistent terminal or subterminal stigmas.

1. Leaves 4–10cm, with 5–6 pairs of veins 2
+ Leaves 10–23cm, with 7–11 pairs of veins 3

2. Leaves obovate or oblanceolate, shortly acuminate, base cuneate; male flowers on pedicels 0.8–1.2mm, stamens 2(–3) .... **1. A. acidum**
+ Leaves broadly oblong, obtuse, base rounded or often subcordate; male flowers sessile, stamens 4–5 .... **2. A. ghaesembilla**

3. Leaves coriaceous, elliptic-obovate, acute or shortly acuminate; stipules early deciduous; male flowers sessile .... **5. A. bunius**
+ Leaves membranous or thinly coriaceous, ovate to elliptic, or oblong-lanceolate, acuminate; stipules mostly persistent; male flowers on pedicels 0.8–1.5mm 4

4. Leaves ovate-lanceolate to elliptic; branchlets glabrous or pubescent; stipules lanceolate 6–15mm; male pedicels 1.2–1.5mm; fruit with terminal style ....**3. A. acuminatum**
+ Leaves oblong-lanceolate; branchlets tomentose; stipules subulate, 3–5mm; male pedicels 0.8–1mm; fruit with subterminal style
4. A. nigricans

1. A. acidum Retzius; *A. diandrum* (Roxb.) Roth, *A. lanceolarium* (Roxb.) Wight. Nep: *Archal* (34). Fig. 48 h–j

Deciduous shrub 2–3m or tree up to 10m, branchlets usually glabrous.

Leaves membranous, obovate or oblanceolate, 5−10×2−3.5cm, shortly acuminate or acute, base cuneate, glabrous or sparsely pubescent on veins, lateral veins 5−6 pairs; petioles 3−4mm; stipules lanceolate, 4−6mm, deciduous. Spikes slender, usually glabrous, simple or little-branched, 2−4cm. Male flowers: pedicels 0.8−1.2mm, calyx cup-shaped, 1mm diameter, shortly 4-lobed, stamens 2(−3). Female flowers: pedicels 1−1.2mm, ovary 1.4mm. Fruit 5−6mm, style terminal.

Bhutan: S—Phuntsholing district (cultivated at Phuntsholing), and Deothang district (near Samdrup Jongkhar), C—Tongsa district (Birti) and Mongar district (Ngasamp); **Sikkim:** Rungit valley, Darjeeling etc. Subtropical and Warm broad-leaved forests, 270−1520m. April−June.

Foliage edible (34), and turns red before falling (16).

2. A. ghaesembilla Gaertner. Nep: *Chipli* (34)

Similar to *A. acidum* but branchlets brownish tomentose; leaves broadly oblong, 4−10×2−6cm, obtuse, base rounded or subcordate, pubescent or tomentose beneath; spikes much branched, 3−7cm, tomentose; male flowers sessile, calyx deeply 5-lobed, stamens 4−5; female flowers on short pedicels 0.8mm; fruit c 4mm.

Bhutan: locally unknown (80); **Sikkim:** Darjeeling foothills. Subtropical forests. April−May.

Fruit edible and the red wood sometimes used (16).

3. A. acuminatum Wight. Nep: *Kalo Bilaune* (34)

Shrub 2−3m or small tree to 10m, branchlets glabrous or pubescent. Leaves membranous, ovate-lanceolate to elliptic, 10−23×4−10cm, acuminate, base cuneate, sparsely pubescent on midrib, lateral veins 8−11 pairs; petioles 4−9mm; stipules lanceolate 6−15mm, pubescent, persistent. Racemes slender, 2−4-branched, 4−11cm, glabrous or pubescent. Male flowers: pedicels 1.2−1.5mm, calyx cup-shaped, c 1.2mm across, 4-lobed, stamens 3−4. Female flowers: pedicels 1−1.6mm, ovary 2−2.4mm. Fruit 4−5mm, style terminal.

Bhutan: S—Sarbhang district (above Sarbhang) and Gaylegphug district (Thewar Khola), C—Tongsa district (Birti, 117); **Sikkim:** foothills. Subtropical forests, 300−1000m. May−June.

Fruit edible (34).

4. A. bunius (L.) Sprengel. Nep: *Himalchari* (34)

Closely allied to *A. acuminatum* but leaves more coriaceous, elliptic-obovate, 10−17×3−6cm, acute or shortly acuminate; lateral veins 7−9 pairs; stipules early deciduous; racemes usually tomentose; male flowers sessile, calyx very shortly lobed, stamens 3; female flowers shortly pedicelled.

84. EUPHORBIACEAE

W. Bengal Duars: Buxa; **Sikkim:** terai and foothills. Subtropical forest, 600–1100m. April–May.
Leaves and fruit edible (34).

5. A. nigricans Tulasne

Similar to *A. acuminatum* and *A. bunius* but shoots tomentose, leaves oblong-lanceolate, 11–18×3–5.5cm, sharply acuminate, lateral veins 8–11 pairs; stipules subulate, 3–5mm, persistent; racemes short, 2–5cm tomentose; male pedicels 0.8–1mm; stamens 4; female pedicels 1–2mm; fruit 7mm, oblique, with subterminal style.

Sikkim: Darjeeling. Subtropical forests, 300m. May.

12. BACCAUREA Loureiro

Evergreen trees. Leaves alternate, pinnately veined; stipules caducous. Dioecious; flowers in simple or branched bracteate racemes borne in clusters on trunk and main branches. Male flowers: sepals 4–5, united at base, petals absent; stamens 5, free, borne around obconic pistillode. Female flowers: calyx as in males but larger, ovary subglobose, 2–3-celled, stigmas 2–3, minute, bilobed, sessile. Fruit leathery, indehiscent, seeds 1–4, red surrounded by white fleshy aril.

1. B. ramiflora Loureiro; *B. sapida* (Roxb.) Mueller. Sha: *Gotham Paisay*; Nep: *Kusum* (34), *Litku, Zat Kusum*. Fig. 48 k–m

Tree 5–15m. Leaves membranous, obovate, 12–23×6–10cm, shortly acuminate, base cuneate, margins entire or sinuate, almost glabrous; petioles 1.5–5cm; stipules lanceolate c8mm. Male racemes 3–10cm, brownish pubescent, bracts ovate 2–4mm, pedicels 1mm, sepals ovate c 2×1.5mm. Females racemes up to 15cm, bracts 3–5mm, pedicels 2mm, sepals oblong 5–7mm, ovary tomentose. Fruiting racemes up to 20cm, fruit globose, 2cm, yellow, on pedicels 8–10mm.

FIG. 48. **Euphorbiaceae.** a–d, *Breynia retusa:* a, portion of male flowering shoot; b, male flower dissected; c, female flower dissected; d, fruit. e–g, *Aporosa octandra:* e, portion of male flowering shoot; f, portion of female flowering shoot; g, female flower. h–j, *Antidesma acidum:* h, portion of male flowering shoot; i, male flower; j, female flower. k–m, *Baccaurea ramiflora:* k, portion of fruiting shoot; 1, fruit; m, transverse section of fruit. n–r, *Croton caudatus:* n, portion of flowering shoot; o, male flower; p, female flower; q, young fruit; r, papilla from surface of fruit. s–u, *Jatropha curcas:* s, portion of flowering shoot; t, male flower dissected, a sepal and two petals removed; u, female flower. Scale: k × ⅕; n, s × ½; e, f × ⅔; 1, m × 3¼; a, h, q × 1; d × 2⅓; t, u × 3; c, o, p × 4; b, g × 6; j × 8; i, r × 10.

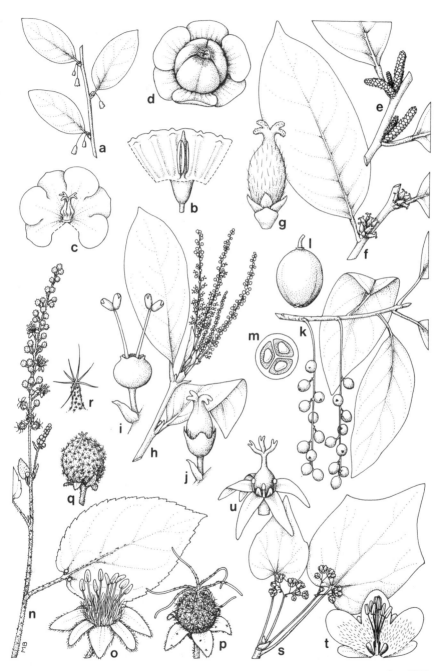

84. EUPHORBIACEAE

Bhutan: S—Samchi, Sarbhang, Gaylegphug and Deothang districts; **Sikkim:** foothills. Subtropical forests, especially on river banks, 300–700m. April–May.

Fruits edible; sometimes cultivated (e.g. at Deothang); bark used as a mordant in dyeing (34).

13. JATROPHA L.

Shrubs, often succulent. Leaves alternate, often palmately lobed, palmately veined at base, peltate or not, stipules pectinate or absent. Monoecious; flowers in terminal cymes with females towards centre. Male flowers: sepals 5, free, petals 5 united in lower half, stamens 6–10, all or inner whorl united at base, disc-glands present. Female flowers similar to males but petals sometimes absent, ovary ellipsoid, 2–4-celled, styles 3, bifid; disc-glands present. Fruit a 2–4-celled capsule, 2–4-seeded.

1. Leaves not peltate; flowers greenish .... 1. **J. curcas**
+ Leaves peltate; flowers red .... 2. **J. podagrica**

1. J. curcas L. Sha: *Nera Khar Shing*; Eng: *Poison nut, Physic nut, Purging nut* (80). Fig. 48 s–u

Shrub 1.5–4m with soapy latex. Leaves ovate or suborbicular 6–17×5–15cm, almost unlobed to deeply 3–5-lobed, apex acute, base broadly cordate, subglabrous, palmately 7-veined at base; stipules absent; petioles 3–11cm. Cymes 4–7cm across, on peduncles 4–6cm. Male sepals elliptic, 4–5mm, green; petals 6–9mm, yellowish-green, hairy within, stamens 10, inner 5 united at base. Female flowers without petals, ovary ellipsoid, 3mm. Capsules 2.5–3×2–2.3cm, 3-lobed.

Bhutan: S—Phuntsholing district (Phuntsholing), Gaylegphug district (Gaylegphug), C—Punakha district (Punakha Dzong) and Tashigang district (S of Tashigang); **Sikkim:** Tista valley. On roadsides, riverbanks and near habitation, cultivated and naturalised, 300–1400m. May–August.

Cultivated as a hedge-plant, sap used by children for bubble-blowing; seeds contain a violent purgative, and an oil used in lamps (146). Sometimes used a food plant for silkworms (48). Probably native to tropical America.

2. J. podagrica Hooker

Differs from *J. curcas* in its very stout swollen stems; leaves deeply 5-lobed, up to 25×20cm, broadly peltate; stipules pectinate; flowers bright red, sepals 2mm, petals 5.5mm, stamens 6–8, filaments united at base; capsules 1.5×1.3cm.

Bhutan: S—Phuntsholing district (Phuntsholing). Cultivated in garden, 300m. June.

Native of tropical America, cultivated as an ornamental.

14. VERNICIA Loureiro

Trees. Leaves alternate, unlobed or 3-lobed, palmately veined at base; petiole 2-glandular at apex; stipules deciduous. Dioecious; flowers in showy terminal panicles. Male flowers: calyx 2-lobed, becoming free, petals 5, spathulate, stamens 8−9, filaments united at base, inner 3−4 longer, disc-glands 5, free. Female flowers: calyx and corolla as in males, ovary ellipsoid, 3-locular; styles 3, free, bifid. Fruit indehiscent fleshy, globose, 3-seeded.

1. V. cordata (Thunberg) Airy Shaw; *Aleurites cordata* (Thunberg) Steudel
Subglabrous tree. Leaves broadly ovate, 14−18×9−15cm, acute, base cordate, entire or palmately 3−5-lobed with round glands at sinuses between lobes, palmately 5-veined, almost glabrous; petioles 7−10cm with 2 glands at apex. Panicles large, 15−20cm across. Calyx c 10mm, petals c 1.7cm, white. Fruit c 4×3cm.
Sikkim: Singtam (109). Cultivated. April.
Native of Japan. Cultivated for its useful seed-oil (*Tung Oil*).

15. CROTON L.

Evergreen or deciduous trees, shrubs, climbers or subshrubs, indumentum stellate. Leaves alternate, opposite or whorled near branch ends, pinnately veined or palmately veined, margins often serrate, with 2 stalked or sessile glands at top of petiole and sometimes marginal glands; stipules minute or absent. Monoecious, flowers solitary or clustered in terminal or axillary, simple or branched racemes. Male flowers: calyx 5-lobed, petals 5−6, stamens 10−12, free, receptacle hairy, disc-glands free. Female flowers: sepals as in males, petals minute or absent, ovary 3-celled, styles long, bifid almost to base. Fruit a 3-lobed, 3-seeded capsule.

1. Subshrub; leaves lanceolate, 2.5−4.5×0.7−2cm ..... **6. C. bonplandianus**
+ Trees or large climbers; leaves ovate, elliptic or ovate-lanceolate, 9−25×3−10cm 2

2. Leaves pinnately veined to base 3
+ Leaves palmately 3- or 5-veined at base 4

3. Leaves ovate or ovate-elliptic, obtusely pointed, margins coarsely crenate-serrate ..... **1. C. roxburghii**
+ Leaves elliptic or lanceolate-elliptic, sharply acuminate, shallowly serrulate or subentire ..... **2. C. joufra**

84. EUPHORBIACEAE

4. Branchlets, petioles and racemes stellate-tomentose; leaf margins with stalked glands between teeth, base cordate or subcordate . **5. C. caudatus**
+ Branchlets, petioles and racemes glabrous or sparsely stellate pubescent; leaf margins without stalked glands, base cuneate or rounded 5

5. Leaves strongly 5-veined at base, stellate-pubescent; glands shortly stalked, borne on apex of petiole .... **3. C. himalaicus**
+ Leaves strongly 3-veined at base (sometimes with 2 additional weaker veins), glabrous or sparsely stellate; glands sessile, borne near base of lamina margin ....**4. C. tiglium**

1. C. roxburghii Balakrishnan; *C. oblongifolius* Roxb. *nom. illeg.*

Deciduous shrub or tree to 8m; young shoots densely stellate-scaly. Leaves coriaceous, alternate below, sometimes opposite or whorled above, ovate or ovate-elliptic, 10−25×4−10cm, obtuse or bluntly pointed, base rounded or broadly cuneate, pinnately veined to base, margins crenate-serrate, sparsely stellate-scaly beneath; petioles 1−3.5cm, scaly; stipules subulate, 2.5mm. Racemes 8−20cm, terminal and in upper leaf axils, stellate-scaly; flowers creamy, females towards base. Male flowers: pedicels slender, 2−6mm, sepals triangular 2.5mm; petals spathulate, equalling sepals, hairy within. Female flowers: pedicels stout, 3mm, calyx as in males, petals smaller, linear, ovary subglobose 3.5mm, styles reflexed. Capsules subglobose, c 11mm, on pedicels 5−7mm.

Bhutan: S—Sankosh district (W of Pinkhua); **Sikkim:** Darjeeling district (34). Subtropical forest, 320−600m. March.

2. C. joufra Roxb.

Similar to *C. roxburghii* but leaves elliptic or elliptic-lanceolate, 10−24×3−7.5cm, sharply acuminate, base cuneate, margins shallowly serrulate or subentire; petioles 2−6cm; racemes 7−15cm, terminal, often branched; male flowers on short pedicels 1mm, petals ovate, longer than sepals; female flowers with petals absent; fruit (80) ovoid, 2.5−3cm.

Bhutan: S—Sarbhang district (Singi Khola), **C**—Tongsa district (below Tama, Mangde Chu). Subtropical and Warm broad-leaved forests, 400−1060m. March.

Some records of *C. oblongifolius* Roxb. from Sikkim and Bhutan (34,117) may belong to this species.

3. C. himalaicus Long; *C. tiglium* auct. p.p. non L. Nep: *Lapche Bis* (34)

Small tree to 6m, shoots sparsely stellate-hairy. Leaves membranous, broadly ovate-elliptic, 7−15×3.5−7cm, finely acuminate; base broadly cuneate or somewhat rounded, margins serrulate, prominently 5-veined at base, stellate pubescent especially beneath; petioles 2.5−5cm, with two shortly stalked glands at apex; stipules 2.5−4mm, caducous. Monoecious;

racemes 10−20cm, terminal, male flowers uppermost. Male flowers: pedicels slender, 2.5−6.5mm, thinly stellate-hairy; sepals ovate 3.3mm, with tuft of hairs at apex; petals oblong, 3mm, stamens 12−18. Female flowers: pedicels stout, 4mm, densely stellate-tomentose; sepals triangular, 4mm, petals oblong 2.5mm; ovary 3−4-celled, densely stellate, styles spreading, 5mm, united in lower ⅙. Fruit globose, c 2cm diameter, thinly stellate-pubescent.

Sikkim: Darjeeling district (Pedong, Poosepong, Kalimpong and Paungaon Forest). Warm broad-leaved forests, 610−1640m. May−June.

A poorly known species restricted to the Darjeeling district, E Nepal and Assam, in the past confused with *C. tiglium*. Young shoots, bark and leaves used to poison fish (34).

4. C. tiglium L.
Similar to *C. himalaicus* but more glabrous throughout; leaves ovate with rounded base, strongly 3-veined (sometimes with 2 additional weaker ones), glabrous or sparsely stellate-pubescent, leaf glands sessile, borne near base of lamina margin; racemes 6−11cm, male pedicels 3.5−5mm, glabrous, female flowers with petals absent or reduced to minute glands; ovary 3-celled; fruit oblong, 1.5−2×1−1.5cm.

Bhutan: locality unknown; **Sikkim:** Darjeeling foothills, cultivated or naturalised (34, 69, 112).

Some or all of the Sikkim records may refer to *C. himalaicus*. Seeds contain an oil used medicinally as a powerful purgative; and powdered seeds thrown into water to stupefy fish (16).

5. C. caudatus Geiseler; *C. malvifolius* Griffith. Sha: *Phikhiroo*; Nep: *Halonre* (34), *Suparey* (34). Fig. 48 n−r.
Similar to *C. himalaicus* and *C. tiglium* but shoots, petioles and racemes stellate-tomentose; leaves broadly ovate, 9−14×3−11cm, base cordate or subcordate, densely stellate beneath especially on veins, 5-veined from base, margin irregularly serrate-dentate, with stalked glands between teeth; petioles 2−5cm, with two long-stalked glands at apex; stipules pinnatifid, c1cm; racemes 12−23cm, sepals stellate-tomentose; capsules subglobose or ovoid 2−2.5×1.5−1.8cm, papillate and stellate-tomentose.

Bhutan: S—Phuntsholing, Chukka and Gaylegphug districts, common, C—Tongsa district (Dakpai); **Sikkim:** foothills. Subtropical and Warm broad-leaved forests, often in secondary scrub and forest margins, 200−1460m. April−May.

Seeds used as a substitute for Betel-Nut (146).

6. C. bonplandianus Baillon; *C. sparsiflorus* Morong. Sha: *Seytsala Ngyon*
Subshrub 30−60cm, sparsely whitish stellate-scaly throughout. Leaves lanceolate, 2.5−4.5×0.7−2cm, tapering to acute or truncate apex, base

rounded, margins serrate, pinnately veined but somewhat 3-veined at base; petioles 2–5mm; with 2 sessile glands at apex; stipules subulate, caducous. Racemes terminal, 5–11cm, males towards apex. Male flowers: pedicels 2mm, sepals ovate, 1mm, petals oblong, whitish, stamens 13–15. Female flowers: pedicels stout, 0.8mm, sepals triangular, 1.4mm, petals absent, ovary densely stellate, styles 3, spreading, united in lower ⅓. Capsules oblong, 6×4mm, thinly stellate.

Bhutan: C—Tashigang district (Cha Zam, Dangme Chu). Weed of roadsides and waste ground, 1000m. June.

Native of tropical America, introduced into Asia.

16. CODIAEUM Jussieu

Evergreen shrubs. Leaves alternate, sometimes shallowly lobed, pinnately veined; stipules minute or absent. Monoecious; flowers in unisexual axillary racemes, males clustered, females solitary. Male flowers: sepals usually 5, free, petals 5, minute, alternating with disc-glands, stamens 15–30, free. Female flowers: sepals as in males, petals absent, ovary 3-celled, styles simple, recurved. Fruit a globose 3-coccous capsule; seeds glossy.

1. C. variegatum (L.) Blume; *C. pictum* (Loddiges) Hooker

Shrub 1–2m, glabrous. Leaves variable in shape, linear, oblong or obovate, 10–35×5–10cm, acute, base cuneate, sometimes with rounded lobes in upper part, often variegated yellow, red and green; petioles 1–3.5cm; stipules absent. Male racemes 15–25cm, pedicels slender, 3–6mm, sepals c 2mm. Female flowers: pedicels 4–7mm, sepals c 2.4mm. Capsules c 8mm.

Bhutan: S—Gaylegphug. Cultivated in garden, 300m. April–July.

Native of Malaysian archipelago cultivated for its showy foliage; the Bhutan plant belongs to the broad, unlobed-leaved var. **variegatum** forma **platyphyllum** Pax.

17. MANIHOT Miller

Glabrous shrubs with large tuberous roots. Leaves alternate, deeply palmately 3–7-lobed; palmately veined; stipules narrowly lanceolate. Monoecious; flowers in axillary panicles, with females towards base. Male flowers: calyx showy, campanulate, deeply 5-lobed, corolla absent, stamens 10, free, inserted between lobes of disc. Female flowers: sepals free, ovary borne on conspicuous disc, 3-celled, style short, stout, 3-branched with swollen, papillate stigmas. Fruit a subglobose, warted, 6-winged, 3-seeded capsule.

17. MANIHOT

1. M. esculenta Crantz; *M. utilissima* Pohl. Kengkha: *Seng Ki;* Nep: *Dori, Simal Tarul* (34); Eng: *Cassava, Tapioca.* Fig. 49 1-o
Shrub 2−3m, branching at apex of main stem; root tubers cylindric, up to 50cm. Leaf lobes elliptic-oblanceolate, 7−16×1.5−4cm, acuminate, pale beneath, united in basal 6−10mm; petioles 8−15cm, swollen at base; stipules 10−13mm. Panicles 4−9cm. Male sepals 12−14mm, yellow green. Female sepals 7−9mm. Fruit 1.5−2cm.
Bhutan: C—Tongsa district (Tama); **Sikkim:** commonly cultivated in terai (34).

Native of S America; two forms are cultivated for their edible roots, the *Sweet Cassava* and the *Bitter Cassava*; the latter is poisonous when uncooked and must be boiled or baked thoroughly before eating. The roots are rich in starch but poor in protein (157).

18. OSTODES Blume

Shrubs or trees. Leaves alternate, pinnately veined but 3-veined at base, 2-glandular at apex of petiole; stipules caducous. Dioecious; flowers in lax panicles (males) or racemes (females) in axils of fallen leaves on older shoots. Male flowers: sepals 3, unequal, petals 5−6, stamens 30−35, free, borne around disc. Female flowers: sepals 3, unequal, petals 8−10, ovary globose, 3-celled, styles 3, stout, bifid at apex. Capsules subglobose, 6-ribbed, 3-seeded.

1. O. paniculata Blume. Nep: *Bepari* (34). Fig. 49 p−r
Shrub to 3m or tree to 15m. Leaves thinly coriaceous, ovate 13−30×6.5−18cm, shortly and abruptly acuminate, base rounded, 3-veined, margins bluntly serrate, glabrous except for tufts of hair in vein-axils beneath; petioles 3.5−20cm, with 2 stalked glands at apex; stipules oblong 3mm. Male panicles 18−30cm, pendulous, pedicels 5−15mm, sepals 4−5mm, petals white tinged pink, 6−7mm. Female racemes 5−9cm, sepals 3−5mm, petals 7−9mm, ovary tomentose. Capsules 2−3cm diameter, woody.
Bhutan: S—Gaylegphug district (Tatapani, Lodrai Khola and Taklai Khola), C—Tongsa district (Shamgong) and Mongar district (Zimgang); **Sikkim:** Kalimpong, Kurseong, etc. Subtropical and Warm broad-leaved forests, 340−1900m. March−June.

A gum which is used in paper-making is extracted from the wood (34); leaves used as cattle fodder (146).

19. CLAOXYLON Jussieu

Evergreen shrubs or small trees. Leaves alternate, pinnately veined; stipules minute. Monoecious, flowers in slender axillary racemes. Male flowers: calyx 3—4-lobed, petals absent, stamens many, free, anther-cells distinct. Female flowers: calyx as in males, ovary 3-celled, disc present, styles 3, short, fimbriate, spreading. Capsules 3-lobed, 3-seeded.

1. C. longipetiolatum Kurz

Shrub 3—5m, finely pubescent throughout. Leaves oblong-lanceolate, 16—26×7—9cm, finely acuminate, base rounded, margin serrulate or crenate-serrate; petioles 7—14cm. Racemes 4—7cm. Male flowers: pedicels 4mm, sepals 1.5mm. Female flowers almost sessile, ovary subglobose, 1.2mm, hirsute. Fruit c1cm, deeply 3-lobed.

Sikkim: Jalpaiguri and Pisu Jhora. Terai forests, 200—300m. December—January.

20. MERCURIALIS L.

Perennial rhizomatous herbs. Leaves opposite, pinnately veined; stipules membranous, persistent. Dioecious or occasionally monoecious; flowers in interrupted axillary spikes. Male flowers: sepals 3, stamens 10—15, free, anther cells distinct, globose. Female flowers: sepals 3, disc-glands 2, ovary 2-locular; styles 2, simple, spreading. Capsules strongly bilobed, 2-seeded.

1. M. leiocarpa Siebold & Zuccarini. Fig. 49 a—c

Stems 30—50cm, glabrous. Leaves ovate 5—12×2—4cm, acute or shortly acuminate, base rounded, margins crenate-serrate, glabrous or sparsely pilose; petioles 1.5—4cm; stipules 2—3mm. Male spikes 6—10cm, flowers green, c 1.5mm across. Female spikes 2—4cm, flowers c 3mm across. Capsules 3—4mm, glabrous.

Bhutan: C—Punakha district (SW of Wangdu Phodrang) and Tongsa district (Tunle La near Kinga Rapden). Shady places in Cool broad-leaved forests, 2740m. April—May.

21. ACALYPHA L.

Shrubs or herbs. Leaves alternate, palmately veined at base, stipulate. Monoecious or dioecious; flowers in long slender axillary unisexual spikes or in short dense bisexual spikes with female flowers towards base; female flowers often with conspicuous accrescent bracts. Male flowers: sepals 4, minute, petals absent, stamens 8, free, anther-cells distinct. Female flowers: sepals usually 4, ovary 3—4-celled, styles 3—4, laciniate. Fruit a 3—4-celled capsule, 3—4-seeded.

1. Annual herb; flower spikes up to 1cm .... **1. A. brachystachya**
+ Cultivated shrubs; flower spikes 2.5–20cm 2

2. Leaves variegated with red; female spikes interrupted, usually shorter than leaves, bracts large, toothed .... **2. A. wilkesiana**
+ Leaves green; female spikes dense, much longer than leaves, bracts small, entire .... **3. A. hispida**

1. A. brachystachya Hornemann. Fig. 49 d&e

Weak annual herb 15–50cm, stems pubescent. Leaves ovate, 3–5×1.5–3cm, acute, base rounded or subcordate, margin crenate, thinly appressed pilose, palmately 3- or 5-veined at base; petioles 1–3cm; stipules minute. Flowers in short axillary spikes up to 1cm, with several female flowers at base, and clusters of male flowers above. Male flowers minute, c 0.25mm. Female flowers 1–3 in axil of deeply 3-lobed bract c 2mm, sepals minute, ovary hairy, 3-celled, styles 3. Capsules 1.5mm, 3-lobed subtended by enlarged bract c 5mm.

Bhutan: C—Thimphu district (Paro and Thimphu), Punakha district (Lometsawa) and Mongar district (Kuru Chu). By streams, ditches and on damp walls, 760–2370m. August–October.

2. A. wilkesiana Mueller

Shrub 1–2m. Leaves ovate 10–20×6–14cm, acute or shortly acuminate, base rounded or broadly cuneate, margins coarsely crenate-serrate, pinnately veined but palmately 3–5-veined at base, variegated red and brownish-red, pubescent on veins; petioles 1–1.5cm; stipules lanceolate, 5mm. Flower spikes unisexual, males slender, 2.5–3cm, in lower axils, females stout, interrupted, 4–8cm, in upper axils. Male flowers subtended by minute ovate bracts c 0.8mm, sepals 0.6mm. Female flowers subtended by broad red 5-toothed bracts 3.5mm long, sepals 1.2mm, styles 3–4, red, 3mm, conspicuously exserted. Fruit not seen.

Bhutan: S—Gaylegphug district (Gaylegphug). Cultivated in garden, 300m. May–June.

Native of Polynesia, cultivated for its colourful foliage.

3. A. hispida Burman f.

Similar to *A. wilkesiana* but leaves green, subcordate at base; petioles 1.5–5.5cm; dioecious, only female plants known; female spikes 7–20cm, dense, uninterrupted; bracts minute, entire, hidden by flowers; sepals c 0.4mm, styles 3, 6.5–7.5mm.

Bhutan: S—Samchi district (Samchi). Cultivated in municipal park, 500m. February–March.

Native possibly of Bismarck Archipelago, cultivated for its ornamental female inflorescences.

22. ALCHORNEA Swartz

Trees or shrubs. Leaves alternate, palmately 3–5-veined at base, with sessile glands between veins at base and stipels at top of petiole; stipules caducous. Monoecious or dioecious; flowers in slender simple or branched axillary or terminal spikes. Male flowers: sepals 2–5, petals absent, stamens 6–8, filaments shortly connate at base. Female flowers: sepals 4–8, sometimes glandular, ovary 3-celled, styles 3, free, simple, often elongate. Capsule 3-valved, 3-seeded.

1. Leaves with base rounded or broadly cuneate, pubescent only on veins beneath; capsules warted .... **1. A. tiliifolia**
+ Leaves with base truncate or cordate, softly pubescent beneath; capsules smooth .... **2. A. mollis**

1. A. tiliifolia (Bentham) Mueller. Nep: *Sanu Malata*
Shrub or small tree, branchlets pubescent. Leaves ovate, 10–20×4–12cm, finely acuminate, base rounded or broadly cuneate, margins serrulate, pubescent on veins beneath, 3-veined at base with discoid glands between veins; petioles 5–15cm, bearing 2 subulate stipels 1.5–2mm at apex; stipules subulate 4–6mm. Dioecious?; male spikes 2–4 clustered in axils of fallen leaves, 3–10cm, flowers clustered, minute, subsessile, sepals c1mm. Female racemes terminal, flowers borne singly, sepals 4–5mm. Capsules ellipsoid, 12–16mm, warted.

Bhutan: S—Sarbhang district (above Sarbhang); **Sikkim:** Darjeeling district. Subtropical forest slopes, 600–1200m. May–July.

2. A. mollis (Bentham) Mueller. Fig. 49 f–h
Similar to *A. tiliifolia* but shoots densely pubescent; leaves broadly ovate or suborbicular, base truncate or shallowly cordate, crenate-serrate, softly pubescent beneath, stipels lanceolate 4–6mm; monoecious; male flowers on short pedicels 2–2.5mm; capsules pubescent but not warted.

Bhutan: C—Mongar district (Shongar Chu valley); **Sikkim:** Tista Valley. Warm broad-leaved forests, 1100m. July.

23. TREWIA L.

Deciduous trees, indumentum of simple and fascicled hairs. Leaves opposite, entire, 3–5-veined at base; stipules caducous. Dioecious; flowers precocious, males numerous in elongate racemes, females few in short racemes or solitary. Male flowers: sepals 3–4, petals absent, stamens numerous, free. Female flowers: sepals 3–5, caducous, ovary large,

3−5-celled, styles 3−5, long, papillose, shortly connate at base. Fruit indehiscent, 3−5-celled, fleshy, 3−5-seeded.

1. T. nudiflora L. Nep: *Ramrita, Aule Kapase* (34), *Pitali*

Tree 10−15m, branchlets woolly. Leaves ovate, 11−20×7−12cm, acuminate, base truncate or cordate, softly pubescent beneath; petioles 2−7.5cm; stipules 4mm. Male racemes slender, 12−15cm, flowers greenish, in clusters of 2−3, pedicels 3−5mm, sepals 3.5mm, concave. Female racemes up to 7cm, stout, sepals and ovary c 7mm, woolly, style branches 15mm. Fruit subglobose, greyish-green, 3.5×3cm.

Bhutan: S—Sarbhang district (Dol Khola) and Gaylegphug district (Gaylegphug and Tori Bari); **Sikkim:** Darjeeling foothills. Subtropical forests, often by streams and on disturbed ground, 300−410m. February−March.

Foliage used as cattle fodder (146); leaves very similar to those of *Gmelina arborea*. Wood soft, used for carving, planking, packing cases and plywood (16).

24. MALLOTUS Loureiro

Evergreen or deciduous trees or shrubs, indumentum of stellate and sometimes simple hairs, usually mixed with sessile gland-dots. Leaves alternate or subopposite, sometimes palmately lobed, palmately 3−9-veined and sometimes peltate at base, usually with discoid glands near base; stipules minute. Dioecious; flowers in terminal or axillary, simple or branched racemes or spikes. Male flowers: sepals 3−4, petals absent, stamens 25−50, free. Female flowers: calyx 3−6-lobed, ovary 2−3-celled, styles 2−3, simple, erect-spreading, papillose or laciniate. Capsules 3-celled, smooth or bristly, 3-seeded.

1. Leaves broadly peltate, petiole inserted 10−25mm above base
 6. M. roxburghianus
+ Leaves not or narrowly peltate, petiole inserted ± basally (2−8mm above base in *M. tetracoccus*) 2

2. Leaves 9−25cm broad 3
+ Leaves 3.5−9cm broad 5

3. Spikes branched, forming panicles .... **3. M. tetracoccus**
+ Spikes simple 4

4. Leaves thinly stellate beneath, not obscuring leaf surface (except when young) .... **1. M. nepalensis**
+ Leaves densely stellate-tomentose beneath, completely obscuring leaf surface .... **2. M. oreophilus**

5. Leaves narrowly ovate, mostly 2× as long as broad; styles 3, capsules 3-lobed .... **4. M. philippensis**
+ Leaves broadly ovate-deltoid, mostly as broad as long; styles 2, capsules 2-lobed .... **5. M. repandus**

1. M. nepalensis Mueller; *M. nepalensis* sensu F.B.I. p.p. Nep: *Malata* (34)
Deciduous tree 4–10m. Leaves membranous, alternate or subopposite, broadly ovate, 12–33×9–25cm, acuminate, base truncate or rounded, not peltate, entire or shallowly 3-lobed in upper half, palmately 3- or 5-veined at base, with sessile discoid glands on basal margin near insertion of petiole, and sometimes on lateral veins near leaf-margins; reddish stellate-tomentose above when young, becoming glabrous, yellow gland-dotted and brownish stellate-pubescent beneath but not obscuring leaf-surface; petioles 8–23cm; stipules minute, caducous. Flowers greenish, clustered, in simple stout stellate racemes 15–25cm. Male flowers; pedicels 2–4mm, sepals 3, ovate, 4mm, stamens c 50. Female flowers subsessile, sepals lanceolate, 4mm, styles 3, stout, exserted. Capsules 8–10mm, with long soft spines, densely stellate tomentose.

Bhutan: S—Chukka district, C—Thimphu, Punakha, Tongsa and Tashigang districts, N—Upper Mo Chu district; **Sikkim.** Cool broad-leaved forests, 1900–2500m. May–July.

2. M. oreophilus Mueller; *M. nepalensis* sensu F.B.I. p.p. non Mueller, *M. nepalensis* Mueller var. *ochraceo-albidus* (Mueller) Pax & Hoffmann. Nep: *Malata*

Similar to *M. nepalensis* but leaves narrowly peltate, petiole attached 1–3mm above base; lower surface completely obscured by dense stellate tomentum; male sepals 4.

Sikkim: Darjeeling and Tista Valley. Warm broad-leaved forests 1200–2100m. July.

FIG 49. **Euphorbiaceae.** a–c, *Mercurialis leiocarpa:* a, portion of male flowering shoot; b, male flower; c, female flower. d & e, *Acalypha brachystachya:* d, portion of flowering shoot; e, close-up of inflorescence. f–h, *Alchornea mollis:* f, portion of flowering shoot; g, male flower; h, fruit. i–k, *Mallotus philippensis:* i, portion of flowering shoot; j, female flower; k, fruit. l–o, *Manihot esculenta:* l, portion of flowering shoot; m, vertical section of male flower; n, vertical section of female flower; o, fruit. p–r *Ostodes paniculata:* p, portion of flowering shoot; q, male flower (one sepal and two petals removed); r, ovary. Scale: $1 \times \frac{1}{4}$; f, p $\times \frac{1}{3}$; a, i $\times \frac{1}{2}$; d $\times \frac{3}{4}$; o $\times 1$; h, k $\times 1\frac{1}{2}$; m $\times 2$; q $\times 2\frac{1}{2}$; n $\times 3$; b, e, j, r $\times 4$; c, g $\times 5$.

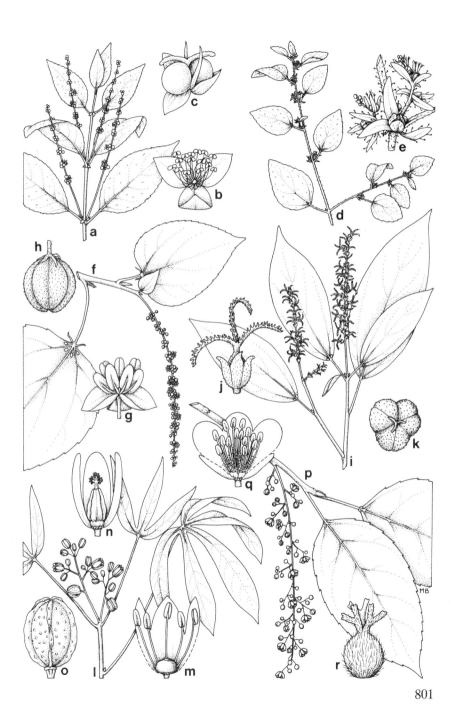

84. EUPHORBIACEAE

3. M. tetracoccus (Roxb.) Kurz; *M. albus* sensu F.B.I. non (Jack) Mueller. Nep: *Jogi Malata* (34), *Kasre Malata*

Similar to *M. nepalensis* and *M. oreophilus* but leaves coriaceous, broadly ovate-deltoid, base peltate, petiole inserted 2−8mm above base, margins often shallowly lobed or dentate, densely brownish-white stellate tomentose beneath, obscuring leaf-surface; flowers in wide branched panicles up to 30×30cm; male flowers smaller, sepals 4, c 2mm, stamens 25−30; female flowers on short pedicels; capsules 1.2−1.5cm, densely tomentose, shortly warted.

Bhutan: S—Deothang district (Samdrup Jongkhar), **C**—Tongsa district (Tama) and Mongar district (Shongar Chu); **W Bengal Duars:** Buxa; **Sikkim:** Darjeeling district. Subtropical forests, 300−1500m. May−July.

4. M. philippensis (Lamarck) Mueller; *M. philippinensis* auct. Dz: *Theshom;* Nep: *Sindure*. Fig. 49 i−k

Small evergreen tree 5−19m. Leaves coriaceous, ovate, 10−20×4−9cm, acuminate, base rounded or cuneate, strongly 3-veined at base, with 2 sessile discoid glands at base of lamina, margins entire, minutely tomentose beneath, mixed with numerous minute red glands; petioles 2−8cm; stipules minute. Spikes slender, erect, 5−14cm, densely stellate-pubescent, males often much-branched at base, forming panicles, females simple. Male flowers clustered, green, subsessile, sepals 3−4, 1.5−2mm, stamens 20−30. Female flowers sessile, not clustered, sepals lanceolate, c 2mm, ovary exserted, styles 3, c 5mm, recurved, papillose, stellate and glandular. Capsules 3-lobed, c 1cm diameter, densely red-glandular.

Bhutan: S—Samchi, Chukka and Gaylegphug districts, **C**—Punakha, Tongsa and Tashigang districts; **Sikkim.** Subtropical and Warm broad-leaved forests and amongst scrub in hot dry valleys, 300−1600m. October−December.

The red glands from the capsules are used to prepare a red dye *(Kamela)* (34).

5. M. repandus (Willdenow) Mueller

Similar to *M. philippensis* but a large climbing shrub; leaves ovate-deltoid, 5−10×3.5−8cm, base truncate, often narrowly peltate, with 6−8 sessile discoid glands near basal margin, stellate-pubescent beneath mixed with yellow gland-dots; male spikes much branched, brown tomentose; female spikes slender, 4−9cm, ovary 2-celled, styles 2, fimbriate; capsules deeply bilobed, c 7mm.

Bhutan: S—locality unknown (80); **Sikkim:** foothills (34). Subtropical forests.

6. M. roxburghianus Mueller. Nep: *Phusre Malata* (34)

Shrub or small tree to 8m. Leaves broadly ovate, 11−21×8−17cm, finely acuminate, base rounded, broadly peltate, petiole inserted 10−25mm above base, margins glandular-toothed, palmately 5−9-veined, with 3−5 sessile discoid glands on veins near base, pubescent with mostly simple hairs above, tomentose with mostly stellate hairs beneath, yellowish gland-dotted on both surfaces; petioles 3.5−10cm; stipules lanceolate, 6−7mm. Flowers in simple racemes, males 10−25cm, females 6−12cm. Male flowers: sepals 4, 2mm, stamens c 35. Female flowers: ovary densely stellate-bristly, styles 3, papillose, recurved. Capsules deeply 3-lobed, 10−12mm broad, stellate-hairy and softly bristly.

Sikkim: Tista and Rungit Valleys etc. Subtropical forests, 300−600m. June−July.

25. CLEIDION Blume

Evergreen shrubs or small trees. Leaves alternate, pinnately veined, with discoid glands near base of lamina and at top of petiole; stipules lanceolate. Dioecious; male flowers in long, simple axillary spikes, females solitary on long axillary pedicels. Male flowers: sepals 3−4, petals absent, stamens numerous, free. Female flowers: sepals 5, ovary 2−3-celled, styles 2−3, bifid with long filiform branches. Fruit a 2−3-celled capsule borne on a long flattened pedicel.

1. C. spiciflorum (Burman f.) Merrill; *C. javanicum* Blume. Nep: *Bepari* (34), *Hare Bepari*

Large climbing shrub or small tree to 12m. Leaves coriaceous, elliptic, 13−25×5−11cm, bluntly apiculate or shortly acuminate, base cuneate, margins serrate, discoid glands 4−6, near base of lamina; petioles 2−7cm with 2 disc-glands at apex; stipules 3mm, caducous. Male spikes 6−25cm; sepals elliptic, 2mm. Female flowers on pedicels 10−14mm, ovary 2mm, styles c 16mm, branches united in lower 4mm. Capsules deeply 2(−3)-lobed, c 1.5×2.5cm, on elongated pedicel 4−5cm.

Bhutan: S—Samchi, Chukka, Sarbhang and Gaylegphug districts; **Sikkim:** Darjeeling, Kalimpong etc. Subtropical forests, 510−910m. February−March.

26. MACARANGA Petit-Thouars

Evergreen trees; indumentum of simple or fascicled (not truly stellate) hairs. Leaves alternate, palmately 5−13-veined and often peltate at base, sometimes pinnately veined to base, minutely gland-dotted beneath, often with discoid glands on veins near base; stipules small or large, often caducous. Dioecious; flowers in axillary racemes or panicles. Male flowers

84. EUPHORBIACEAE

minute, clustered in axil of bract; calyx 2−4-lobed, petals absent, stamens 2−20, free, anthers 3−4-celled. Female flowers few in axil of bract, calyx 2−4-lobed, petals absent, ovary 1−3-celled, styles 1−3, simple. Capsules simple or deeply 2−3-lobed.

1. Leaves broadly peltate or occasionally deeply cordate; petioles inserted 1−6.5cm above base 2
+ Leaves neither peltate nor cordate; petiole inserted at base of lamina ... 4

2. Leaves broadly ovate-deltoid, base truncate, petiole inserted 1−2.5cm above base; basal lateral veins straight or weakly curved
1. M. denticulata
+ Leaves suborbicular, base rounded; petiole inserted 2.5−6.5cm above base; basal lateral veins strongly arched upwards 3

3. Leaves with 1−4 large glands on veins near base; panicles slender, lax, 7−20×2−6cm, caducous bracteoles small, lanceolate, entire, often gland-tipped; capsules 3−4mm .... **2. M. indica**
+ Leaves without basal glands; panicles stout, narrow, racemiform, 3−5×1cm; caducous bracteoles conspicuous, ovate, serrate; capsules 4−6mm .... **3. M. peltata**

4. Leaves broadly ovate, distinctly palmately 5−7-veined at base
4. M. pustulata
+ Leaves oblong-lanceolate, pinnately veined but obscurely 3-veined at base .... **5. M. gamblei**

1. M. denticulata (Blume) Mueller; *M. gummiflua* (Miquel) Mueller. Dz: *Bomchu Shi;* Nep: *Malata* (34). Fig. 50 a−c

Tree 3−10m, shoots striate, tomentose when young. Leaves broadly ovate-deltoid, 12−30×11−24cm, shortly acuminate, base peltate, truncate or cordate, petiole inserted 1−2.5cm above base, margins sinuate-denticulate, palmately 9−11-veined at base, with 5−11 sessile discoid glands on lowermost veins on upper surface, minor veins parallel, prominent beneath; subglabrous, lower surface densely gland-dotted; petioles 8−17cm; stipules lanceolate, c 7mm, caducous. Panicles axillary, 5−10cm. Male flowers minute, sessile, c 1.5mm, sepals 2−3, stamens c 12. Female flowers on short pedicels, calyx c 1mm, ovary deeply bilobed, styles 2−3, recurved. Capsules deeply bilobed, c 4mm, gland-dotted.

Bhutan: S—Phuntsholing, Chukka, Sarbhang and Deothang districts; **Sikkim:** foothills and duars. Secondary subtropical forests, 550−1000m. March−May.

Foliage used as cattle fodder (146). A fast-growing tree rapidly colonising cleared forests.

2. M. indica Wight. Nep: *Malata* (34)

Similar to *M. denticulata* but leaves suborbicular, 12.5−23×12.5−22cm, base rounded, broadly peltate, petiole inserted 2.5−6.5cm above base, 1−4 basal veins often bearing a conspicuous linear discoid gland, margins entire; stipules ovate, c 15mm, recurved; panicles slender, lax, 7−20×2−6cm, with minute caducous, lanceolate, sometimes gland-tipped bracteoles; stamens 3−8; ovary subglobose, 1-celled, with single sub-basal style; capsules 3−4mm, in broad panicles.

Sikkim: Darjeeling foothills and Terai. Subtropical forests, 910−1670m. September−October.

Milky sap used medicinally to help healing of sores (131).

3. M. peltata (Roxb.) Mueller; *M. roxburghii* Wight.

Similar to *M. denticulata* and particularly *M. indica*, differing from the latter in its leaves without glands near base, and flowers in stout racemiform panicles 3−5×1cm, bracteoles conspicuous at first, ovate, serrate, stamens 2−5; capsules 4−6mm.

Sikkim: locality unknown. Subtropical forests, 910m.

4. M. pustulata Hook. f.; *M. gmelinifolia* Hook. f. Nep: *Malata* (34)

Large shrub or tree to 12m. Leaves broadly ovate, 10−17×10−17cm, shortly acuminate, base truncate or weakly cordate, margins shallowly sinuate-dentate, palmately 5−7-veined at base, with 2 conspicuous discoid glands on base of lamina, glabrous but densely gland-dotted beneath; petioles 5−15cm; stipules ovate, 5−7mm, recurved. Male panicles 4−5cm, flowers c 1mm diameter, stamens c 20. Females panicles 3−8cm; flowers on short thick pedicels; calyx c 1.5mm, 4-lobed; ovary 2-lobed, with bands of gland dots and brown tomentum, styles 2, recurved. Capsules 2-lobed, c 5×9mm, brown tomentose, styles persistent.

Bhutan: S—Gaylegphug district (Sham Khara) and Deothang district (Morong), **N**—Upper Mo Chu district (Tamji); **Sikkim.** Warm broad-leaved forests, 1450−2130m. November−March.

5. M. gamblei Hook. f.

Small glabrous tree. Leaves oblong-lanceolate, 7−13×3.5−5cm, acuminate, base rounded, margins obscurely sinuate-crenate, pinnately veined but obscurely 3-veined with 2 discoid glands above base, obscurely gland-dotted beneath; petioles 1−2.5cm; stipules minute, subulate. Male panicles slender, pubescent, flowers c1mm, stamens 15−20. Female flowers and fruit unknown.

Sikkim: Dulka Jhar, Darjeeling terai. Terai forests. April−September.

27. HOMONOIA Loureiro

Shrubs or small trees, indumentum of simple hairs and peltate scales. Leaves alternate, narrow, pinnately veined, aromatic; stipules subulate, subpersistent. Dioecious; flowers in slender axillary spikes. Male flowers: calyx 3-lobed, petals absent, stamens many forming a globose mass, filaments forming fascicles and united into a column at base. Female flowers: sepals 5–8, unequal, ovary 3-celled, styles 3, simple, spreading. Capsules 3-lobed, 3-seeded.

1. H. riparia Loureiro. Nep: *Khola Ruis* (34). Fig. 50 d–f

Rigid shrub or small tree to 6m. Leaves erect, narrowly linear-elliptic, 8–20×0.8–1.7cm, acute or glandular-mucronate, base cuneate, margins remotely serrulate, sparsely pubescent on veins, minutely papillose and densely peltate-scaly beneath; petioles 4–12mm; stipules subulate, 4–10mm. Male spikes 4–6cm, sepals rounded, reddish, c 3mm, puberulous. Female spikes 4–10cm, sepals lanceolate, 3mm, pubescent, ovary subglobose, styles c 4.5mm, papillose. Capsules c 3mm, pubescent.

Bhutan: C—Tongsa district (Mangde Chu); **Sikkim**: Rungit and Tista valleys. On river banks, 300–600m. January–May.

28. LASIOCOCCA Hook. f.

Evergreen trees, indumentum of simple often gland-tipped hairs. Leaves alternate or subopposite, often becoming ± whorled in 3's at branch ends, entire, pinnately veined, eglandular; stipules narrow lanceolate, caducous. Monoecious; males in axillary racemes, females solitary on slender pedicels in upper leaf axils. Male flowers: calyx 3-lobed, petals absent, stamens many, filaments united below. Female flowers: sepals 5–7, unequal, persistent, ovary 3-celled, styles 3, simple, united at base. Capsule 3-seeded.

FIG. 50. **Euphorbiaceae** and **Daphniphyllaceae**. **Euphorbiaceae**. a–c, *Macaranga denticulata*: a, portion of male flowering shoot; b, male flower; c, portion of infructescence. d–f, *Homonoia riparia*: d, portion of male flowering shoot; e, male flower; f, portion of female inflorescence. g–j, *Ricinus communis*: g, portion of flowering shoot; h, male flower; i, fascicle of stamens; j, immature fruit. k & l, *Suregada multiflora*: k, portion of flowering shoot; l, male flower. m–o, *Baliospermum densiflorum*: m, portion of flowering shoot; n, male flower (two sepals removed); o, female flower (two sepals removed). p–s, *Sapium insigne*: p, portion of fruiting shoot; q, portion of male inflorescence; r, male flower; s, female flower. **Daphniphyllaceae**. t–v, *Daphniphyllum chartaceum*: t, portion of female flowering shoot with young leaves; u, male flower; v, fruit. Scale: m × ¼; a, d, g, k, p × ⅓; t × ½; c, j, q, v × 1; h × 1½; u × 2; f, i, l, o × 3; e × 4; n, s × 5; r × 6; b × 8.

84. EUPHORBIACEAE

1. L. symphylliifolia (Gamble) Hook.f.; *Homonoia symphylliifolia* Gamble. Nep: *Jhankhri-Kath* (34).

Tree to 12m, branchlets with white bark. Leaves elliptic-oblanceolate, 7−13×2.5−4cm, acuminate, base narrowly cordate, glabrous; petioles 3−8mm, pubescent; stipules 4mm. Male racemes pendulous 2−5cm, sepals rounded, pubescent. Female flowers erect, on pedicels 8−25mm, sepals ovate, 2.5−3mm, acuminate, glandular-pubescent. Capsules (immature) 12mm diameter, subtended by enlarged persistent sepals 6.5mm or more.

Sikkim: Tista valley and terai. Terai forests. April−May.

29. RICINUS L.

Glabrous evergreen herbs, shrubs or small trees. Leaves alternate, peltate, deeply palmately 5−10-lobed, glandular-serrate, palmately veined; petiole with a large discoid gland at apex; stipules sheathing, caducous. Monoecious; flowers in terminal panicles with females uppermost. Male flowers clustered, sepals 3−6, petals absent, stamens many, filaments united into fascicles. Female flowers: sepals 2, ovary 3-celled, styles 3, spreading, bifid, papillose. Capsules 3-lobed, softly prickly, 3-seeded.

1. R. communis L. Med: *Denrog;* Sha: *Chamling Shing;* Nep: *Reri* (34); Eng: *Castor-oil Plant.* Fig. 50 g−j

Shrub 2m or small tree to 5m. Leaves orbicular, 10−30cm long and broad, lobes lanceolate, acuminate, coarsely serrate; petioles 7−15cm; stipules 1.4−1.8cm, early caducous. Panicles 8−17cm. Male flowers c 1cm diameter, sepals ovate, 5−7mm. Female flowers: sepals ovate, 6−7mm, ovary globose, densely prickly. Capsules ellipsoid-globose 1.5−2.5cm.

Bhutan: S—Phuntsholing district, **C**—Thimphu, Punakha and Tongsa districts; **Sikkim.** Waste ground in towns and villages, 200−3000m. January−March.

Often cultivated for its seed oil used as fuel for lamps and sometimes used medicinally as castor-oil. Seeds poisonous. Sometimes used as a food-plant for silkworms (48). Probably native to tropical Africa (16).

30. ENDOSPERMUM Bentham

Evergreen trees, indumentum of stellate and simple hairs. Leaves coriaceous, alternate, entire, palmately 3-veined and with 2 large glands at base; stipules minute, caducous. Dioecious; flowers in simple or branched racemes. Male flowers clustered, calyx 4-toothed, stamens c 10, filaments free, anthers peltate, 4-celled. Female flowers solitary, calyx 5-toothed, ovary 2−3-celled, styles united into a disc. Fruit 2−3-lobed, indehiscent, 2−3-seeded.

30. ENDOSPERMUM

1. E. chinense Bentham. Nep: *Seti-Kath* (34)
Large tree. Leaves broadly ovate, 8−14×6−9cm, subacute, base rounded with 2 conspicuous globose glands; petioles 6−9cm. Male racemes branched, up to 21×8cm, calyx 2mm diameter. Female racemes simple, tomentose, calyx 3mm diameter. Fruits ellipsoid, c 12×8mm, tomentose.
Sikkim: foothills, rare (34). Subtropical forests.

31. SUREGADA Rottler

Evergreen shrubs or small trees. Leaves coriaceous, alternate, entire, pinnately veined, eglandular but with pellucid dots between reticulations; stipules minute, early caducous. Dioecious; flowers in short leaf-opposed cymes or fascicles. Male flowers: sepals 5, stamens many, exserted, free, with numerous small glands at base of filaments. Female flowers: sepals 5, disc present, ovary 3-celled, styles 3, short, bifid, spreading. Fruit a fleshy subglobose capsule, shallowly 3-lobed, 3-seeded.

1. S. multiflora (Jussieu) Baillon; *Gelonium multiflorum* Jussieu. Fig. 50 k&l
Shrub or tree to 12m. Leaves elliptic, 12−30×4−9cm, acute or shortly and bluntly acuminate, base cuneate, glabrous. Cymes on short peduncles up to 1cm; sepals 4−5mm, finely pubescent, yellow. Fruit subglobose, 2cm diameter.
Sikkim: terai at Mahanuddi. Subtropical forests. February−April.

32. BALIOSPERMUM Blume

Subshrubs. Leaves alternate, often crenate or serrate, pinnately veined or palmately veined at base, often with 2 discoid glands at base of lamina; stipules minute, persistent. Dioecious or monoecious; flowers clustered, in racemes or panicles, sometimes with long naked peduncle. Male flowers: sepals 5, petals absent, disc annular or as free glands, stamens 12−20, free, anther cells united, dehiscing laterally. Female flowers: sepals 5−6, often enlarging and persistent in fruit, ovary 3-celled, styles free, bifid. Capsule 3-celled, 3-seeded.

1. Leaves coarsely crenate-dentate; monoecious; inflorescences leafy, without naked peduncles .... **4. B. montanum**
+ Leaves serrate or crenate-serrate; dioecious; inflorescences not leafy, peduncles naked 2

2. Leaves ovate-deltoid, base cordate or occasionally truncate, strongly 5-veined; male panicles usually dense, corymbose, sepals c 3mm
1. B. densiflorum
+ Leaves broadly ovate to elliptic, base rounded or cuneate, sometimes

truncate, weakly to strongly 3-veined; male panicles lax or racemiform, sepals 1.5–2.5mm 3

3. Leaves elliptic or ovate-elliptic, 4.5–8.5cm broad, base round or cuneate; male panicles narrow, racemiform .... **2. B. corymbiferum**
+ Leaves broadly ovate, 8–10cm broad, base rounded or sometimes truncate; male panicles lax, corymbose .... **3. B. nepalense**

1. B. densiflorum Long. Fig. 50 m–o

Erect subshrub 1–3m. Leaves ovate-deltoid, 13–20×9–13cm, acuminate, base cordate or occasionally truncate, strongly 5-veined, margins coarsely crenate-serrate, teeth gland-tipped, subglabrous or pubescent beneath; petioles 8–13cm, with 2 sessile round glands at apex on some leaves; stipules oblong, c 2mm, subpersistent. Dioecious; male panicles densely corymbiform, 1.5–4×1.5–4cm, on naked peduncles 10–28cm; bracts lanceolate, 2–3mm. Male sepals rounded, c 3mm, stamens 18–20. Female panicles 2–3cm, on peduncles 10–14cm; bracts linear, 3–6mm, with 1–2 stalked glands near base. Female sepals lanceolate, c 4mm, with 1–2 glandular marginal teeth. Capsules unknown.

Bhutan: C—Mongar district (Saleng and Kori La) and Tashigang district (E side of Kori La). Warm broad-leaved and Evergreen oak forests, 910–2130m. June–August.

Records of *B. calycinum* Mueller (a Mishmi species) from Bhutan (117) have not been verified and possibly refer to *B. densiflorum* or some other species.

2. B. corymbiferum Hook. f.

Similar to *B. densiflorum* but leaves ovate-elliptic, 11–19×4.5–8.5cm, caudate acuminate, narrowed and rounded at base, remotely serrate with rounded gland-tipped teeth, venation ± pinnate, weakly 3-veined at base; petioles 3–9cm; male panicles narrow, 3–13×1–2cm, on peduncles 6–15cm, sepals c 2mm; female panicles and fruit not seen.

Bhutan: S—Chukka district (Marichong); **Sikkim:** Darjeeling, Mamrung and Selim. Warm broad-leaved forests, 1200–1600m. August–September.

3. B. nepalense Hurusawa & Tanaka

Similar to *B. densiflorum* but leaves broadly ovate, 10–15 (–25)×8–10cm, base truncate, rounded or shallowly cordate, 3-veined, margins coarsely glandular-dentate; male panicles widely-branched, 4–7×4–6cm, on peduncles 10–30cm, flowers small, sepals c 1.6mm, stamens 12–15; female inflorescences on peduncles 5–25cm, with bracts 5–10mm, sepals c 5mm.

Bhutan: S—Chukka district (Giengo); **Sikkim:** Tista valley. Warm broad-leaved forests, 600–1500m.

4. B. montanum (Willdenow) Mueller; *B. axillare* Blume. Nep: *Harital* (34)
Subshrub 1−2m. Leaves ovate-elliptic 8−18×4−11cm, acute, base rounded or cuneate, margins coarsely crenate-dentate, 3-veined at base, with 2 discoid glands on margin near petiole, pubescent beneath; petioles 2−10cm. Monoecious, inflorescences leafy, flowers in axillary clusters or short racemes, males and females on the same or different shoots. Male flowers: sepals 2.5mm, stamens c 20. Female flowers: sepals c 2mm, ovary pubescent, sometimes with a few stamens. Capsules 3-lobed, c 1cm diameter.
Bhutan: locality unknown; **Sikkim:** Tista valley, terai and foothills (34). Subtropical forests, 100−900m. February−March.

33. PTEROCOCCUS Hasskarl

Twining shrubs. Leaves alternate, serrate, palmately 3−5-veined at base, 2-glandular at apex of petiole; stipules persistent. Monoecious; flowers in lateral or axillary racemes, female flowers towards base. Male flowers: calyx 4−5-lobed, petals absent, stamens 8−13, free, with short thick filaments. Female flowers: sepals 4−5, ovary 4-celled, styles united into a column, shortly 4-lobed. Capsule depressed, stellately 4-lobed, lobes winged at apex, 4-seeded.

1. P. corniculatus (Smith) Pax & Hoffmann; *Plukenetia corniculata* Smith
Twining shrub, subglabrous. Leaves oblong, 7−18×3−10cm, caudate-acuminate, base truncate, margins serrate; petioles 2−6cm; stipules lanceolate 1−2mm. Racemes shortly pedunculate, 2−4cm. Male flowers: sepals c 1mm. Female flowers: sepals 1.5mm, ovary 4-winged. Capsules on pedicels 4−5cm, c 1.5×2.5cm; seeds c 1cm.
Sikkim: Rishap, Darjeeling. Subtropical forests, 600m. August.

34. SAPIUM Browne

Evergreen or deciduous trees or shrubs, usually glabrous. Leaves alternate, pinnately veined, entire or serrate; petioles mostly 2-glandular at apex; stipules persistent or caducous. Monoecious or dioecious; flowers in terminal simple or branched spikes or racemes, when monoecious females in lower part. Male flowers: calyx 2-or 3-lobed, petals absent, stamens 2−3, free, anther cells distinct. Female flowers: calyx 3-lobed, ovary 2−3-celled, styles free or united at base, spreading, recurved. Fruit a 3-lobed, 3-seeded capsule or 1−2-seeded and fleshy.

84. EUPHORBIACEAE

1. Deciduous tree; leaves serrulate; dioecious .... **1. S. insigne**
+ Evergreen trees or shrubs; leaves entire; monoecious 2

2. Leaves broadly rhombic-ovate or suborbicular .... **4. S. sebiferum**
+ Leaves ovate or ovate-elliptic 3

3. Leaves acuminate; racemes forming terminal panicles; fruit fleshy
   2. S. baccatum
+ Leaves acute; racemes simple, terminal; fruit capsular
   3. S. eugeniifolium

1. S. insigne (Royle) Hook. f. Dz: *Sho Shi.* Fig. 50 p−s

Robust glabrous deciduous tree 7−12m, branches stout, with large leaf-scars. Leaves ovate-elliptic 10−25×6−11cm, acuminate, base cuneate, margins serrulate, minor veins reticulate beneath; petioles 4−8cm; stipules lanceolate, 1.5mm, fimbriate. Dioecious; flower spikes terminal, erect, rigid, males 15−20cm, deciduous, females stouter, 7−15cm. Male flowers minute, green, sessile in clusters of 10 in axil of bract, calyx 2-lobed, stamens 2. Female flowers solitary, subsessile, calyx 2-lipped, ovary ovoid, styles short, recurved. Fruit subglobose, glossy, dark crimson, fleshy c 8mm diameter, dehiscing to leave bristle-like vascular strands between the 2 ovoid seeds 5−6mm.

Bhutan: S—Sarbhang district (Sarbhang), **C**—Punakha district (Wangdu Phodrang, Lobesa and Punakha) and Tongsa district (Wangde Khola near Shamgong); **Sikkim:** foothills. Amongst scrub in dry valleys, 460−1400m. January−April.

2. S. baccatum Roxb. Nep: *Ankhataruwa* (34)

Glabrous evergreen tree 8−12m. Leaves orange-red when young, ovate 9−18×4−7cm, finely acuminate, base rounded, margins entire, lateral veins sometimes with discoid glands near leaf base and margins; petioles 3.5−7cm, eglandular; stipules oblong, c 12mm, early caducous. Monoecious; flowers shortly pedicellate, in terminal paniculately-branched spikes 10−13cm long, rachis bearing clusters of glands between flowers. Male flowers minute, calyx irregularly dentate, stamens 2. Female flowers: calyx 3-lobed, ovary ovoid. Fruit subglobose or 2-lobed, 8−12mm, fleshy, purple, 1−2-seeded.

Bhutan: S—Sarbhang district (E of Sarbhang); **W Bengal Duars:** Buxa; **Sikkim:** terai and foothills. Subtropical and terai forests, 300−600m. April.

Two sterile specimens with leaves peltate may represent juvenile foliage of this or some other species.

3. S. eugeniifolium Hook. f. Nep: *Pipalpate* (34), *Phirphire* (34)

Similar to *S. baccatum* but leaves ovate-elliptic, 8−11×3−5.5cm, acute;

petioles 3.5—5cm, with 2 large glands near apex; flowers pedicellate, in a simple dense terminal raceme 4—8cm, females at base; styles 2—3; fruit a globose woody capsule c 1cm.
Sikkim: Darjeeling foothills. Secondary subtropical forests, 600—1000m. May—June.

4. S. sebiferum (L.) Roxb. Dz: *Ja Shing;* Eng: *Chinese Tallow Tree*
Evergreen shrub 2—3m or tree to 8m. Leaves broadly rhombic-ovate or suborbicular, 4—8×3—6cm, acuminate, base cuneate or rounded, margins entire, glabrous; petioles mostly 2—6cm, 2-glandular at apex; stipules c1.5mm. Monoecious; flowers in simple or little-branched terminal racemes 6—12cm, females at base. Male flowers small, clustered, pedicels 1.5mm, calyx 3-lobed, c 0.5mm, stamens 2—3. Female flowers solitary, sepals 3, ovate, c 2mm, ovary ovoid 1.5mm, styles 3. Capsules rhomboid-globose, 1—1.5cm, 3-valved, seeds depressed-globose, 5—6mm, enclosed in whitish wax.

Bhutan: C—Punakha district (Mo Chu near Punakha); **Sikkim:** cultivated in terai and foothills (34). Cultivated and naturalised, 500—1500m. May—September.

A native of China, widely cultivated for the wax that surrounds the seeds which is used to make candles; leaves are used to produce a black dye (48).

Family 85. DAPHNIPHYLLACEAE

by D. G. Long

Dioecious evergreen trees or shrubs. Leaves alternate but crowded at branch-ends, simple, entire, pinnately veined, exstipulate. Flowers in axillary racemes, unisexual, actinomorphic. Perianth absent (in Bhutan). Male flowers: stamens 5—10, free, in a whorl, rudimentary ovary absent. Female flowers: ovary superior, 2-celled, surrounded by 5—10 staminodes; styles 2, short, thick, recurved; ovules 2 per cell. Fruit a 1-seeded drupe.

1. DAPHNIPHYLLUM Blume

Description as for Daphniphyllaceae.

1. Leaves oblong-elliptic, 13—25×4—8cm, base broadly cuneate, lateral veins prominent beneath; petioles 2.5—6cm; anthers broad c 2.5× 1.5mm, emarginate; female pedicels ± erect .... **1. D. chartaceum**
+ Leaves oblanceolate, 9—20×3—6.5cm, base attenuate, lateral veins

scarcely prominent beneath; petioles 1.5−4(−5)cm; anthers narrow, c3×1mm, apiculate; female pedicels strongly recurved .. **2. D. himalense**

1. D. chartaceum Rosenthal; *D. bengalense* Rosenthal, *D. himalense* (Bentham) Mueller var. *chartaceum* (Rosenthal) Huang. Dz: *Juroo Shing;* Nep: *Lal Chandan* (34). Fig. 50 t−v

Tree 5−12m. Scales of terminal buds apiculate. Leaves thinly coriaceous, oblong-elliptic, 13−40×4−11cm, shortly acuminate, base broadly cuneate, glossy above, pale but obscurely papillate beneath; lateral veins 10−14 pairs, prominent beneath; petioles crimson, 2.5−6cm. Racemes crimson, 10−15-flowered, bearing ovate, caducous bracts 5−6×3−4mm. Male racemes 2−4cm; pedicels 5−7mm, anthers broad, c 2.5×1.5mm, emarginate, dark purple. Female racemes 3−6cm, erect, pedicels 4−10mm, erect or weakly recurved, ovary ellipsoid c 2mm. Fruiting racemes up to 13cm; drupes black, ellipsoid c 14×9mm, stigmas persistent, becoming sub-terminal.

Bhutan: S—Chukka district (Tala), **C**—Punakha district (Gangtokha) and Mongar district (Ngasamp and Rudo La), **N**—Upper Mo Chu district (Kencho); **Sikkim.** Warm broad-leaved and Evergreen oak forests, 1600−2290m. April−June.

2. D. himalense (Bentham) Mueller

Similar to *D. chartaceum* but a shrub or small tree to 6m; scales of terminal buds obtuse; leaves thickly coriaceous, oblanceolate, 9−20× 3−6.5cm, acute, base attenuate, pale but densely and conspicuously papillate beneath; lateral veins scarcely prominent beneath; petioles 1.5−4(−5)cm; anthers narrow c 3×1mm, apiculate; female racemes recurved at apex, pedicels strongly recurved.

Bhutan: C—Punakha district (Mara Chu), **N**—Upper Mo Chu district (Gasa) and Upper Kuru Chu district (Denchung); **Sikkim:** Rimbick, Zemu Chu, etc.; **Chumbi.** Cool broad-leaved and moist conifer forests, 2270−3050m. May−September.

Wood streaked red and of ornamental value; red wood used to make dye for caste-marks (48).

INDEX OF BOTANICAL NAMES

Abarema clypearia (Jack) Kostermans, 650
 monadelpha (Roxb.) Kostermans, 650
Abrus Adanson, 655
 fruticulosus Wight & Arnott, 655
 precatorius L., 655
 pulchellus Thwaites, 655
Acacia Miller, 640
 caesia (L.) Wight & Arnott, 642
 catechu (L.f.) Willdenow, 642
 catechuoides (Roxb.) Bentham, 642
 concinna (Willdenow) DC., 641
 decurrens Willdenow, 643
 farnesiana (L.) Willdenow, 641
 gageana Craib 642
 intsia (L.) Willdenow, 642
 lenticularis Bentham, 642
 melanoxylon R. Brown, 642
 pennata (L.) Willdenow, 641
 rugata (Lamarck) Voigt, 641
Acalypha L., 796
 brachystachya Hornemann, 797, 801
 hispida Burman f., 797
 wilkesiana Mueller, 797
Acomostylis elata (G. Don) Bolle, 580
 sikkimensis (Prain) Bolle, 581
Acrocarpus Arnott, 620
 fraxinifolius Arnott, 620, 637
Adenanthera L., 636
 pavonina L., 636
Aeschynomene L., 710
 aspera L., 710
 indica L., 710
Agrimonia L., 582
 pilosa Ledebour, 582, 583
  var. *nepalensis* (D.Don) Nakai, 582
  var. *zeylanica* (Hook. f.) Purohit & Panigrahi, 582
Albizia Durazzini, 643
 chinensis (Osbeck) Merrill, 646
 gamblei Prain, 637, 645
 julibrissin Durazzini, 645
  var. *mollis* (Wall.) Bentham, 646
 lebbeck (L.) Bentham, 644
 lucida (Roxb.) Bentham, 644
 lucidior (Steudel) Hara, 644
 myriophylla Bentham, 645
 odoratissima (L.f.) Bentham, 644
 procera (Roxb.) Bentham, 645
 sherriffii E. G. Baker, 646
 stipulata (Roxb.) Boivin, 646
Alchornea Swartz, 798
 mollis (Bentham) Mueller, 798, 801
 tiliifolia (Bentham) Mueller, 798

Aleurites cordata (Thunberg) Steudel, 791
Altingia Noronha, 471
 excelsa Noronha, 469, 471
Alysicarpus Desveaux, 681
 rugosus (Willdenow) DC., 681
 vaginalis (L.) DC., 671, 681
Amphicarpaea ferruginea Bentham, 695
Amygdalus communis L., 543
Anisadenia Meisner, 752
 khasyana Griff., 752
 pubescens Griff., 747, 753
 saxatilis Meisner, 752
Antidesma L., 786
 acidum Retzius, 786, 789
 acuminatum Wight, 787
 bunius (L.) Sprengel, 787
 diandrum (Roxb.) Roth, 786
 ghaesembilla Gaertner, 787
 lanceolarium (Roxb.) Wight, 786
 nigricans Tulasne, 788
Apios Fabricius, 689
 carnea (Wall.) Baker, 649, 689
Aporosa Blume, 785
 dioica (Roxb.) Mueller, 785
 octandra (D. Don) Vickery, 785, 789
 roxburghii Baillon, 785
Arachis L., 712
 hypogaea L., 712
Archidendron F. v. Mueller, 648
 clypearia (Jack) Nielsen, 650
 monadelphum (Roxb.) Nielsen, 649, 650
Armeniaca vulgaris Lamarck, 542
Aruncus L., 536
 dioicus (Walter) Fernald, 536
  subsp. *triternatus* (Maximowicz) Hara, 536
Astilbe D. Don, 488
 rivularis D. Don, 488
 rubra Hook. f. & Thomson, 487, 488
Astragalus L., 714
 acaulis Baker, 716
 balfourianus Simpson, 719
 bhotanensis Baker, 717, 719
 chlorostachys Lindley, 716
 concretus Bentham, 716
 donianus DC., 719
 floridus Bunge, 718
 kongrensis Baker, 718
 larkyaensis auct. p.p. non Kitamura, 714
 lessertioides Bunge, 718
 pycnorhizus Bentham, 719
 rigidulus Bunge, 720
 sikkimensis Bunge, 721

INDEX OF BOTANICAL NAMES

stipulatus Sims, 715
strictus Bentham, 720
tenuicaulis Bunge, 721
tongolensis Ulbrich, 718
vicioides Baker non Ledebour, 716
xiphocarpus auct. non Bunge, 716
zemuensis W.W. Smith, 720
Atylosia elongata Bentham, 704
grandiflora Baker, 704
mollis Bentham, 704
scarabaeoides (L.) Bentham, 703
villosa Baker, 705
Averrhoa L., 739
carambola L. 740, 741

Baccaurea Loureiro, 788
ramiflora Loureiro, 788, 789
sapida (Roxb.) Mueller, 788
BALANITACEAE, 750
Balanites Delile, 750
aegyptiaca (L.) Delile, 750
roxburghii Planchon, 750
Baliospermum Blume, 809
axillare Blume, 811
calycinum Mueller, 810
corymbiferum Hook. f., 810
densiflorum Long, 807, 810
montanum (Willenow) Mueller, 811
nepalense Hurasawa & Tanaka, 810
Bauhinia L., 632
anguina Roxb., 634
glabrifolia (Bentham) Baker, 635
macrostachya Baker non Bentham, 635
malabarica Roxb., 633
purpurea L., 633
scandens L., 634
vahlii Wight & Arnott, 634
variegata L., 625, 634
wallichii MacBride, 635
Bergenia Moench, 491
ciliata (Haworth) Sternberg, 492
forma ligulata (Wall.) Yeo, 492
purpurascens (Hook. f. & Thomson) Engler, 492
Biophyum DC., 743
reinwardtii (Zuccarini) Klotzsch, 741, 743
Brachycaulos Dixit & Panigrahi, 579
simplicifolius Dixit & Panigrahi, 579
Breynia J. R. & J. G. A. Forster, 781
patens (Roxb.) Bentham, 782
retusa (Dennstedt) Alston, 782, 789
Bridelia Willdenow, 767
montana sensu F.B.I. p.p. non Willdenow, 768

pubescens Kurz, 770
retusa (L.) Sprengel, 763, 769
sikkimensis Gehrmann, 768
stipularis (L.) Blume, 769
tomentosa Blume, 770
verrucosa Haines, 768
Bryophyllum calycinum Salisbury, 473
pinnatum (Lamarck) Oken, 473
Bucklandia populnea Griffith, 470
Butea Roxb., 687
buteiformis (Voigt) Grierson & Long, 688
frondosa Roxb., 688
minor Baker, 687
monosperma (Lamarck) Kuntze, 661, 688
parviflora Roxb, 688

Caesalpinia L., 622
bonduc (L.) Roxb., 623, 625
bonducella (L.) Fleming, 623
cinclidocarpa Miquel, 624
crista L., 623
cucullata Roxb., 623, 625
decapetala (Roth) Alston, 624, 625
nuga (L.) Aiton, 623
pulcherrima (L.) Swartz, 626
sepiaria Roxb., 624
tortuosa Roxb., 624
Caesalpinioideae, 608
Cajanus DC., 702
cajan (L.) Huth, 671, 703
elongatus (Bentham) van der Maesen, 704
grandiflorus (Baker) van der Maesen, 704
indicus Sprengel, 703
mollis (Bentham) van der Maesen, 704
scarabaeoides (L.) Wall., 703
villosus (Baker) van der Maesen, 705
Calliandra Bentham, 647
haematocephala Hasskarl, 647
Campylotropis Bunge, 681
eriocarpa (Maximowicz) Schindler, 681
griffithii Schindler, 682
macrostyla (D. Don) Miquel, 682
var. *griffithii* (Schindler) Ohashi, 682
speciosa (Schindler) Schindler, 681
Canavalia DC., 690
gladiata (Jacquin) DC., 690
gladiolata Sauer, 690
Caragana Fabricius, 713
chumbica Prain, 713
crassicaulis sensu F.B.I. p.p. non Baker, 714
gerardiana Royle, 713
jubata (Pallas) Poiret, 713, 717
nepalensis Kitamura, 713

INDEX OF BOTANICAL NAMES

spinifera Komarov, 714
sukiensis Schneider, 713
Cassia L., 627
  alata L., 629, 637
  fistula L., 625, 628
  floribunda Cavanilles, 631
  grandis L.f., 629
  hochstetteri Ghesquiere, 631
  javanica L., 629
    subsp. nodosa (Roxb.) K. & S. Larsen, 629
  laevigata Willdenow, 631
  lechenaultiana DC., 625, 630
  mimosoides L., 630
    var. *auricoma* Bentham, 630
    var. *dimidiata* (Roxb.) Baker, 631
    var. *wallichiana* DC., 630
  occidentalis L., 631
  sophera L., 631
  spectabilis DC., 629
  surattensis Burman f., 632
  tora L., 632
Chamaesyce hirta (L.) Millspaugh, 766
  hypericifolia (L.) Millspaugh, 766
  prostrata (Aiton) Small, 767
  thymifolia (L.) Millspaugh, 766
Chesneya nubigena auct. p.p. non (D. Don) Ali, 714
  polystichoides (Handel-Mazzetti) Ali, 714
    subsp. *bhutanica* Ohashi, 714
  purpurea Li, 714
Choenomeles Lindley, 602
  lagenaria (Loiseleur) Koidzumi, 602
    var. lagenaria, 603
    var. wilsonii Rehder, 603
Christia Moench, 680
  vespertilionis (L.f.) van Meeuwen, 661, 680
Chrysosplenium L., 489
  adoxoides (Griffith) Maximowicz, 490
  alternifolium sensu F.B.I. non L., 491
  carnosum Hook. f. & Thomson, 490
  forrestii Diels, 491
  griffithii Hook. f. & Thomson, 487, 491
  lanuginosum Hook. f. & Thomson, 490
  nepalense D. Don, 489
  nudicaule Bunge, 491
    var. intermedium Hara, 491
    var. nudicaule, 491
  singalilense Hara, 490
  tenellum Hook. f. & Thomson, 490
Cicer L., 727
  arietinum L., 661, 727

Cladrastis Rafinesque, 651
  sinensis Hemsley, 649, 651
Claoxylon Jussieu, 795
  longipetiolatum Kurz, 796
Cleidion Blume, 803
  javanicum Blume, 803
  spiciflorum (Burman f.) Merrill, 803
Clitoria L., 696
  mariana L., 697
  ternatea L., 696
Cochlianthus Bentham, 689
  gracilis Bentham, 689
Codariocalyx gyroides (Link) Hasskarl, 670
  motorius (Houttuyn) Ohashi, 670
Codiaeum Jussieu, 794
  pictum (Loddiges) Hooker, 794
  variegatum (L.) Blume, 794
    forma platyphyllum Pax, 794
Coluria longifolia Maximowicz, 581
Connarus L., 607
  paniculatus Roxb., 605, 607
CONNARACEAE, 607
Corylopsis Siebold & Zuccarini, 468
  himalayana Griff., 469, 470
Cotoneaster L., 588
  acuminatus Lindley, 591
  adpressus Bois, 591
  bacillaris Lindley, 590
  bakeri Klotz, 591
  cavei Klotz, 590
  congestus Baker, 589
  cooperi Marquand, 590
  distichus Lange, 590
  frigidus Lindley, 590
  gamblei Klotz, 590
  griffithii Klotz, 590
  horizontalis Decaisne, 591
  integrifolius (Roxb.) Klotz, 589
  ludlowii Klotz, 589
  microphyllus Lindley, 589
  mucronatus Franchet, 591
  nitidus Jacques, 590
  nummularia sensu F.B.I. non Fischer & Meyer, 590
  obtusus Lindley, 590
  racemiflorus (Desfontaines) Koch, 583, 590
  rotundifolius Lindley, 589
  rubens W. W. Smith, 591
  sandakphuensis Klotz, 590
  sanguineus Yu, 591
  sherriffii Klotz, 589
  simonsii Baker, 591

INDEX OF BOTANICAL NAMES

thymifolius Baker, 589
wallichianus Klotz, 591
Cotyledon spathulata Poiret, 473
Crassula L., 474
   indica Decaisne, 473
   pentandra (Edgeworth) Schoenland, 474
   schimperi Fischer & Meyer, 474
CRASSULACEAE, 471
Crataegus L., 592
   crenulata (D. Don) Roxb., 592
   sp., 592
Crotalaria L., 731
   acicularis Bentham, 736
   alata D. Don, 735
   albida Roth, 735
   bhutanica Thothathri, 735
   bialata Schrank, 735
   bracteata Roxb., 732
   calycina Schrank, 734
   capitata Baker, 734
   cytisoides DC., 717, 732
   ferruginea Bentham, 735
   humifusa Bentham, 736
   juncea L., 733
   leschenaultii DC., 733
   linifolia L.f., 735
   montana Roth, 735
   mucronata Desveaux, 732
   occulta Bentham, 734
   pallida Aiton, 732
   psoralioides D. Don non Lamarck, 732
   sericea Retzius non Burman f., 733
   sessiliflora L., 734
   spectabilis Roth, 733
   striata DC., 732
   tetragona Andrews, 733
   trifoliastrum Willdenow, 733
Croton L., 791
   bonplondianus Baillon, 793
   caudatus Geiseler, 789, 793
   himalaicus Long, 792
   joufra Roxb., 792
   malvifolius Griff., 793
   oblongifolius Roxb., 792
   roxburghii Balakrishnan, 792
   sparsiflorus Morong, 793
   tiglium L., 793
   tiglium auct. p.p. non L., 792
Cyclostemon assamicus Hook. f., 784
   griffithii Hook. f., 785
   indicus Mueller, 785
   lancifolius Hook. f., 785
   subsessilis Kurz, 785
Cydonia cathayensis Hemsley, 602

Dalbergia L.f., 652
   assamica Bentham, 654
   bhutanica Thothathri, 654
   hircina Bentham, 654
   latifolia Roxb., 653
   mimosoides Franchet, 654
   pinnata (Loureiro) Prain, 653
   rimosa Roxb., 653
   sericea G. Don, 654
   sissoo DC., 652, 671
   stenocarpa Kurz, 654
   stipulacea Roxb., 653
   tamarindifolia Roxb., 653
   volubilis Roxb., 655
DAPHNIPHYLLACEAE, 813
Daphniphyllum Blume, 813
   bengalense Rosenthal, 814
   chartaceum Rosenthal, 807, 814
   himalense (Bentham) Mueller, 814
      var. *chartaceum* (Rosenthal) Huang, 814
Delonix Rafinesque, 621
   regia (Hooker) Rafinesque, 622
Dendrolobium triangulare (Retzius) Schindler, 669
Derris Loureiro, 656
   acuminata Bentham, 656
      var. *sikkimensis* Thothathri, 657
   cuneifolia Bentham, 657
      var. longipedicellata Thothathri, 657
   discolor Bentham, 657
   ferruginea (Roxb.) Bentham, 657
   microptera Bentham, 657
   monticola (Kurz) Prain, 657
   polystachya Bentham, 656
Desmodium Desveaux, 667
   capitatum (Burman f.) DC., 675
   caudatum (Thunberg) DC., 677
   cephalotes (Roxb.) Wight & Arnott, 669
   concinnum DC., 676
   confertum DC., 677
   duclouxii Pampanini, 676
   elegans DC., 674
   floribundum (D. Don) G. Don, 676
   gangeticum (L.) DC., 672
   gyrans (L.f.) DC., 670
      var. *roylei* (Wight & Arnott) Baker, 670
   gyroides (Link) DC., 670
   heterocarpon (L.) DC., 674
      var. heterocarpon, 675
      var. strigosum van Meeuwen, 675
   khasianum Prain, 674
   kulhaitense Prain, 674
   laburnifolium (Poiret) DC., 677

INDEX OF BOTANICAL NAMES

latifolium DC., 672
laxiflorum DC., 678
laxum DC., 676
microphyllum (Thunberg) DC., 673
motorium (Houttuyn) Merrill, 670, 671
multiflorum DC., 676
oblongum Bentham, 671, 672
oojeinense (Roxb.) Ohashi, 673
oxyphyllum DC., 675
oxyphyllum sensu F.B.I., 674
parvifolium DC., 673
podocarpum DC., 671, 675
var. *laxum* (DC.) Baker, 676
subsp. oxyphyllum (DC.) Ohashi, 676
polycarpum (Poiret) DC., 674
pulchellum (L.) Baker, 669
renifolium (L.) Schindler, 673
reniforme (L.) DC., 673
retroflexum (L.) DC., 675
sequax Wall., 677
sinuatum (Miquel) Baker, 677
styracifolium (Osbeck) Merrill, 675
tiliifolium (D. Don) Wall., 674
triangulare (Retzius) Merrill, 669
triflorum (L.) DC., 673
triquetrum (L.) DC., 669
subsp. pseudotriquetrum (DC.) Prain, 669
subsp. triquetrum, 669
velutinum (Willdenow) DC., 672
williamsii Ohashi, 676
Deutzia Thunberg, 526
bhutanensis Zaikonnikova, 526
compacta auct. non Craib, 526
corymbosa G. Don, 519, 526
crenata Siebold & Zuccarini, 527
hookeriana (Schneider) Airy Shaw, 526
staminea Wall., 526
Dichroa Loureiro, 521
febrifuga Loureiro, 522
Docynia Decaisne, 603
griffithiana Decaisne, 603
indica (Wall.) Decaisne, 603, 605
Dolichos L., 698
biflorus sensu F.B.I. non L., 699
falcatus sensu F.B.I. p.p. non Willdenow, 698
tenuicaulis (Baker) Craib, 698
Drypetes Vahl, 784
assamica (Hook. f.) Pax & Hoffmann, 784
griffithii (Hook. f.) Pax & Hoffmann, 785
indica (Mueller) Pax & Hoffmann, 785
lancifolia (Hook. f.) Pax & Hoffmann, 785

subsessilis (Kurz) Pax & Hoffmann, 785
Duchesnea Smith, 579
indica (Andrews) Focke, 579, 583
Dumasia DC., 695
villosa DC., 696
var. leiocarpa (Bentham) Baker, 696
var. villosa, 696
Dunbaria Wight & Arnott, 705
circinalis (Bentham) Baker, 705
conspersa Bentham, 706
debilis Baker, 706
pulchra Baker, 704
rotundifolia (Loureiro) Merrill, 706
Dysolobium (Bentham) Prain, 697
grande (Bentham) Prain, 697
pilosum (Willdenow) Marechal, 700

Emblica officinalis Gaertner, 772
Endospermum Bentham, 808
chinense Bentham, 809
Entada Adanson, 638
"*entity C*" Brenan, 638
laotica Gagnepain, 638
pusaetha DC., 638
subsp. *sinohimalensis* Grierson & Long, 638
rheedii Sprengel, 638
subsp. sinohimalensis (Grierson & Long) Panigrahi, 638
scandens sensu F.B.I. p.p. non (L.) Bentham, 638
Enterolobium saman (Jacquin) Prain, 647
Eriobotrya Lindley, 601
bengalensis (Roxb.) Hook.f., 602
dubia (Lindley) Decaisne, 601
hookeriana Decaisne, 601
petiolata Hook.f., 602
Eriococcus hamiltonianus (Mueller) Hurusawa & Tanaka, 774
Eriolobus indicus (Wall.) Schneider, 603
Eriosema (DC.) G. Don, 709
chinense sensu F.B.I. non Vogel, 709
himalaicum Ohashi, 709
tuberosum (D. Don) Wang & Tang non Richter, 709
Erythrina L., 683
arborescens Roxb., 649, 684
indica Lamarck, 684
stricta Roxb., 683
suberosa Roxb., 684
variegata L., 684
Euchresta Bennett, 736
horsfieldii (Leschenault) Bennett, 736
var. *bhutanica* Ohashi, 736

819

INDEX OF BOTANICAL NAMES

Euphorbia L., 759
  antiquorum L., 762
  griffithii Hook. f., 763, 764
    var. bhutanica (Fischer) Long, 764
    var. griffithii, 764
  himalayensis (Klotzsch) Boissier, 765
  himalayensis sensu F.B.I. non (Klotzsch) Boissier, 764
  hirta L., 766
  hypericifolia L. agg., 766
  indica Lamarck, 766
  leucocephala Lotsy, 764
  longifolia D. Don, 765
  luteo-viridis Long, 764
  milii Des Moulins, 762
  parviflora L., 766
  pilulifera L., 766
  prostrata Aiton, 767
  pulcherrima Klotzsch, 762
  royleana Boissier, 761
  sikkimensis Boissier, 764
    subsp. bhutanica Fischer, 764
  splendens Hooker, 762
  stracheyi Boissier, 765
  thymifolia L., 766
EUPHORBIACEAE, 754
Exbucklandia R. W. Brown, 470
  populnea (Griffith) R. W. Brown, 469, 470

Federovia glauca (Wall.) Yakovlev, 650
Flemingia Aiton, 706
  bhutanica Grierson, 708
  bracteata (Roxb.) Wight, 706
  congesta Aiton, 707
  fruticulosa Bentham, 707
  involucrata Bentham, 707
  macrophylla (Willdenow) Merrill, 707
  paniculata Bentham, 707
  semialata Roxb., 707
  stricta Roxb., 708
  strobilifera (L.) Aiton, 706
  wightiana Wight & Arnott, 707
Flueggea Willdenow, 775
  microcarpa Blume, 775
  virosa (Willdenow) Voigt, 775
    subsp. himalaica Long, 776
    subsp. virosa, 776
Flueggiopsis glauca (Mueller) A. Das, 774
Fragaria L., 577
  daltoniana Gay, 578
  indica Andrews, 579
  nilgerrensis Gay, 578
  nubicola (Hook. f.) Lacaita, 578
    rubiginosa Lacaita, 578
    sikkimensis Kurz, 578
  vesca L., 578
    var. nubicola Hook. f., 578

Gelonium multiflorum Jussieu, 809
GERANIACEAE, 744
Geranium L., 744
  chumbiense Knuth, 746
  collinum sensu F.B.I. p.p. non Willdenow, 748
  donianum Sweet, 748
  grevilleanum Wall., 746
  grevilleanum sensu F.B.I. p.p. non Wall., 746
  lambertii Sweet, 746
    var. backhousianum (Regel) Hara, 746
  nakaoanum Hara, 745
  nepalense Sweet, 748
  polyanthes Edgeworth & Hook. f., 745
  procurrens Yeo, 746, 747
  refractum Edgeworth & Hook. f., 748
  stenorrhizum Stapf, 748
  wallichianum D. Don, 746
Geum L., 580
  aleppicum Jacquin, 581
  elatum G. Don, 580
    var. elatum, 580
    var. humile (Royle) Hook. f., 580
  macrosepalum Ludlow, 581
  sikkimense Prain, 581
  urbanum L., 582
Gliricidia Humboldt, Bonpland & Kunth, 660
  sepium (Jacquin) Walpers, 660
Glochidion J. R. & J. G. A. Forster, 776
  acuminatum Mueller, 779
  assamicum (Mueller) Hook. f., 780
    var. brevipedicellatum Hurusawa & Tanaka, 780
  bhutanicum Long, 780
  daltonii (Mueller) Kurz, 781
  gamblei Hook. f., 781
  heyneanum (Wight) Beddome, 779
  hirsutum (Roxb.) Voigt, 778
  khasicum (Mueller) Hook. f., 780
  lanceolarium (Roxb.) Voigt, 780
  multiloculare (Willdenow) Mueller, 778
  nubigenum Hook. f., 779
  oblatum Hook. f., 779
  sphaerogynum (Mueller) Kurz, 781
  thomsonii Hook. f., 781
  velutinum Wight, 763, 779
Glycine Willdenow, 693
  hispida (Moench) Maximowicz, 693

INDEX OF BOTANICAL NAMES

max (L.) Merrill, 693
soja sensu F.B.I. non Siebold & Zuccarini, 693
GROSSULARIACEAE, 522
Gueldenstaedtia Fischer, 722
himalaica Baker, 723
santapauii Thothathri, 723

HAMAMELIDACEAE, 468
Hedysarum L., 723
sikkimense Baker, 723
Hemicicca glauca (Mueller) Hurusawa & Tanaka, 774
Homonoia Loureiro, 806
riparia Loureiro, 806, 807
symphylliifolia Gamble, 808
Hydrangea L., 518
altissima Wall. 520
anomala D. Don, 519, 520
aspera D. Don, 520
  subsp. aspera, 520
  subsp. robusta (Hook. f. & Thomson) McClintock, 520
heteromalla D. Don, 520
macrophylla (Thunberg) Seringe, 521
  subsp. *stylosa* (Hook. f. & Thomson) McClintock, 521
robusta Hook. f. & Thomson, 520
stylosa Hook. f. & Thomson, 521
vestita sensu. F.B.I. non Wall., 520
HYDRANGEACEAE, 517
Hydrobryum Endlicher, 738
griffithii (Griff.) Tulasne, 738, 741
Hylotelephium spectabile (Boreau) Ohba, 483

Indigofera L., 662
argentea L. non Burman f., 664
astragalina DC., 663
atropurpurea Hornemann, 667
bracteata Baker, 667
caerulea Roxb., 664
cassioides DC., 664
cylindracea Baker, 666
dosua D. Don, 664
  var. dosua, 664
  var. tomentosa Baker, 664
exilis Grierson & Long, 665
gerardiana Baker, 665
hebepetala Baker, 666
  var. glabra Ali, 666
  var. hebepetala, 666
heterantha Brandis, 665
  var. *longipedicellata* Thothathri, 666
hirsuta sensu F.B.I. non L., 663

leptostachya sensu F.B.I. non DC., 665
linifolia (L.f.) Retzius, 663
pseudoreticulata Grierson & Long, 666
pulchella sensu F.B.I. non Roxb., 664
stachyodes Lindley, 664
tinctoria L., 665
trifoliata L., 663
zollingeriana Miquel, 665
Itea L., 527
macrophylla Roxb., 519, 528
ITEACEAE, 527

Jatropha L., 790
curcas L., 789, 790
podagrica Hooker, 790

Kalanchoe Adanson, 472
integra (Medikus) Kuntze, 469, 473
pinnata (Lamarck) Persoon, 473
spathulata DC., 473
Kirganelia microcarpa (Bentham) Hurusawa & Tanaka, 773
reticulata (Poiret) Gamble, 773
Lablab Adanson, 697
purpureus (L.) Sweet, 698, 717
Lasiobema scandens (L.) de Wit, 634
Lasiococca Hook. f., 806
symphylliifolia (Gamble) Hook. f., 808
Lathyrus L., 726
odoratus L., 726
Laurocerasus acuminata (Wall.) Roemer, 539
undulata Roemer, 539
LEGUMINOSAE, 607
Lespedeza Michaux, 682
cuneata (Dumont de Courset) G. Don, 682
eriocarpa sensu F.B.I. non DC., 681
gerardiana Maximowicz, 683
juncea (L.f.) Persoon, 682
  var. juncea, 683
  var. sericea (Thunberg) Lace & Hemsley, 683
  var. variegata (Cambessedes) Ali, 683
sericea Miquel, 682
Leucaena Bentham, 639
glauca Bentham non L., 640
leucocephala (Lamarck) de Wit, 640
LINACEAE, 751
Linum L., 753
nutans Maximowicz, 753
usitatissimum L., 753
Lotus L., 724
corniculatus L., 724
Lourea vespertilionis (L.f.) Desveaux, 680

INDEX OF BOTANICAL NAMES

Macaranga Petit-Thouars, 803
  denticulata (Blume) Mueller, 804, 807
  gamblei Hook. f., 805
  gmelinifolia Hook. f., 805
  gummiflua (Miquel) Mueller, 804
  indica Wight, 805
  peltata Hook. f., 805
  pustulata Hook. f., 805
  roxburghii Wight, 805
Macrotyloma (Wight & Arnott) Verdcourt, 698
  uniflorum (Lamarck) Verdcourt, 699
Maddenia Hook. f. & Thomson, 537
  himalaica Hook. f. & Thomson, 537
    var. glabrifolia Hara, 538
  hypoleuca Koehne, 537
  pedicellata Hook. f., 540
Mallotus Loureiro, 799
  albus sensu F.B.I. non (Jack) Mueller, 802
  nepalensis Mueller, 800
  nepalensis sensu F.B.I. p.p. non Mueller, 800
    var. ochraceo-albidus (Mueller) Pax & Hoffmann, 800
  oreophilus Mueller, 800
  philippensis (Lamarck) Mueller, 801, 802
  philippinensis auct., 802
  repandus (Willdenow) Mueller, 802
  roxburghianus Mueller, 802
  tetracoccus (Roxb.) Kurz, 802
Malus L., 603
  baccata (L.) Borkhausen, 604
  pumila Miller, 604
  sikkimensis (Wenzig) Koehne, 604
Manihot Miller, 794
  esculenta Crantz, 795, 801
  utilissima Pohl, 795
Mastersia Bentham, 694
  assamica Bentham, 694
  cleistocarpa Bentham, 694
Melilotus Miller, 728
  indica (L.) Allioni, 728
  parviflora Desfontaines, 728
Mercurialis L., 796
  leiocarpa Siebold & Zuccarini, 796, 801
Mezoneuron cucullatum (Roxb.) Wight & Arnott, 623
Micromeles ferruginea (Wenzig) Schneider, 596
  griffithii Decaisne, 595
  rhamnoides Decaisne, 596
  thomsonii (Hook. f.) Schneider, 596
Millettia Wight & Arnott, 658

  auriculata Brandis, 658
  cinerea Bentham, 658
  extensa (Bentham) Baker, 658
  glaucescens Kurz, 659
  .pachycarpa Bentham, 658
  piscidia (Roxb.) Wight, 659
  prainii Dunn, 659
Mimosa L., 638
  himalayana Gamble, 637, 639
  pudica L., 639, 649
    var. hispida Brenan, 639
  rubicaulis Lamarck, 639
    subsp. *himalayana* (Gamble) Ohashi, 639
Mimosoideae, 608
Moghania fruticulosa (Bentham) Mukerjee, 707
  involucrata (Bentham) Kuntze, 707
  macrophylla (Willdenow) Kuntze, 707
  paniculata (Bentham) Li, 707
  stricta (Roxb.) Kuntze, 708
  strobilifera (L.) Kuntze, 706
Mucuna Adanson, 685
  capitata Wight & Arnott, 687
  imbricata Baker, 671, 686
  imbricata sensu F.B.I. p.p. non Baker, 686
  interrupta Gagnepain, 686
  macrocarpa Wall., 671, 686
  nigricans (Loureiro) Steudel, 686
  pruriens (L.) DC., 671, 687
    var. pruriens, 687
    var. utilis (Wight) Burck, 687
  sempervirens Hemsley, 686

Neillia D. Don, 536
  rubiflora D. Don, 537
  thyrsiflora D. Don, 536

Ormosia Jackson, 650
  glauca Wall., 650
Ostodes Blume, 795
  paniculata Blume, 795, 801
Ougeinia dalbergioides Bentham, 673
OXALIDACEAE, 739
Oxalis L., 740
  acetosella sensu F.B.I. p.p. non L., 742
  acetosella L., 742
    subsp. *griffithii* (Edgeworth & Hook. f.) Hara, 742
  corniculata L., 742
  corymbosa DC., 743
  griffithii Edgeworth & Hook. f., 742
  latifolia Humboldt, Bonpland & Kunth, 743

INDEX OF BOTANICAL NAMES

leucolepis Diels, 741, 742
martiana Zuccarini, 743
Oxytropis DC., 721
humifusa Karelin & Kirilow, 722
lapponica (Wahlenberg) Gay, 717, 721
var. lapponica, 722
var. xanthantha Baker, 722
microphylla (Pallas) DC., 722
sericopetala Fischer, 722
sulphurea Ledebour, 722

Pachyrhizus DC., 690
angulosus DC., 691
erosus (L.) Urban, 691
Papilionoideae, 608
Parnassia L., 515
affinis Hook. f. & Thomson, 516
chinensis Franchet, 516
cooperi Evans, 517
delavayi Franchet, 517, 519
mysorensis sensu F.B.I. p.p. non Wight & Arnott, 516
nubicola Royle, 516, 519
ornata Arnott, 517
ovata sensu F.B.I. non Ledebour, 516
pusilla Arnott, 516
tenella Hook. f. & Thomson, 517
wightiana Wight & Arnott, 517, 519
PARNASSIACEAE, 515
Parochetus D. Don, 728
communis D. Don, 649, 728
Pedilanthus Poiteau, 767
tithymaloides (L.) Poiteau, 763, 767
Pelargonium Aiton, 748
zonale (L.) Aiton, 749
Peltophorum (Vogel) Bentham, 621
ferrugineum Bentham, 621
pterocarpum (DC.) Heyne, 621
Persica vulgaris Miller, 542
Phanera vahlii (Wight & Arnott) Bentham, 634
Phaseolus L., 701
aureus Roxb., 700
calcaratus Roxb., 701
coccineus L., 701
lunatus L., 702
mungo L., 701
mungo sensu F.B.I. non L., 700
tenuicaulis Baker, 698
trilobus Aiton, 700
velutinus Baker, 697
vulgaris L., 649, 702
PHILADELPHACEAE, 525
Philadelphus L., 527

coronarius L., 527
var. *tomentosus* (G. Don) Clarke, 527
tomentosus G. Don, 527
Photinia Lindley, 599
arguta Lindley, 601
var. hookeri (Decaisne) Vidal, 601
beauverdiana Schneider, 600
griffithii Decaisne, 600
integrifolia Lindley, 600
mollis Hook. f., 601
Phyllanthus L., 770
acidus (L.) Skeels, 773
clarkei Hook. f., 774
dalbergioides J. J. Smith, 773, 774
debilis Willdenow, 772
distichus (L.) Mueller, 773
emblica L., 763, 772
glaucus Mueller, 774
griffithii Mueller, 775
hamiltonianus Mueller, 774
leschenaultii Mueller, 775
parvifolius D. Don, 775
pulcher Mueller, 774
reticulatus Poiret, 763, 773
sikkimensis Mueller, 774
simplex Retzius, 772
urinaria L., 772
virgatus Forster f., 772
Phyllodium pulchellum (L.) Desveaux, 669
Piliostigma malabaricum (Roxb.) Bentham, 633
Piptanthus Sweet, 737
bombycinus Marquand, 737
laburnifolius (D. Don) Stapf, 737
nepalensis (Hook.) Sweet, 737
tomentosus Franchet, 737
Pisum L., 726
sativum L., 727
var. arvense (L.) Poiret, 727
Pithecellobium Martius, 648
clypearia (Jack) Bentham, 650
dulce (Roxb.) Bentham, 648
heterophyllum (Roxb.) MacBride, 650
monadelphum (Roxb.) Kostermans, 650
Pithecolobium angulatum Bentham, 650
bigeminum sensu F.B.I. p.p., 650
saman (Jacquin) Bentham, 647
PITTOSPORACEAE, 528
Pittosporum Solander, 528
floribundum sensu F.B.I. p.p. non Wight & Arnott, 528
glabratum sensu F.B.I. non Lindley, 529
napaulense (DC.) Rehder & Wilson, 519, 528

823

INDEX OF BOTANICAL NAMES

var. *rawalpindiense* Gowda, 528
podocarpum Gagnepain, 529
Plukenetia corniculata Smith, 811
PODOSTEMACEAE, 738
Poinciana regia Hooker, 622
Poinsettia pulcherrima (Klotzsch) Graham, 762
Potentilla L., 562
   albifolia Hook. f., 576
   ambigua Cambessedes, 566
   anserina L., 571
   arbuscula D. Don, 565
      var. arbuscula, 566
      var. pumila (Hook. f.) Handel-Mazzetti, 566
      var. unifoliolata Ludlow, 566
   argyrophylla Lehmann, 569
   armerioides (Hook. f.) Grierson & Long, 566
   atrosanguinea Loddiges, 569
      var. argyrophylla (Lehmann) Grierson & Long, 569
      var. atrosanguinea, 569
   bhutanica Ludlow, 567
   bifurca L., 569
      var. moorcroftii (Lehmann) Wolf, 569
   bryoides Sojak, 573
   caliginosa Sojak, 570
   commutata Lehmann, 573
   coriandrifolia D. Don, 573
   cuneata Lehmann, 566
   eriocarpa Lehmann, 567
      var. eriocarpa, 567
      var. dissecta Marquand, 567
      var. *tsarongensis* W. E. Evans, 567
   eriocarpoides Krause, 567
   forrestii W. W. Smith, 569
      var. caespitosa (Wolf) Sojak, 569
      var. forrestii, 569
      var. segmentata Sojak, 569
   fragarioides L. agg., 568
   fruticosa sensu F.B.I. p.p. non L., 565
   fruticosa L, 566
      var. *armerioides* Hook. f., 566
      var. *pumila* Hook. f., 566
   fulgens Hooker, 571
   griffithii Hook. f., 570
   indica (Andrews) Wolf, 579
   kleiniana Wight, 567
   latipetiolata Fischer, 566
   leschenaultiana Seringe, 570
   leuconota D. Don, 572
   lineata Treviranus, 571
      var. intermedia (Hook. f.) Dixit & Panigrahi, 571
   microphylla D. Don, 572
      var. achilleifolia Hook. f., 572
      var. *commutata* (Lehmann) Hook. f., 573
      var. *depressa* Lehmann, 572
      var. latifolia Lehmann, 573
      var. latiloba Lehmann, 573
      var. microphylla, 572
   monanthes Lehmann, 568
      var. monanthes, 568
      var. sibthorpioides Hook. f., 568
   mooniana Wight, 571
   moorcroftii Lehmann, 569
   multifida L., 568
      var. *saundersiana* (Royle) Hook. f., 568
   nivea sensu F.B.I. p.p. non L., 568, 569
   peduncularis D. Don, 571
      var. clarkei Hook. f., 572
      var. peduncularis, 572
   perpusilla Hook. f., 576
   perpusilloides W. W. Smith, 575
   polyphylla Lehmann, 571
   potaninii Wolf, 568
   purpurea (Royle) Hook. f., 576
   ribui Gandoger, 565
   saundersiana Royle, 568
      var. caespitosa Wolf, 569
      var. potaninii (Wolf) Handel-Mazzetti, 568
      var. saundersiana, 568
   sericea L., 577
      var. *compacta* W. W. Smith & Cave, 577
   sibbaldii sensu F.B.I. non Haller f., 574
   siemersiana Lehmann, 571
   sikkimensis Wolf non Prain, 570
   spodiochlora Sojak, 570
   sundaica (Blume) Kuntze, 567
   supina L., 570
   tetrandra (Bunge) Hook. f., 575
   trullifolia Hook. f., 576
   wallichiana Lehmann non Seringe, 567
Poterium diandrum Hook. f., 584
   filiforme Hook. f., 585
Pourthiaea arguta (Lindley) Decaisne, 601
Prinsepia Royle, 543
   utilis Royle, 543
Priotropis cytisoides (DC.) Wight & Arnott, 732
Prunus L., 538
   acuminata (Wall.) Dietrich, 539
   amygdalus Batsch, 543

INDEX OF BOTANICAL NAMES

arborea (Blume) Kalkman, 539
  var. montana (Hook. f.) Kalkman, 540
armeniaca L., 542
carmesina Hara, 541
cerasoides D. Don, 540, 605
  var. *rubea* Ingram, 541
cerasus L., 541
ceylanica (Wight) Miquel, 539
communis Hudson, 542
cornuta (Royle) Steudel, 540
domestica L., 542
dulcis (Miller) Webb, 543
glauciphylla Ghora & Panigrahi, 540
jenkinsii Hook. f. & Thomson, 539
napaulensis (Seringe) Steudel, 540
padus sensu F.B.I. non L., 540
persica (L.) Batsch, 542
puddum (Seringe) Brandis, 540
racemosa Lamarck, 540
rufa Hook. f., 541
  var. trichantha (Koehne) Hara, 541
trichantha Koehne, 541
undulata D. Don, 539
undulata sensu F.B.I. non D. Don, 540
venosa Koehne, 540
wallichii Steudel, 539
wattii Ghora & Panigrahi, 540
Pterococcus Hasskarl, 811
  corniculatus (Smith) Pax & Hoffmann, 811
Pterolobium Wight & Arnott, 626
  hexapetalum (Roth) Santapau, 626
  indicum sensu F.B.I. non A. Richard, 626
  indicum A. Richard, 626
    var. *macropterum* (Kurz) Baker, 626
  macropterum Kurz, 625, 626
Pueraria DC., 691
  edulis Pampanini, 692
  lobata (Willdenow) Ohwi, 692
    var. thomsonii (Bentham) van der Maesen, 692
  peduncularis (Bentham) Bentham, 692
  phaseoloides (DC.) Bentham, 693
    var. subspicata (Bentham) van der Maesen, 693
  quadristipellata, W. W. Smith, 692
  sikkimensis Prain, 691 '
  thomsonii Bentham, 692
  thunbergiana (Siebold & Zuccarini) Bentham, 692
  wallichii DC., 692
Pygeum acuminatum Colebrooke, 539
  ceylanicum sensu F.B.I. non Gaertner, 539
  glaberrimum Hook. f., 539

montanum Hook. f., 539
Pyracantha Roemer, 592
  crenulata (D. Don) Roemer, 592
Pyrus L., 606
  baccata L., 604
  bhutanica W. W. Smith, 596
  communis L., 606
  ferruginea (Wenzig) Hook. f., 596
  foliolosa Wall., 598
  griffithii (Decaisne) Hook. f., 595
  insignis (Hook. f., 599
  kurzii Prain, 599
  malus L., 604
  microphylla (Wenzig) Hook. f., 597
  pashia D. Don, 606
  rhamnoides (Decaisne) Hook. f., 596
  serrulata Rehder, 606
  sikkimensis (Wenzig) Hook. f., 604
  thomsonii Hook. f., 596
  ursina G. Don, 598
  vestita G. Don, 595
  wallichii Hook. f., 598

Reidia hamiltoniana (Mueller) Cowan & Cowan, 774
Reinwardtia Dumortier, 751
  cicanoba (D. Don) Hara, 752
  indica Dumortier, 747, 752
  tetragyna (Bentham) Planchon, 752
  trigyna Planchon, 752
Rhodiola L., 474
  amabilis (Ohba) Ohba, 480
  asiatica D. Don, 477
  atsaensis (Froderstrom) Ohba, 481
  bupleuroides (Hook. f. & Thomson) Fu, 469, 479
  chrysanthemifolia (Leveille) Fu, 476
  coccinea (Royle) Borissova, 477
    subsp. coccinea, 477
    subsp. scabrida (Franchet) Ohba, 477
  crenulata (Hook. f. & Thomson) Ohba, 478
  cretinii (Hamet) Ohba, 481
  fastigiata (Hook. f. & Thomson) Fu, 478
  himalensis (D. Don) Fu, 477
  hobsonii (Hamet) Fu, 479
  humilis (Hook. f. & Thomson) Fu, 480
  imbricata Edgeworth, 478
  ludlowii Ohba, 481
  marginata Grierson, 476
  nobilis (Franchet) Fu, 477
  prainii (Hamet) Ohba, 476
  purpureoviridis (Praeger) Fu, 479
    subsp. *phariensis* (Ohba) Ohba, 479

825

INDEX OF BOTANICAL NAMES

sherriffii Ohba, 479
smithii (Hamet) Fu, 480
stapfii (Hamet) Fu, 476
wallichiana (Hooker) Fu, 478
Rhynchosia Loureiro, 708
  harae Ohashi & Tateishi, 709
  minima (L.) DC., 709
Ribes L., 522
  acuminatum G. Don, 524
  alpestre Decaisne, 523
  desmocarpum Hook. f. & Thomson, 524
  emodense Rehder, 524
  glaciale Wall., 525
  griffithii Hook. f. & Thomson, 519, 523
  grossularia sensu F.B.I. non L., 523
  himalense Decaisne, 524
  laciniatum Hook. f. & Thomson, 525
  luridum Hook. f. & Thomson, 525
  orientale Poiret, 524
  rubrum sensu F.B.I. non L., 524
  takare D. Don, 524
  tenue Janczewski, 525
Ricinus L., 808
  communis L., 807, 808
Rodgersia Gray, 486
  nepalensis Cullen, 486, 487
Rosa L., 585
  brunonii Lindley, 586
  hybrida Hortorum, 588
  involucrata Roxb., 587
  lyellii Lindley, 587
  macrophylla Lindley, 583, 587
  microphylla Lindley non Desfontaines, 587
  moschata sensu F.B.I. non Herrmann, 586
  roxburghii Trattinik, 587
  sericea Lindley, 586
    var. hookeri Regel, 586
    var. omeiensis (Rolfe) Rowley, 586
    var. pteracantha (Franchet) Bean, 586
    var. sericea, 586
ROSACEAE, 529
Rubus L., 543
  acaenocalyx Hara, 557
  acuminatus Smith, 550
  alexeterius Focke, 557
  alpestris sensu F.B.I. p.p. non Blume, 556
  andersonii Hook. f. non Lefevre, 555
  arcticus L., 550
    var. *fragarioides* (Bertoloni) Focke, 550
  arcuatus Kuntze, 554
  asper D. Don non Presl, 562
  barbatus Rehder non Fritsch, 550

bhotanensis Kuntze, 553
biflorus Smith, 558, 605
calophyllus Clarke, 552
calycinoides Kuntze, 553
calycinus D. Don, 549
cooperi Long, 555
cordifolius D. Don non Noronha, 552
coronarius (Sims) Sweet, 562
darschilingensis Kuntze, 553
diffisus Focke, 553
distans D. Don, 560
efferatus Craib, 553
ellipticus Smith, 557
  var. *denudatus* Hook. f., 557
excurvatus Kuntze, 551
fallax Kuntze, 553
ferox Focke non Trattinik, 553
flavus D. Don, 557
fockeanus Kurz, 549
fragarioides Bertoloni, 550
gigantiflorus Hara, 559
glandulifer Balakrishnan, 553
gracilis DC. non Presl, 559
griffithii Hook. f., 551
hamiltonii Hook. f., 551
himalaicus Kuntze, 553
hookeri Focke non Koch, 559
horridulus Hook. f. non Mueller, 561
hypargyrus Edgeworth, 559
  var. hypargyrus, 560
  var. niveus (G. Don) Hara, 560
indotibetanus Koidzumi, 562
inopertus (Focke) Focke, 561
insignis Hook. f., 553
  var. *ochraceus* Focke, 553
irritans Focke, 558
kumaonensis Balakrishnan, 554
kurzii Balakrishnan, 553
lanatus Hook. f. non Focke, 553
lasiocarpus Smith, 560
  var. *micranthus* (D. Don) Hook. f., 561
  var. *rosifolius* Hook. f., 561
latifolius Kuntze, 550
lineatus Blume, 555
  var. *andersonii* Kuntze, 555
  var. *angustifolius* Kuntze, 555
  var. *intermedius* Kuntze, 555
  var. *pulcherrimus* Kuntze, 555
macilentus Cambessedes, 560
macrocarpus Clarke non Bentham, 559
mesogaeus Focke, 560
micranthus D. Don, 561
moluccanus L., 554
  var. *calycinoides* (Kuntze) Kuntze, 553

INDEX OF BOTANICAL NAMES

var. *ferox* Kuntze, 553
var. *insignis* (Hook. f.) Kuntze, 553
var. *paniculatus* (Smith) Kuntze, 552
var. *tiliaceus* (Smith) Kuntze, 552
var. *treutleri* (Hook. f.) Kuntze, 554
moluccanus auct. p.p. non L., 553, 554
nepalensis (Hook. f.) Kuntze, 550
niveus Thunberg, 560
 var. *hypargyrus* (Edgeworth) Hook. f., 560
 subsp. *inopertus* Focke, 561
 var. *micranthus* (D. Don) Hara, 561
 var. *microcarpus* Hook. f., 560
 var. *niveus,* 561
 var. *rosifolius* (Hook. f.) Hara, 561
niveus G. Don non Thunberg, 559
nutans G. Don non Vest, 550
 var. *fockeanus* (Kurz) Kuntze, 549
 var. *nepalensis* Hook. f., 550
nutantiflorus Hara, 550
 var. *nepalensis* (Hook. f.) Balakrishnan, 550
paniculatus Smith, 552
 forma *tiliaceus* (Smith) Hara, 552
paniculatus sensu F.B.I. p.p. non Smith, 552
parapungens Hara, 561
pectinaroides Hara, 549
pedunculosus auct. non D. Don, 559
pentagonus Focke, 556
phengodes Focke, 555
pilocalyx Kuntze, 550
poliophyllus Kuntze, 552
preptanthus Focke, 551
pungens Cambessedes, 561
 var. *horridulus* Hara, 562
purpureus sensu F.B.I. p.p. non Bunge, 558
reticulatus Hook. f. non Kerner, 554
rosifolius Smith, 562
 forma *coronarius* (Sims) Kuntze, 562
rosifolius sensu F.B.I. p.p. non Smith, 562
rosulans Kuntze, 554
rugosus Smith, 554
rugosus auct. p.p. non Smith, 553
senchalensis Hara, 556
sengorensis Grierson & Long, 550
sikkimensis Hook. f., 558
 var. sikkimensis, 558
 var. *canescens* Long, 558
sikkimensis Kuntze non Hook. f., 562
sorbifolius Maximowicz, 562
splendidissimus Hara, 555
sterilis Kuntze, 553

subherbaceus Kuntze, 549
sumatranus Miquel, 562
thomsonii Focke, 556
tiliaceus Smith, 552
tonglooensis Kuntze, 554
treutleri Hook. f., 554
tridactylus Focke, 556
wardii Merrill, 559

Samanea (DC.) Merrill, 646
 saman (Jacquin) Merrill, 647
Sanguisorba L., 584
 diandra (Hook. f.) Nordborg, 583, 584
  var. diandra, 585
  var. *villosa* Purohit & Panigrahi, 585
 filiformis (Hook. f.) Handel-Mazzetti, 585
Sapium Browne, 811
 baccatum Roxb., 812
 eugeniifolium Hook. f., 812
 insigne (Royle) Hook. f., 807, 812
 sebiferum (L.) Roxb. 813
Sauropus Blume, 782
 albicans Blume, 783
 androgynus (L.) Merrill, 783
 compressus Mueller, 783
 macranthus Hasskarl, 783
 macrophyllus Hook. f., 783
 pubescens Hook. f., 783
 quadrangularis (Willdenow) Mueller, 783
  var. *compressus* (Mueller) Airy Shaw, 783
  var. *puberulus* Kurz, 783
 repandus Mueller, 784
 stipitatus Hook. f., 783
 trinervius sensu F.B.I. p.p. non Mueller, 784
Saxifraga L., 492
 alookthangensis Wadhwa, 511
 andersonii Engler, 487, 497
 aristulata Hook. f. & Thomson, 510
 asarifolia Sternberg, 502
 bergenioides Marquand, 504
 brachypoda D. Don, 503
 brunoniana Sternberg, 499
 brunonis Seringe, 487, 499
 bumthangensis Wadhwa, 500
 caveana W. W. Smith, 510
 cernua L., 503
 chumbiensis Engler & Irmscher, 512
 clivorum H. Smith, 497
 coarctata W.W. Smith, 503
 contraria H. Smith, 514
 cordigera Hook. f. & Thomson, 509, 510

827

INDEX OF BOTANICAL NAMES

corymbosa Hook. f. & Thomson non Boissier, 508
deminuta H. Smith, 513
diapensia H. Smith, 510
diversifolia Seringe, 507
diversifolia sensu F.B.I. p.p. non Seringe, 507
dungbooii Engler & Irmscher, 502
elliptica Engler & Irmscher, 511
engleriana H. Smith, 514
erinacea H. Smith, 505
exigua H. Smith, 512
filicaulis Seringe, 504
fimbriata Seringe, 503
finitima W. W. Smith, 513, 514
flagellaris Sternberg, 499
    var. *mucronulata* (Royle) Clarke, 499
flavida H. Smith, 498
gageana W. W. Smith, 501
georgei Anthony, 498
glabricaulis H. Smith, 511
gouldii Fischer, 503, 505
    var. *eglandulosa* H. Smith, 503
    var. *gouldii*, 505
granulifera H. Smith, 502
haematochroa H. Smith, 509
harry-smithii Wadhwa, 506
hemisphaerica Hook. f. & Thomson, 487, 503
himalaica Balakrishnan, 501
hirculoides Decaisne, 508
hirculus L., 508, 511
    var. *indica* Hook. f., 511
    var. *subdioica* Clarke, 508
hispidula D. Don, 500
    var. *doniana* Engler, 500
    var. *hispidula*, 500
hookeri Engler & Irmscher, 508
    var. *glabrisepala* Engler & Irmscher, 509
humilis Engler & Irmscher, 503
imbricata Royle non Lamarck, 498
inconspicua W. W. Smith, 512
indica (Hook. f.) Wadhwa, 511
jacquemontiana Decaisne, 514
    var. *stella-aurea* (Hook. f. & Thomson) Clarke, 513
kinchingingae Engler, 512
kingiana Engler & Irmscher, 506
lamninamensis Ohba, 508
latiflora Hook. f. & Thomson, 506
lepida H. Smith, 510
lepidostolonosa H. Smith, 510
ligulata Wall., 492

llonakhensis W. W. Smith, 510
lychnitis Hook. f. & Thomson, 504
matta-florida H. Smith, 498
matta-viridis H. Smith, 513
megalantha Marquand, 505
melanocentra Franchet, 502
micrantha Edgeworth non Fischer, 501
microphylla Hook. f. & Thomson, 515
montana H. Smith, 511
montanella H. Smith, 511
moorcroftiana (Seringe) Sternberg, 487, 507
mucronulata Royle, 499
    subsp. *sikkimensis* (Hulten) Hara, 499
nigroglandulifera Balakrishnan, 505
nigroglandulosa Engler & Irmscher, 512
nutans Hook. f. & Thomson non D. Don nec Adams, 505
odontophylla Hook. f. & Thomson non Sternberg, 502
pallida Seringe, 501, 502
palpebrata Hook. f. & Thomson sensu F.B.I. p.p., 509, 510
parnassifolia D. Don, 507, 508
parva Hemsley, 509, 511
perpusilla Hook. f. & Thomson, 513
petrophila Franchet, 509
pilifera Hook. f. & Thomson, 500
pluviarum W. W. Smith, 501
przewalskii Engler, 508
pseudohirculus Engler & Irmscher, 505
pseudopallida Engler & Irmscher, 502
pulvinaria H. Smith, 498
punctulata Engler, 514
purpurascens Hook. f. & Thomson, 492
rubiflora H. Smith, 501
saginoides Hook. f. & Thomson, 513
sarmentosa Schreber, 499
saxorum H. Smith, 498
serrula H. Smith, 506
sherriffii H. Smith, 496
sikkimensis Engler, 509
sphaeradena H. Smith, 511
    subsp. *dhwojii* H. Smith, 507
stella-aurea Hook. f. & Thomson, 513
    var. ciliata Marquand & Shaw, 514
    var. polyadena H. Smith, 514
stolitzkae Engler & Irmscher, 497
stolonifera Curtis, 499
strigosa Seringe, 500
subaristulata Engler, 510
subsessiliflora Engler & Irmscher, 497
subspathulata Engler & Irmscher, 512
tangutica Engler, 487, 508

INDEX OF BOTANICAL NAMES

tentaculata Fischer, 500
thiantha H. Smith, 496
   var. citrina H. Smith, 496
   var. thiantha, 496
thimpuana Wadhwa, 512
tsangchanensis Franchet, 512
umbellulata Hook. f. & Thomson, 508
vacillans H. Smith, 497
viscidula Hook. f. & Thomson, 505
wardii W. W. Smith, 505
SAXIFRAGACEAE, 485
Securinega suffruticosa (Pallas) Rehder, 776
virosa (Willdenow) Baillon, 775
Sedum L., 481
   amabile Ohba, 480
   asiaticum (D. Don) DC., 477
   atsaense Froderstrom, 481
   barnesianum Praeger, 480
   bhutanense Praeger, 479
   bhutanicum Praeger, 479
   bupleuroides Hook. f. & Thomson, 479
   cavei Hamet, 485
   chrysanthemifolium Léveille, 476
   coccineum Royle, 477
   cooperi Praeger non Clemenceau, 479
   correptum Hamet, 484
   crassipes Hook. f. & Thomson, 478
   crenulatum Hook. f. & Thomson, 478
   cretinii Hamet, 481
   discolor Franchet, 479
   elongatum Hook. f. & Thomson non Ledebour, 479
   fastigiatum Hook. f. & Thomson, 478
   filipes Hemsley, 483
   fischeri Hamet, 483
   gagei Hamet, 484
   griffithii Clarke, 485
   henrici-robertii Hamet, 484
   himalense D. Don, 477
   hobsonii Hamet, 479
   hookeri Balakrishnan, 479
   humile Hook. f. & Thomson, 480
   indicum (Decaisne) Hamet, 473
   jaeschkei Kurz, 485
   levii Hamet, 480
   linearifolium Royle, 476
      var. *ovatisepalum* Hamet, 476
   mirabile Ohba, 479
   multicaule Lindley, 469, 484
   oreades (Decaisne) Hamet, 485
   ovatisepalum (Hamet) Ohba, 476
   perpusillum Hook. f. & Thomson, 483
   phariense Ohba, 479
   praegerianum W. W. Smith, 479

   prainii Hamet, 476
   przewalskii Maximowicz, 483
   pseudostapfii Praeger, 483
   pseudosubtile Hara, 485
   quadrifidum sense F.B.I. p.p. non Pallas, 477, 478
      var. *himalense* (D. Don) Froderstrom, 477
   quevae Hamet, 485
   rhodiola sensu F.B.I. non DC., 478
   rotundatum Hemsley, 478
   smithii Hamet, 480
   spectabile Boreau, 483
   stapfii Hamet, 476
   triactina Berger, 482
   trifidum Hook. f. & Thomson, 476
   trullipetalum Hook. f. & Thomson, 484
   venustum Praeger p.p., 478
   verticillatum (Hook. f. & Thomson) Hamet non L., 482
   wallichianum Hook. f. & Thomson, 478
Sesbania Scopoli, 660
   aegyptiaca Persoon, 662
   sesban (L.) Merrill, 662
      var. bicolor Wight & Arnott, 662
      var. picta Prain, 662
Shuteria Wight & Arnott, 694
   ferruginea (Bentham) Baker, 695
   hirsuta Baker, 695
   involucrata (Wall.) Wight & Arnott, 695
      var. glabrata (Wight & Arnott) Ohashi, 659
   vestita Wight & Arnott, 695
Sibbaldia L., 573
   byssitecta Sojak, 576
   compacta (Smith & Cave) Dixit & Panigrahi, 577
   cuneata Kuntze, 574
   macropetala Muravjeva, 577
   melinotricha Handel-Mazzetti, 575
   micropetala (D. Don) Handel-Mazzetti, 576
   parviflora Willdenow, 574
      var. micrantha (Hook. f.) Dixit & Panigrahi, 575
   pentaphylla Krause, 577
   perpusilla (Hook. f.) Chatterjee, 576
   perpusilloides (W. W. Smith) Handel-Mazzetti, 575
   purpurea Royle, 576
   sikkimensis (Prain) Chatterjee, 575
   tetrandra Bunge, 575
   trullifolia (Hook. f.) Chatterjee, 576

INDEX OF BOTANICAL NAMES

Sinocrassula Berger, 473
  indica (Decaisne) Berger, 473
Smithia Aiton, 710
  ciliata Royle, 711
  grandis Baker, 711
  sensitiva Aiton, 711
Sophora L., 651
  acuminata sensu F.B.I. non Desveaux, 651
  benthamii van Steenis, 651
  bhutanica Ohashi, 651
  glauca DC. non Salisbury, 652
  velutina Lindley, 652
  wightii Baker, 651
    subsp. bhutania (Ohashi) Grierson & Long, 651, 671
Sorbus L., 592
  arachnoidea Koehne, 597
  bhutanica (W. W. Smith) Balakrishnan, 596
  cuspidata (Spach) Hedlund, 595
  ferruginea (Wenzig) Rehder, 596
  foliolosa (Wall.) Spach, 598
  foliolosa auct. p.p. non (Wall.) Spach, 597, 598
  griffithii (Decaisne) Rehder, 595
  hedlundii Schneider, 595
  himalaica Gabrielian, 598
  hupehensis Schneider, 599
    var. oligodonta (Cardot) Yu, 599
  insignis (Hook. f.) Hedlund, 599
  kurzii (Prain) Schneider, 599
  microphylla Wenzig, 597
  oligodonta (Cardot) Handel-Mazzetti, 599
  prattii Koehne, 597
  rhamnoides (Decaisne) Rehder, 596
  rufopilosa Schneider, 597
  sikkimensis Wenzig 595, 596
    var. ferruginea Wenzig, 596
    var. microcarpa Wenzig, 595
    var. oblongifolia Wenzig, 596
  thibetica (Cardot) Handel-Mazzetti, 594
  thomsonii (Hook. f.) Rehder, 596
  ursina (G. Don) Schauer, 598
    var. wenzigiana Schneider, 598
  verticillata Merrill, 596
  vestita (G. Don) Loddiges, 595
  wallichii (Hook. f.) Yu, 598
  wattii Koehne, 599
Spartium L., 738
  junceum L., 738
Spatholobus roxburghii Bentham, 688
Spenceria Trimen, 584
  parviflora Stapf, 584
  ramalana Trimen, 584
Spiraea L., 534
  alpina Pallas, 535
  arcuata Hook. f., 535, 605
  aruncus L., 536
  bella Sims, 534
  canescens D. Don, 535
  hemicryptophyta Grierson, 534
  micrantha Hook. f., 534
  ulicina Prain, 535
  x vanhouttei (Briot) Zabel, 535
Spongiocarpella Yakovlev & Ulzijkhutag, 714
  purpurea (Li) Yakovlev, 714
Stracheya Bentham, 723
  tibetica Bentham, 717, 724
Suregada Rottler, 809
  multiflora (Jussieu) Baillon, 807, 809
Symingtonia populnea (Griff.) van Steenis, 470

Tamarindus L., 635
  indicus L., 636, 637
Tadehagi triquetrum (L.) Ohashi, 669
Tephrosia Persoon, 659
  candida DC., 659
Teramnus P. Browne, 694
  flexilis Bentham, 694
Thermopsis Brown, 737
  barbata Royle, 737
Tiarella L., 488
  polyphylla D. Don, 487, 489
Tillaea pentandra Edgeworth, 474
Tithymalus griffithii (Hook. f.) Hara, 764
  himalayensis Klotzsch, 765
  longifolius (D. Don) Hurusawa & Tanaka, 765
  sikkimensis (Boissier) Hurusawa & Tanaka, 764
  stracheyi (Boissier) Hurusawa & Tanaka, 765
Trewia L., 798
  nudiflora L., 799
Triactina verticillata Hook. f. & Thomson, 482
Tribulus L., 750
  terrestris L., 747, 750
Trifolium L., 730
  dubium Sibthorp, 730
  minus Smith, 730
  pratense L., 731
  repens L., 731

INDEX OF BOTANICAL NAMES

Trigonella L., 729
   corniculata (L.) L., 730
   emodi Bentham, 729
   foenum-graecum L., 729
   incisa Bentham, 729
   polycerata sensu F.B.I. non L., 729
TROPAEOLACEAE, 749
Tropaeolum L., 749
   majus L., 747, 749

Uraria Desveaux, 678
   hamosa (Roxb.) Wight & Arnott, 680
   lagopodioides (L.) Desveaux, 679
   lagopoides (Burman f.) DC., 679
   lagopus DC., 679
   picta (Jacquin) Desveaux, 678
   prunellifolia Baker, 679
   rufescens (DC.) Schindler, 680
   sinensis Franchet, 680

Vernicia Loureiro, 791
   cordata (Thunberg) Airy Shaw, 791

Vicia L., 725
   angustifolia L., 725
   hirsuta (L.) Gray, 725
   sativa L., 661, 725
      var. angustifolia L., 725
      var. sativa, 725
   tetrasperma Moench, 725
   tibetica Fischer, 726
Vigna Savi, 699
   clarkei Prain, 701
   mungo (L.) Hepper, 701
   pilosa (Willdenow) Baker, 700
   radiata (L.) Wilczek, 700
   trilobata (l..) Verdcourt, 700
   umbellata (Thunberg) Ohwi & Ohashi, 701
   vexillata (L.) A. Richard, 661, 700

Zornia Gmelin, 712
   diphylla sensu F.B.I. p.p. non Persoon, 712
   gibbosa Spanoghe, 661, 712
ZYGOPHYLLACEAE, 750

INDEX OF COMMON NAMES

Ainselu, 557
Almond, 543
Amala, 772
Amil Tanki, 633
Amrul, 742
Ankhataruwa, 812
Apple, 604
Apricot, 542
Arare, 639, 642
Arare Khanra, 641
Arari Khanra, 642
Archal, 786
Arere Khanra, 639
Aru, 542
Arupate, 540
Asare, 785
Aselu, 553, 554, 557
Aule Kapase, 799

Baiari, 657
Baldengra, 686, 688
Bandarlata, 628
Bandre Siris, 654
Bangikath, 780
Barikaunli, 785
Basak, 522
Bashak, 522
Basuri, 522
Batwasi, 707
Benanglebee, 732
Bepari, 795, 803
Beymangrobu, 674
Bhogote, 520
Bhokote, 520
Bhorla, 634
Bin Aselu, 549
Biraley Kara, 550
Birali, 692
Birali Lahara, 688, 691
Bird Cherry, 540
Birdsfoot Trefoil, 724
Bitter Cassava, 795
Bjake Tshalu, 556
Black Gram, 701
Bohari Jhar, 639
Bokshi Khanra, 623
Bomchu Shi, 804
Bonavist Bean, 698
Bonga Karpo, 744
Booarey, 639
Brumzey, 582
Brush jhar, 589
Bun Mara, 659
Burmese Storax, 471

Buro Okhate, 488
Buru Okhati, 488

Carambola Apple Tree, 740
Cassava, 795
Castor-oil Plant, 808
Chamling Shing, 808
Chamze Nam, 524
Chana, 727
Charu, 590
Chasokey, 594
Chassee, 684
Chenju Shing, 470
Cherry, 541
Chhinpashosha, 686
Chhorgeng, 772
Chickpea, 727
Chinese Tallow Tree, 813
Chipli, 787
Chipli Khari, 785
Chiringi Jhar, 664
Chiriya Phal, 565
Choga Sey Sey, 562
Churoo, 772
Crown of Thorns, 762

Darim Pate, 775
Debri Lahara, 688
Denrog, 808
Dharani, 539
Domaytsalu, 552
Donga, 628
Donka Sey, 628
Donko Shing, 628
Dori, 795
Dosem, 773, 774
Dosem Metog, 521
Dreeta Sazin, 499
Durjit, 764, 765
Dushi Tsang, 543

Fever Nut, 623
Field Pea, 727
Flamboyant, 622
Flame Tree, 622
Flax, 753

Gadur, 746
Gahate-phul, 737
Garden pea, 727
Gaunji Lahara, 688
Gayo, 768, 769
Gempe Aselu, 555, 562
Geykang Shing, 775

Gold Mohur, 622
Gongsey, 557
Gotham Paisay, 788
Green Gram, 700
Grong Grongmo Shing, 470
Ground Nut, 712

Haldi Kath, 780
Halonre, 520, 793
Hare, 600, 785
Hare Bepari, 803
Hare Kusum, 785
Harital, 811
Harra Siris, 645
Hawthorn, 592
Heyduk, 631
Hil Bashak, 522
Himalchari, 787
Himli, 636
Hindo Nam, 522
Hogena, 522
Hyacinth Bean, 698

Indian Cress, 749
Indian Hemp, 733

Jadum, 584
Japho Tsi Tsi, 589
Ja Shing, 813
Jhankhri-Kath, 808
Jhikre, 536
Jogi Malata, 802
Juroo Shing, 814

Kakur, 642
Kakushbish, 658
Kalo Aselu, 560
Kalo Bilaune, 787
Kalo Siris, 644, 646, 650
Kamaranga, 740
Kamrak, 740
Kaochir, 687
Kaoso, 686
Karkur Siris, 644
Kasre Malata, 802
Kasreto, 769
Kauso, 687
Kautcho, 687
Kentakare, 558
Khair, 642
Khangtsalu, 534
Khelsho, 683
Khenzem, 536
Khola Ruis, 806

INDEX OF COMMON NAMES

Khomang Shing, 602, 604
Khorsane, 528
Kimbu, 525
Koerlo, 634
Koiralo, 634
Kolokpo, 638
Komatsang, 550
Kotokmo Shing, 779, 780
Kuach, 687
Kuhir, 769
Kumchingma Shing, 664
Kumchumo Shing, 659
Kurku, 658
Kurkus, 658
Kursimla, 600
Kusum, 788

Laha Siris, 654
Lahara Gaijo, 768
Lahara Gayo, 769
Lahara Pasi, 596
Lahara Siris, 653
Lahare Phul, 534
Lahari Siris, 646
Lajunia, 639
Lal Chandan, 814
Lalgeri, 655
Lali, 539
Lam Shing, 646
Lapche Bis, 792
Latikath, 779, 781
Lati Mauwa, 780
Lebbek Siris, 646
Lee, 606
Lekh Arupate, 539
Lekh Mehel, 595
Lekh Paiyun, 541
Lem Shing, 470
Lengsey, 542
Ligadur, 746
Lima Bean, 702
Linseed, 753
Lise, 606
Lishi, 606
Litku, 788
Litong, 606
Losi Metok, 737
Lu Shing, 761

Malata, 800, 804, 805
Malchina, 781
Mandane, 620
Mardum Kumchimo Shing, 670

Mauwa, 688
Maya, 601, 602
Mehel, 603
Migme Sangey, 744
Mindu Shing, 604
Monkey Nut, 712
Mung Bean, 700
Muse Gayo, 770

Nagbaele, 634
Naspati, 606
Nasturtium, 749
Nectarine, 542
Neptang Shing, 674
Nera Khar Shing, 790
Ngingshosha, 638
Nicker Tree, 623
Nimthomozoo, 761

Olla Sema, 653
Omla, 772
Orey, 698

Paiyu, 540
Paiyun, 540
Pakhanbet, 492
Palas, 688
Pangra, 638
Pangro, 638
Pan-Kara, 553
Parke Siris, 644
Pasi, 596, 597
Patpate Siris, 645
Pchang, 654
Peach, 542
Pea Nut, 712
Pear, 606
Pegpeyposhing, 633
Pekpekpu, 732
Penma, 565
Phalame, 600, 775
Phaledo, 683, 684
Phikhiroo, 793
Phirphire, 812
Phrogpa Laga, 688
Phurke, 528
Phusre Asaelu, 561
Phusre Aselu, 555
Phusre Malata, 802
Physic Nut, 790
Pigeon Pea, 703
Pipalpate, 812
Pipli, 470
Pitali, 799

Plum, 542
Poinsettia, 762
Poison Nut, 790
Portka Siris, 644
Prang Seng, 774
Purging Nut, 790

Quince, 602

Rain Tree, 647
Raj Birse, 628
Rajbriksh, 628
Ramrita, 799
Rato Siris, 644, 646
Reri, 808
Re Sheng, 662
Rokte, 781
Roringa, 684

Sandan Pipli, 673
Sanu Aselu, 550
Sanu Malata, 798
Sansu Pasi, 597
Sanu Tenga, 595
Sarkinu, 669, 674
Saro Kat, 600
Satisal, 653
Scarlet Runner Bean, 701
Sedgno, 721
Sedkar, 722
Semchu, 698
Semchung Robjay, 704
Seng Ki, 795
Sergong, 557
Seti, 471
Seti-kath, 809
Seto Siris, 645
Sewai Metog, 586
Sew Shing, 586
Seytsala Ngyon, 793
Sheonri, 761
Shimi, 698
Sho Shi, 812
Simal Tarul, 795
Sindure, 802
Siris Lahara, 653
Sishi, 541
Sisi Chungchung, 586
Sissau, 652
Sissoo, 652
Slipper-flower, 767
Soybean, 693
Star Gooseberry, 773
Sunalo, 628

INDEX OF COMMON NAMES

Sungmulagu Shing, 520
Sunn, 733
Suparey, 793
Sweet Cassava, 795
Sweet Pea, 726
Sword Bean, 690

Taki, 634
Taktokhajim, 638
Taktse Metog, 558
Taktsher, 586
Tamarind, 636
Tanki, 633
Tapioca, 795
Tata, 644
Tatebiri, 653
Tatse Tsang, 624

Tatur Shi, 674
Tenga, 595
Tengsey, 636
Thalumbo, 554
Theshom, 802
Thulu Gong, 560
Thulu Gongsey, 558
Thung Kakpa, 603
Tikpi-Kung, 650
Tilki, 528
Titiri, 636
Titri, 636
Tong Shing, 603
Tonsar Gugay, 488
Totom, 536
Totuma, 537
Tsang Guma, 560

Tsangi Metog, 624
Tsaulane, 528
Tse Hein, 623
Tsema Shing, 597
Tshema, 557
Tshema Tshelu, 557
Tshenmar, 481
Tung Oil, 791

Urd, 701

Yakima, 491
Yuguli, 562

Zalibi, 687
Zat Kusum, 788